Respiratory
Pharmacology and
Pharmacotherapy

Airways Smooth Muscle: Peptide Receptors, Ion Channels and Signal Transduction

Edited by
D. Raeburn
M. A. Giembycz

Birkhäuser Verlag
Basel · Boston · Berlin

Editors:

Dr. David Raeburn
Department Head
Discovery Biology
Rhône-Poulenc Rorer Ltd
Dagenham Research Centre
Dagenham
Essex RM10 7XS
England

Dr. Mark A. Giembycz
Lecturer
Department of Thoracic Medicine
Royal Brompton National Heart and Lung Institute
Dovehouse Street
London SW3 6LY
England

Library of Congress Cataloging-in-Publication Data

Airways smooth muscle : peptide receptors, ion channels and signal
transduction / edited by D. Raeburn ; M. A. Giembycz.
    p. cm. – (Respiratory pharmacology and pharmacotherapy)
Includes bibliographical references and index.
ISBN-13: 978-3-0348-7364-2   e-ISBN-13: 978-3-0348-7362-8
DOI: 10.1007/978-3-0348-7362-8
1. Respiratory Muscles – Cytochemistry. 2. Smooth muscle –
Cytochemistry. I. Raeburn, D. (David), 1953– . II. Giembycz,
M. A. (Mark A.), 1961– . III. Series.
    [DNLM: 1. Respiratory Muscles – physiology. 2. Muscle, Smooth –
– physiology. 3. Receptors, Peptide – physiology. 4. Ion Channels –
– physiology. 5. Signal Transduction – physiology. WF 102 A29863
1995]
QP121.A425 1995
612.2 – dc20
DNLM/DLC
for Library of Congress

Die Deutsche Bibliothek - CIP - Einheitsaufnahme

Airways smooth muscle : Peptide receptors, ion channels and signal transduction. - 1995
/ ed. by D. Raeburn ; M. A. Giembycz.
Basel ; Boston ; Berlin : Birkhäuser, 1995
    (Respiratory pharmacology and pharmacotherapy)
    ISBN-13: 978-3-0348-7364-2

NE: Raeburn, David [Hrsg.]

© 1995 Birkhäuser Verlag
Softcover reprint of the hardcover 1st edition 1995
P.O. Box 133
CH-4010 Basel/Switzerland

ISBN-13: 978-3-0348-7364-2

9 8 7 6 5 4 3 2 1

Contents

List of Contributors..................................... VII

1. Endothelins
   D. W. P. Hay.. 1

2. Bradykinin
   S. G. Farmer... 51

3. Tachykinins and Calcitonin Gene-Related Peptide
   C. A. Maggi.. 67

4. Vasoactive Intestinal Peptide
   S. I. Said... 87

5. Atrial Natriuretic Peptides
   N. C. Thomson.. 115

6. Platelet-Derived Growth Factor, Transforming Growth
   Factor-β and Connective Tissue Growth Factor
   J. Kelley 131

7. Voltage-Dependent and Receptor-Operated Calcium
   Channels
   I. W. Rodger... 155

8. High Conductance Calcium-Activated Potassium Channels
   G. J. Kaczorowski and T. R. Jones...................... 169

9. Adenosine Triphosphate-Activated Potassium Channels
   S. L. Underwood and D. Raeburn........................ 199

10. Sodium/Potassium/Chloride Co-Transport
    A. J. Knox 217

11. Sodium/Hydrogen Exchange
    R. Bose 233

12. Sodium/Potassium-Dependent Adenosine Triphosphatase
    S. J. Gunst 255

Index... 273

Contributors

Ratna Bose, Department of Pharmacology and Therapeutics,
University of Manitoba, Winnipeg, Manitoba, Canada

Stephen G. Farmer, Pulmonary Pharmacology Section, Zeneca
Pharmaceuticals Group, Wilmington, Delaware, USA

Susan J. Gunst, Department of Physiology and Biophysics, Indiana
University School of Medicine, Indianapolis, Indiana, USA

Douglas W. P. Hay, Department of Pulmonary Pharmacology,
SmithKline Beecham Pharmaceuticals, King of Prussia,
Pennsylvania, USA

Thomas R. Jones, Department of Pharmacology, Merck Frosst
Centre for Therapeutic Research, Pointe Claire-Dorval, Canada

Gregory J. Kaczorowski, Department of Membrane Biochemistry and
Biophysics, Merck Research Laboratories, Rahway, New Jersey,
USA

Jason Kelley, University of Vermont College of Medicine, Given,
Burlington, Vermont, USA

Alan J. Knox, Respiratory Medicine Unit, City Hospital,
Nottingham, UK

Carlo Alberto Maggi, Department of Pharmacology, A. Menarini
Pharmaceuticals, Florence, Italy

David Raeburn, Department of Inflammation, Rhône-Poulenc Rorer,
Inc., Collegeville, Philadelphia, USA

Ian W. Rodger, Merck Frosst Centre for Therapeutic Research,
Pointe Claire-Dorval, Quebec, Canada

Sami I. Said, Department of Veterans Affairs, Medical Center at
Northport, New York, and University Medical Center, Stony
Brook, New York, USA

Neil C. Thomson, Department of Respiratory Medicine, Western
Infirmary, Glasgow, Scotland, UK

Stephen L. Underwood, Rhône-Poulenc Rorer Ltd., Dagenham
Research Centre, Dagenham, Essex, UK

Airways Smooth Muscle: Peptide Receptors, Ion Channels and Signal Transduction
ed. by D. Raeburn and M. A. Giembycz
© 1995 Birkhäuser Verlag Basel/Switzerland

CHAPTER 1
Endothelins

Douglas W. P. Hay

*Department of Pulmonary Pharmacology, SmithKline Beecham
Pharmaceuticals, King of Prussia, Pennsylvania, USA.*

1   Introduction
2   Distribution, Synthesis, Release, Metabolism, Uptake and Clearance
2.1   Distribution
2.2   Synthesis
2.3   Release
2.4   Metabolism
2.5   Uptake and Clearance
3   Receptors
3.1   Background
3.2   Biochemical and Molecular Biological Studies
3.3   Functional Studies
3.4   Receptor Localization
4   Biological Effects in the Pulmonary System
4.1   *In Vitro* Studies
4.1.1   Contractile Activity
4.1.2   Relaxant Activity
4.1.3   Airway Epithelium
4.1.4   Mucous Glands
4.1.5   Smooth Muscle and Fibroblast Proliferation
4.1.6   Mediator Release
4.1.7   Microvascular Permeability
4.1.8   Inflammatory Cell Function
4.1.9   Modulation of Neurotransmission
4.2   *In Vivo* Studies
4.2.1   Actions on Bronchoconstrictor Tone
4.2.2   Actions on Vasomotor Tone
4.2.3   Other *In Vivo* Effects
5   Signal Transduction Mechanisms
6   Potential Pathophysiological Role
6.1   Background
6.2   Asthma
6.3   Pulmonary Hypertension
6.4   Other Pulmonary Disorders
6.5   Animal Studies
7   Conclusions
        Acknowledgements
        References

1. Introduction

The endothelium is a critical cellular layer in the cardiovascular system
which releases a variety of biologically active substances [1, 2]. In the

1980s considerable attention in the field of endothelial cell biology was directed towards the isolation and characterization of endothelium-derived modulators of the responsiveness of the underlying vascular smooth muscle. The focus of these studies was on relaxant substances released from the endothelium, such as prostacyclin and, in particular, an unknown material which was designated endothelium-derived relaxing factor or EDRF. Recent research has provided convincing evidence that EDRF is nitric oxide or a nitric oxide-containing moiety [3]. During this period there was limited research and information on endothelium-derived vasoconstrictor substances. However, evidence was provided that thromboxane and superoxide elicited contraction of vascular smooth muscle following their release from the endothelium [1]. Furthermore, two groups of researchers reported that a protease-sensitive material in the supernatant of cultured bovine aortic endothelial cells contracted vascular smooth muscle [4–6]. This suggested the presence of a vasoconstrictor peptide that was released from the endothelium.

The research area of endothelium-derived constrictor substances was irreversibly transformed in 1988 when Yanagisawa and co-workers published a landmark paper which described the isolation, purification, cloning and expression, and initial pharmacological characterization of a potent vasoconstrictor 21-amino acid peptide, named endothelin, which was released from porcine aortic endothelial cells [7]. It was demonstrated subsequently that this material was a member of a mammalian family of vasoconstrictor peptides, the endothelins (ETs), designated ET-1, ET-2 and ET-3, which are encoded by three similar but distinct ET-related genes [8]. ET-1 is the original porcine/human ET, ET-2 differs by two amino acid substitutions from ET-1, and ET-3 differs by six amino acids. The genes encoding ET-1, ET-2 and ET-3 are localized to chromosomes 6, 1 and 20, respectively [9–11]. Interestingly, the ETs possess a close structural and functional homology to a group of snake venom toxins, the sarafotoxins, which are found in the venom of the Middle Eastern burrowing asp, *Atractaspis engaddensis* [12–14].

An illustration of the massive amount of research conducted on the biology of the ETs and their putative physiological and pathophysiological roles is the realization that there have been over 3300 publications on this peptide family since the discovery of ET-1 nearly 6 years ago. Perhaps not surprisingly, in light of its recognized potent vasoconstrictor properties and the fact that it was originally isolated from endothelial cells, initial interest on the ETs focused on their activity and potential pathophysiological relevance in the cardiovascular system. However, research quickly revealed that the ETs possess a spectrum of diverse biological activities in a variety of other systems and tissues [12–14]. For example, there has been a significant increase in information in recent years on the actions of the ETs in the pulmonary system.

In this chapter a comprehensive review of our current knowledge of the influence of the ETs in various cells in the pulmonary system will be presented, with particular attention being paid to ET receptors. In addition, the evidence for a potential pathophysiological role of the ETs in respiratory tract disorders will be summarized and critiqued. Before the topics which are the major focus of this chapter are addressed, a brief synopsis of some of the general characteristics of the ETs and their relation to the pulmonary system, namely, distribution, synthesis, release, metabolism, uptake and clearance, will be given. Note, that most of the material relates to ET-1, which, compared to ET-2 and ET-3, has been the focus of research.

2. Distribution, Synthesis, Release, Metabolism, Uptake and Clearance

2.1. Distribution

Of the tissues and organs examined, ET levels in the lung are among the highest detected [15–18]. ET-1 mRNA was detected in human lung homogenates [19]. The cellular origins of ET have still to be clarified, although evidence indicates that they include the endothelium, epithelium, endocrine cells and some inflammatory cells. ET-like immunoreactivity (ir-ET) is present in most epithelial cells in the conducting airways of the rat and mouse [20]. In rabbit trachea, ET-1 immunoreactivity was detected in cells scattered throughout the epithelium, and was also observed in cultured tracheal epithelial cells [21]. Furthermore, in rat foetal lung significant quantities of ET mRNA are detected, localized in respiratory epithelial cells of bronchioles and also blood vessels [22]. In this study Northern blot analysis detected two forms of ET mRNA (2.5-kilobase and 3.7-kilobase forms), both of which were present in lung.

Immunocytochemical analysis and molecular biological techniques have also demonstrated ir-ET and mRNA, respectively, for the three ET isoforms in human airway epithelial and endocrine cells. Immunoreactivity was observed predominantly in pulmonary endocrine cells and was less evident in the airway epithelium (in about 50% of human adults) [23]. The amounts of immunoreactivity and mRNA present in vascular endothelial cells was highest in the developing lung, started to decrease before birth and was minimal in adults, and, accordingly, it was proposed that ET may play a role in growth regulation.

The expression and release of ET-1 has been demonstrated from human macrophages, a predominant inflammatory cell in the lung; the expression but not the release of ET-3 was noted [24]. Furthermore, ir-ET-1, but not ir-ET-3, is present in mouse primary bone marrow mast cells in large quantities [25].

2.2. Synthesis

It was proposed originally by Yanagisawa and co-workers that ET-1 was synthesized *via* an unusual, two-step, proteolytic process [7]. The initial stage involved the formation of a 39-amino acid residue interme- diate, designated "big endothelin", from a 203-residue preproendothelin *via* the activity of an endopeptidase(s) which is specific for the paired dibasic amino acid residues [7]. Note, the human form of preproen- dothelin consists of 212-amino acid residues and big endothelin pos- sesses 38 amino acids [8, 26]. Big endothelin then undergoes a previously unknown type of cleavage between Trp^{73} and Val^{74}, *via* the activity of a putative endopeptidase with chymotrypsin-like activity, which was called "endothelin-converting enzyme" or "ECE" [7]. The biosynthetic pathway for ET-1, in addition to the two potential molecu- lar targets for therapeutic intervention, namely the ECE and also the ET receptors, are shown in Figure 1.

ET-1, Big ET-1 and also the carboxyterminal of Big ET-1 (big ET^{22-39}) are detected in the supernatant of cultured endothelial cells [27,

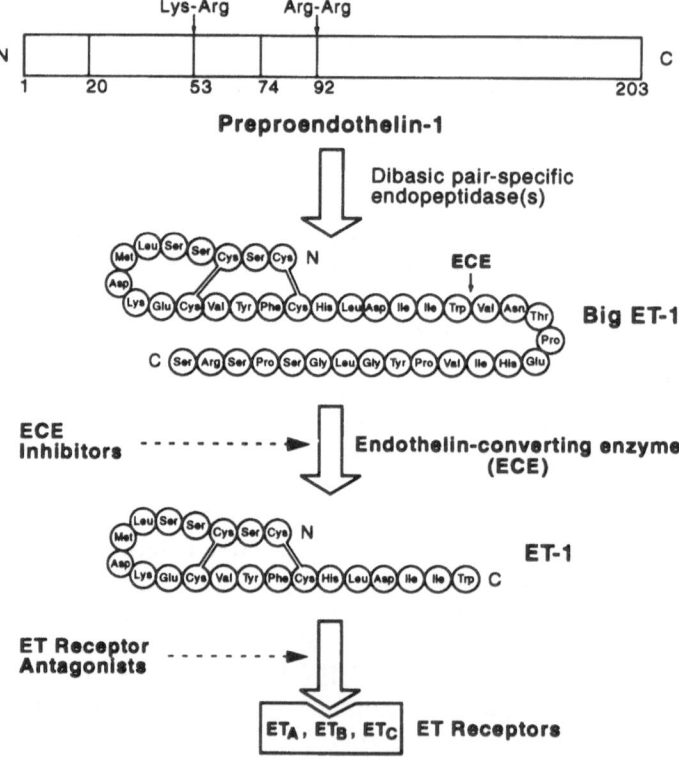

Figure 1. Endothelin-1 (ET-1) biosynthetic pathway and potential molecular targets for novel therapeutic agents (ET receptor antagonists and ECE inhibitors). Adapted from [7].

28] and in human plasma [29]. The vasoconstrictor potency of big ET-1 in isolated blood vessels is appreciably less than ET-1 [30, 31], suggesting that the formation of ET-1 is critical for optimal biological activity of the ET system. Furthermore, evidence indicates that ET is not stored preformed but is released constitutively by *de novo* synthesis, with the regulation of synthesis predominantly at the transcription level [12, 14, 32].

Until recently, there was considerable constroversy as to the identity, location and characteristics of the putative ECE, with evidence for the existence of at least three ECE-like enzymatic activities, two of which are cytosolic and the other membrane-associated. Most of the information favoured a membrane-bound, neutral metalloprotease, which is sensitive to phosphoramidon, as the most likely candidate as the physiologically relevant ECE [14, 32]. A phosphoramidon-sensitive ($IC_{50} =$ 0.5 µM) neutral protease which was able to convert big ET-1 to ET-1 was identified in rat lung, with a relative abundance in the membrane versus the cytosol of 4:1 [33]. Furthermore, the partial solubilization and purification of ECE from porcine lung was reported recently [34] and successful purification to homogeneity of ECE from rat lung microsomes was demonstrated [35]. The purified enzyme had a molecular weight of 130 kD and specifically catalyzed the conversion of big ET-1 to ET-1 *via* a mechanism which was inhibited by phosphoramidon and metal chelators [35]. Sucrose-gradient ultracentrifugation analysis of rat lung microsomal fractions revealed that a major portion of the phosphoramidon-sensitive ECE activity appears to be associated with the Golgi apparatus, where it co-exists with endogenous ET-1 [36]. It was proposed that in rat lung most of the big ET-1 is converted to ET-1 during its passage *via* intracellular secretory pathways, although a smaller component of ECE may exist as an ecto-enzyme on the plasma membrane [36], accounting for the rapid one-pass conversion of intravenous big ET-1 to ET-1 in perfused lungs [37].

There is also evidence *in vivo*, and using perfused lungs, that the ECE responsible for the conversion of big ET-1 to ET-1 in airways is sensitive to phosphoramidon [38–40].

The presence of an aspartic protease with ECE activity in rat lung was reported [41]. Furthermore, evidence was presented for a novel, serine protease enzyme, which is present in the soluble fraction of porcine lung which degraded big ET-1 by cleavage at the bond between Val[22] and Asn[23]. This enzyme, which was sensitive to diisopropylfluorophosphate, was designated ET-Val-generating endopeptidase [42]. It was also suggested that lung mast cell-derived chymase produces extracellular processing of big ET-1 to ET-1 in the perfused rat lung [43]. The physiological significance of these putative ECEs in the pulmonary system has still to be ascertained and most of the evidence indicates that, as in other systems, the relevant ECE in the lung is a phosphoramidon-sensitive metalloprotease. In support of this assertion, and as

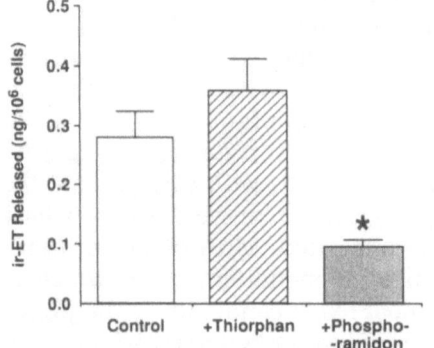
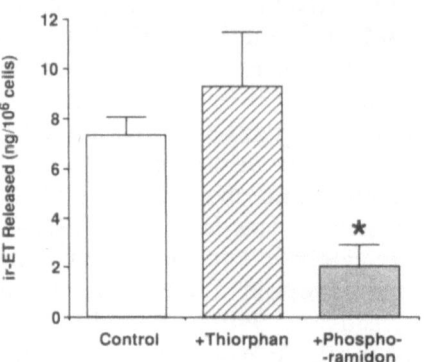

Figure 2. Effect of thiorphan (10 μM) or phosphoramidon (10 μM) on basal release of ir-ET-1, over a 24 hour period, from (A) cultured guinea-pig tracheal epithelial cells and (B) cultured human umbilical vein endothelial cells (HUVECs). Results are the Mean ± S.E.M. of 4 experiments. * indicates signficant compared to control, $P < 0.05$. The results indicate the ET release in both guinea-pig tracheal epithelial cells and HUVECs is sensitive to phosphoramidon (metalloprotease inhibitor; NEP inhibitor) but not thiorphan (NEP inhibitor) suggesting that the ECE responsible for the conversion of big ET-1 to ET-1 in both systems is a phosphoramidon-sensitive metalloprotease. Note the different scales for the y-axes in the two graphs indicating that the release of the ETs from HUVECs is much greater (ca. 30-fold) than from guinea-pig tracheal epithelial cells.

outlined in Figure 2, experiments performed in the author's laboratory indicate that basal release of ir-ET from guinea-pig cultured tracheal epithelial cells, and also human umbilical vein endothelial cells (HUVECs), is inhibited by phosphoramidon (100 μM) but not thiorphan (10 μM), another inhibitor of neutral endopeptidase (NEP; EC 3.4.24.11) [44]. The basal release of ir-ET from HUVECs was about 30-fold higher than from guinea-pig tracheal epithelial cells.

Considerably less is known about the processing of the precursors for ET-2 or ET-3 than for ET-1, and it is a controversial issue as to whether different ECEs exist which are specific for the individual ET peptides [45, 46].

2.3. Release

As indicated above, ET mRNA and ir-ET have been detected in airway epithelial cells and ECE is present in the lung. In support of these observations, ET-1 is released basally from porcine, canine [47] and human cultured bronchial epithelial cells [48] and guinea-pig [49, 50] and rabbit tracheal epithelial cells [51]. In addition, ET-3 was detected in supernatants from canine and porcine cultured tracheal epithelial cells [47]. ET-1 is also released from murine cultured bone marrow-derived mast cells following long-term incubation with IgE (10-fold increase after 20 hr) [25]. The release of ET from cultured tracheal epithelial cells is increased by endotoxin, thrombin and various cytokines [49–51], whereas in perfused guinea pig lungs capsaicin, antidromic nerve stimulation or anoxia did not increase the outflow of ir-ET [52]. The regulatory role of cytokines on ET-1 synthesis and release may be important in inflammation and tissue repair associated with many pulmonary disorders, processes in which the cytokines are thought to play a key role.

2.4. Metabolism

The epithelium inhibits ET-1-induced contractile responses in guinea-pig trachea via a mechanism which is sensitive to phosphoramidon, an inhibitor of NEP [54]. The epithelium is a rich source of NEP [54] and it was hypothesized that this modulatory role was indicative of ET-1 being metabolized by epithelium-derived NEP [53]. The ETs were subsequently observed to be good substrates for NEP [55, 56]. Further functional data in support of a modulatory role of epithelium-derived NEP on ET-1-indiced contraction in guinea-pig trachea has been provided [57, 58], including the observation that recombinant human NEP decreased ET-1-induced contraction [58]. In vivo studies in guinea pigs have also demonstrated that phosphoramidon potentiates ET-1-induced bronchoconstriction [59]. In human bronchus, evidence indicated that NEP is involved in the local metabolism of ET-3 but not ET-1; similar findings were observed in rabbit bronchus [60]. However, other studies demonstrated that phosphoramidon potentiated contraction induced by ET-1 [61, 62], and also ET-2 and ET-3 [61], in human bronchus. It was concluded that the epithelium modulated ET-induced contraction in human bronchus, in part because of the removal of the metabolic influence of NEP, but also via other mechanisms [61]. In pig lung membrane fraction, NEP was identified as the principal metabolic pathway for ET-1 [63].

These and other data suggest strongly that two phosphoramidon-sensitive enzymes have opposing effects in the control of ET-1 levels in airways: thus, phosphoramidon inhibits the putative ECE responsible for the formation of ET-1 from big ET-1 and also NEP which appears to be involved in the breakdown of ET-1.

Although, NEP appears to be the major enzyme involved in ET-1 degradation in the airways, other pathways may also contribute. For example, soybean trypsin inhibitor essentially abolished the degradation of ET-1 induced by activated human polymorphonuclear neutrophils (PMNs), whereas phosphoramidon was without effect, suggesting that cathepsin G, rather than NEP, is involved in ET-1 metabolism by these cells [64].

2.5. Uptake and Clearance

Intravenously administered ET-1 is rapidly eliminated from the circulation in rats, with accumulation predominantly in the lungs and kidneys [65–67]; for example, it was calculated that 82% of $[^{125}I]$-ET-1 was taken up by the lungs [67]. Substantial removal of ET-1 by the pulmonary circulation (> 50% in a single passage) also occurred in guinea-pig and rat isolated perfused lungs [68] and in rabbits significant uptake of ET-1 was demonstrated by the pulmonary but not the coronary circulation [69]. In contrast, during infusion of ET-1 in pig no evidence was observed for clearance in the lungs [17]. In two studies in humans, involving intravenous infusion of ET-1, it was estimated that the lung is the main organ responsible for its removal (approximately 50% of the elimination) [70, 71]. In contrast, another study concluded that ET was not extracted by the pulmonary system in humans [72]. Thus, marked species, in addition to regional, differences appear to exist in the contribution of the pulmonary circulation to the uptake and clearance of ET.

3. Receptors

3.1. Background

It quickly became apparent that quantitative and qualitative differences existed in the pharmacological profiles of the ET isoforms [12–14, 73]. For example, ET-1 is more potent than ET-2 in eliciting vasoconstriction whereas they have equivalent potencies as vasodilators [8, 74, 75]. In addition, it has been observed that administration of the ETs *in vivo* in animals elicits an initial transient depressor response followed by a sustained increase in blood pressure [7, 8, 74, 76]. This formed part of the initial indirect evidence to suggest that the ETs exert their diverse biological effects *via* multiple ET receptors which possess different tissue distributions [13, 14].

3.2. Biochemical and Molecular Biological Studies

Results of biochemical and molecular biological studies subsequently provided unequivocal evidence for the existence of distinct ET receptor subtypes. It is noteworthy that the lung, largely because of the abundance of ET receptors, has been used extensively in studies investigating ET receptor characterization, subtyping, isolation and purification.

For example, cross-linking, affinity labeling studies in rat lung membranes provided evidence for two ET receptors, a 44 kD type which had higher affinity for ET-1 and ET-2 than ET-3 and another with a molecular weight of 32 kD which had a higher affinity for ET-3 [77]. Ligand binding and affinity labeling studies in porcine tissues, including lung, also suggested the presence of two distinct ET receptor subtypes, an ET-1-specific receptor and a receptor that was common to the ET/sarafotoxin family [78]. In this study, in both porcine and rat lung membranes three major bands, with molecular weights of about 120 000, 47 000 and 35 000, were labeled with [^{125}I]-ET-1 in cross-linking experiments. ET-1 prevented the labeling in all bands whereas ET-3 abolished only the labeling in the 35 000 molecular weight band. In another study photoaffinity labeling of the ET-1 receptor in bovine and rat lung membranes yielded only one labeled band with a molecular weight of 34 000 [79].

Using several systems, including rat and bovine lung, the cloning and expression of cDNA for a selective and also for a non-selective ET receptor, designated ET$_A$ and ET$_B$, respectively, has been accomplished [14, 73, 74, 80–82]. The ET$_A$ receptor has a higher affinity for ET-1 or ET-2 compared with ET-3, whereas, the ET$_B$ receptor has equal affinity for the various ET isoforms. Evidence has also been provided in rat cultured anterior pituitary cells and rat PC12 phaeochromocytoma cells, for an ET receptor subtype, designated ET$_C$ [14], which possesses selectivity for ET-3 [83, 84]. This receptor has recently been cloned from *Xenopus laevis* dermal melanophores [85]. All the ET receptors belong to the superfamily of G-protein-linked, seven transmembrane-spanning receptors [14, 73, 74, 80–82, 85]. A comparison of the characteristics of the three ET receptors is given in Table 1.

The human ET$_A$ and the ET$_B$, but not ET$_C$, receptors have been cloned [82, 86–90]. For the ET$_A$ or ET$_B$ receptor there is considerable homology between species ($\geq 88\%$), although significant differences exist between the ET$_A$ and ET$_B$ receptors. The ET$_A$ receptor was localized to chromosome 4 [87]. High or moderate distribution of mRNA for ET$_A$ and ET$_B$ receptors in human tissues has been detected in various tissues including lung (Table 1) [82, 88, 90]. There is no information on the distribution and function of ET$_C$ receptors in humans.

The solubilization of ET$_A$ and ET$_B$ receptors from rat lung [91] and the purification of the ET$_B$ receptor from bovine lung [92, 93] have been reported. In bovine lung, two forms of the ET$_B$ receptor were identified, a 34 kDa species which was proposed to be a proteolytic compound of the native form (52 kDa molecular weight). Both forms exhibited nearly identical ligand affinities and specificities indicating that the binding activity of the receptor is located within the 34 kDa structure [92]. Cross-linking of [^{125}I]-ET-1 and [^{125}I]-ET-3 labeled solubilized ET receptors both produced a major band of 48 kDa and a minor band, 37 kDa, which was regarded as a proteolytic product [91].

Table 1. Properties of the ET_a, ET_B and ET_C receptor[1]

Parameter	ET_A	ET_B	ET_C	References
Relative agonist affinities	ET-1 ⩾ ET-2 ≫ ET-3	ET-1 = ET-2 = ET-3	ET-3 ≫ ET-1 = ET-2	14, 73, 74, 82, 85, 86, 88, 90
Seven transmembrane, G-protein linked receptor	yes	yes	yes	14, 73, 74, 80–82, 85, 88, 90
Homology	—	64% to ET_A	47% to ET_A 52% to ET_B	73, 82, 85, 86, 90
Amino acid number	427	442	424	82, 85, 86, 88–90
Potential glycosylation sites	2	1	2	82, 85, 86–90
Prominent distribution	heart, lung, colon, blood vessels, brain	endothelium, brain, lung, kidney, blood vessels, duodenum, adrenal, colon	unknown	82, 88, 90
Selective agonists	?	S6c, [Ala1,3,11,15]-ET-1 IRL 1620, BQ-3020	ET-3	85, 99, 110, 120, 121
Selective antagonists	BQ-123 FR 139 317	IRL 1038	None	100, 101, 122

[1]ET_A and ET_B are the human receptors whereas ET_C was cloned from *Xenopus laevis* dermal melanophores.

3.3. Functional Studies

There is accumulating data from functional studies in support of ET receptor subtypes in the pulmonary system. In the initial series of studies examining this phenomenon Maggi and co-workers comprehensively investigated the contractile effects of several members of the ET and sarafotoxin families and peptide analogs in isolated tissues including guinea-pig airways [94–96]. For example, they noted that the hexapeptide ET-(16-21) was a full agonist in guinea-pig bronchus but was without effect in rat aorta, whereas ET-1 effectively contracted both tissues. Based on these observations they provisionally designated two ET receptor subtypes as ET_A (representing the aorta) and ET_B (for bronchus) [95]. However, a subsequent study indicated that ET-(16-21) was a much less effective agonist in guinea-pig trachea and was without marked effect on [^{125}I]-ET-1 binding [97]. These data suggests that ET-(16-21) may be of limited usefulness for ET receptor classification. Following studies comparing the relative contractile activities of ET-1, ET-2 and ET-3, in addition to their ability to cause cross-sensitization, it was proposed that distinct ET receptors mediate contraction in guinea-pig pulmonary artery and trachea [98].

More direct evidence for ET receptor subtypes in the pulmonary system has been provided utilizing selective ligands, in particular, sarafotoxin S6c, the ET_B-selective agonist [99] and peptide ET_A-selective antagonists such as the cyclic pentapeptide BQ-123 [100] and FR 139317 [101]. Using these experimental tools, functional evidence was provided for distinct ET receptors mediating contraction in guinea-pig pulmonary artery and rat aorta (ET_A-subtype) compared with guinea-pig trachea and bronchus (non-ET_A, probably ET_B) [102, 103]. In human bronchus a non-ET_A (ET_B?) receptor population is the predominant one mediating ET-1-induced contraction whereas in human pulmonary artery contraction is meditated *via* the ET_A receptor [104] (Figure 3). In human bronchus ET_A receptors may be involved in smooth muscle proliferation [105].

In addition, regional differences in the relative distribution of ET_A and non-ET_A, probably ET_B, receptors occurs in guinea-pig airways. Thus, the contribution of ET_B receptors, assessed functionally using sarafotoxin S6c, predominated over ET_A receptors in guinea pig bronchus, and increased from the upper trachea (little contribution) to the lower trachea (marked contribution) [104]. In a more recent study utilizing various ligands, including antagonists, which interact with ET_A and ET_B receptors, it was concluded that ET_B receptors mediate contractions produced by ET-1 in guinea-pig trachea, bronchus and parenchyma and ET_A receptor activation also contributes to the response in guinea-pig trachea and parenchyma [106]. *In vivo* experiments indicate that ET-induced bronchoconstriction in guinea-pigs was not sensitive

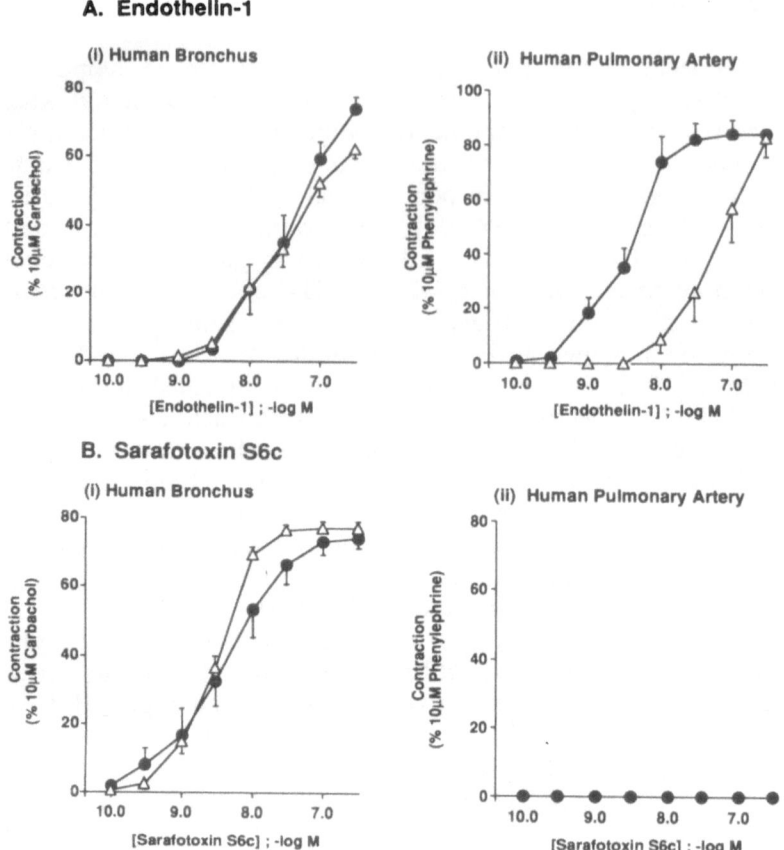

Figure 3. Effect of BQ-123 (10 μM), the ET_A receptor antagonist, on responses to (A) ET-1 and (B) sarafotoxin S6c (S6c), the ET_B-selective agonist, in (i) human bronchus and (ii) human pulmonary artery. Results are expressed as a percentage of the reference contraction and are the Mean ± S.E.M. of 5–7 experiments; ● = Control; △ = +10 μM BQ-123. BQ-123 antagonized ET-1-induced contraction in human pulmonary artery (pK_B = 6.2) but not in human bronchus. Furthermore, S6c did not contract human pulmonary artery. The results provide evidence that ET-1 induced contraction in human pulmonary artery is mediated *via* ET_A receptor activation whereas in human bronchus it is due to stimulation of a non-ET_A receptor, probably ET_B. Adapted from [104].

to BQ-123 and, therefore, does not appear to be mediated by ET_A receptor activation [107]. Interestingly, evidence was provided for different receptors mediating ET-1-induced contraction (non-ET_A) and prostanoid release in human bronchus (ET_A) [104]. In rat perfused lungs ET-1-induced PGI_2 release was antagonized by BQ-123 [108].

It was proposed initially that ET_A receptors on vascular smooth muscle elicit contraction whereas ET_B receptors located on endothelial cells produce vasodilatation [14, 73]. In support of this hypothesis, stimulation of the ET_A receptor subtype appears to mediate contraction elicited by the ETs in most vascular smooth muscle [109–112]. ET_B-selective agonists have been shown to elicit endothelium-dependent relaxation in isolated vascular smooth muscles [113] and ET-1-induced relaxation in rat aorta is antagonized by the ET_B receptor antagonist, IRL 1038 [114]. However, ET_B-receptor activation mediates contraction in some vascular smooth muscles [113, 115–117]. It was proposed, based on binding studies in rat brain and atrium, that there are two subtypes of ET_B receptors, called ET_{B1}, which were designated as super-high affinity sites (in the pM range) and ET_{B2}, high affinity sites [118]. It was hypothesized that ET_{B1} receptors are involved in the vasodilator effects of the ETs, whereas, ET_{B2} receptors mediate the vasoconstriction.

There is *in vivo* functional evidence, obtained by exploring the responses and cross-tachyphylaxis induced by ET-1, ET-2 and ET-3, for the existence of ET_A-like and ET_C-like receptors in the cat pulmonary vascular bed [119].

Selective ligands for the ET_A, ET_B and ET_C receptor have been identified (Table 1). These include sarafotoxin S6c, [Ala^{1,3,11,15}]-ET-1 IRL 1620 and BQ-3020 (ET_B receptor agonists) [99, 110, 120, 121], BQ-123 and FR 139317 (peptide ET_A receptor antagonists) [100, 101] and IRL 1038 (peptide ET_B receptor antagonists) [122]. These compounds will continue to be invaluable experimental tools with which to decipher the receptors mediating the diverse biological effects of the ETs. In addition, the recently identified non-peptide receptor antagonists, such as Ro 46-20058 [111], SB 209670 [112] and CGS 27830 [123], may also be of therapeutic benefit in the various diseases in which the ETs have been implicated. Ro 46-20058 has about equivalent potency as an ET_A and ET_B receptor antagonist [111], whereas SB 209670 and CGS 27830 have higher affinity for the ET_A versus the ET_B receptor (ca. 20-fold) [112, 123]. A natural product isolated from the bayberry *Myria cenfera*, 27-0-caffeoyl myricerone (50-235) has been proposed to a selective ET_A receptor antagonist ($K_i = 51$ nM) with minimal affinity for the ET_B receptors [109, 124]. There are no known ET_C receptor antagonists.

It is likely that the present classicfiation of ET receptors is incomplete and there is increasing evidence from functional and binding studies for the existence of further subtypes of ET receptors [116, 125–128]. For example, evidence was presented for a novel subtype of isopeptide-nonselective ET_B receptor which mediates contraction to sarafotoxin S6c, ET-1, ET-2, ET-3 and IRL 1620 in swine pulmonary vein (not inhibited by the purported ET_B receptor antagonist, IRL 1038), and which is distinct from the ET_B receptor mediating ET-3-induced relaxation on precontracted swine pulmonary artery (antagonized by IRL 1038) [116].

3.4. Receptor Localization

Binding and autoradiographic studies indicate that significant quantities of a single class of high-affinity ET-1 binding sites are differentially located in various regions of the respiratory tract of several species including humans [97, 129–134]. In human bronchial tissues, labeling of [^{125}I]-ET-1 was localized largely to airways and vascular smooth muscle, with little or no binding on cartilage, connective tissues, the submucosal layer (including glandular cells) and the epithelium [129, 132]. Similar results were observed in mouse, rat and guinea-pig tracheal sections [132]. In contrast, recent studies in sheep trachea have revealed high densities of specific binding sites for ET-1 in cells associated with submucosal glands and in the submucosa immediately below the epithelium (lamina propria). In addition, in human, and also rat and guinea-pig airways, it was observed that there was significant binding associated with alveolar septae and parasympathetic ganglia, and also with paravascular nerves and nerves in the connective tissues [129, 130, 132, 134].

A single, specific high affinity binding site for [^{125}I]-ET-1 was detected on human cultured bronchial smooth muscle cells, with an apparent binding affinity (K_d) of 0.11 nM, and a maximum binding capacity (B_{max}) of 22.1 fmol/10^6 cells [48].

In contrast to some of the above studies, binding sites in airway epithelium have been reported. For example, some diffuse binding in the guinea-pig tracheal epithelium was noted, although the density of binding was much less (16%) than that detected in the smooth muscle and especially in the submucosal region (6%) [97]. Two binding sites for ET-1 were observed, with ET-3 able to interact with both sites. Similarly, significant labeling of [^{125}I]-ET has been detected in rat airway epithelium [65] and feline tracheal epithelial cells (two sites) [135]. A single class of binding sites for [^{125}I]-ET-1 (K_d = 390 pM and B_{max} = 307 fmol/mg protein) was detected in rat alveolar type II cells [136]. Furthermore, in situ hybridization techniques have revealed strongly positive staining for ET_B receptor probes in ciliated and nonciliated rat and rabbit bronchial epithelial cells [137]. In addition, endothelium of large-calibre vessels, type II alveolar epithelial cells and visceral pleural mesothelium were stained. The presence of ET receptors on the same cells that are able to synthesize ET suggests that there may be autocrine regulation of its secretion [137].

Following in vivo labeling studies in rats the highest density of labeling of [^{125}I]-ET-1 was observed in the lung and kidney [65]. Furthermore, in a comparison of [^{125}I]-ET binding to various tissue membrane fractions, the highest density of binding was detected in trachea followed by lung parenchyma and vas deferens [138].

Both ET_A and ET_B receptors are abundant in mammalian lung. Using immunohistochemical and immunoprecipitation techniques with an ET_B-specific antiserum, it was calculated that about 70% of the ET receptors in bovine lung were of the ET_B subtype, with the estimated amount of ET_B receptors $= 385 \pm 58$ fmol/mg protein [139]. A preliminary study indicated that the proportions of ET_B and ET_A receptors in human bronchial smooth muscle from both non-asthmatic and asthmatic lung was about 88% ET_B:12% non-ET_B [140]. However, the relative regional distribution of ET receptor subtypes in the pulmonary system appears to be species-dependent. For example, in porcine pulmonary tissues, binding studies indicated that the ratio of ET_A:ET_B receptors in bronchus appeared to be about 7:3 whereas, in lung parenchyma it was the reverse [141]. Autoradiographic analysis supported the differential predominance of ET receptor subtypes in rat pulmonary tissues, with the ET_A receptor prominent in the bronchi and vasculature, and the ET_B receptor more abundant in the parenchyma [141].

Two separate binding studies using rat lung membranes provided evidence that the ratio of ET_A:ET_B was 6:4 [108] and 7:3 [142]. Recently, using the ET_B-selective radioligand, [^{125}I]-BQ-3020, autoradiographic studies in porcine lung revealed significant binding to parenchyma, parasympathetic ganglia, pulmonary and submucosal plexuses, but minimal binding to circular smooth muscle layers or airway epithelium. Interestingly, the binding of [^{125}I]-BQ-3020 in blood vessels paralleled the acetylcholinesterase activity suggesting that the ET_B receptors in blood vessels may be located on parasympathetic nerves [143]. Autoradiographic data have recently provided evidence that the ET_A and ET_B receptors exist in similar amounts in rat tracheal smooth muscle [144]. In contrast, ovine tracheal smooth muscle appears to possess an almost homogeneous population of ET_A receptors [145, 146].

Specific binding sites for [^{125}I]-ET-1 ($K_d = 38$ pM; $B_{max} = 810$ fmol/mg) and [^{125}I]-ET-3 ($K_d = 5.1$ pM; $B_{max} = 360$ fmol/mg) were detected in guinea-pig tracheal smooth muscle preparations and were concluded to represent ET_A and ET_B receptors; similar data were obtained in primary cultures of tracheal smooth muscle cells [147]. In agreement with previous reports [104, 106], analysis of functional responses, involving Ca^{2+} mobilization in cultured cells and contraction in whole tissues, provided further evidence for the presence of ET_A and ET_B receptors which contribute to the contractile response in guinea-pig trachea [147].

4. Biological Effects in the Pulmonary System

It is recognized that the ETs are potent and effective bronchoconstrictor agonists and this was the initial focus of research on their activity in the respiratory tract. However, information is accumulating on the actions

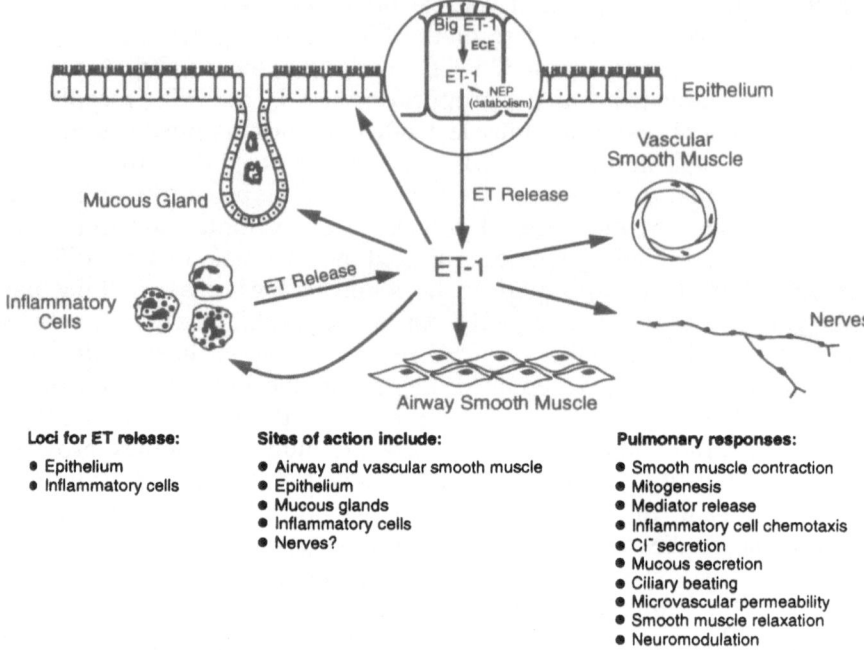

Figure 4. Summary of loci for release, potential sites of action and effects of ET-1 in the lung.

of the ETs in cells and tissues other than airways smooth muscle. A review is given of the effects of the ETs in the different components of the pulmonary system, with emphasis on material that may be relevant to a potential pathophysiological role (Figure 4).

4.1. In Vitro *Studies*

4.1.1. *Contractile Activity*

4.1.1.1. Airways Smooth Muscle. The ETs are potent and effective contractile agonists in isolated airways smooth muscle preparations from a variety of species including humans, with EC_{50}s in the $1-30$ nM range [131, 132, 134, 148–152]. The contraction is relatively slow to develop, well maintained and only slowly reversed by washing. A correlation exists between the density of ET-1 binding sites and the contractile effects induced by ET-1 in human, rat and guinea-pig preparations [132]. In human bronchus, evidence suggests that non-ET_A, probably ET_B, receptors mediate predominantly, if not exclusively, ET-1-induced contraction [104], whereas in guinea-pig airways both ET_A and non-ET_A receptors are involved [104, 106, 147].

In human bronchus, ET-1 appears to be more potent than ET-2 or ET-3 [131, 134, 150]; however, this may be related to differences in susceptibility to degradation by NEP [60, 62]. ET-1-induced contraction of human bronchus is not modulated by cyclooxygenase products [134, 150, 153]. Similar findings were observed using ferret [154] and bovine airways [155]. In guinea-pig trachea there is conflicting information on the ability of cyclooxygenase products to modulate ET-1-induced contraction [132, 149, 151, 152, 156, 157].

In addition, there are discrepancies in the results of studies exploring the role of secondary mediators in contractions elicited by ET-1. For example, in human bronchus, use of specific receptor antagonists and direct measurement of the release of histamine and peptidoleukotrienes provided evidence that ET-1 produced contraction *via* a direct mechanism which did not involve the significant contribution of acetylcholine, PAF, histamine, peptidoleukotrienes or tachykinins [153]. Similar findings and conclusions were reached for the response in guinea-pig trachea [151, 158]. In contrast, ET-1-induced contraction in guinea-pig trachea was reported to be antagonized by PAF receptor antagonists [159]. Furthermore, ET-1-induced contraction and thromboxane release was concluded to be mediated, in part, by PAF and peptidoleukotriene release [160]. It was reported also that a histamine receptor antagonist antagonized ET-1-induced response in guinea-pig trachea [161]. It is noteworthy that ET-1 stimulated the release of various prostanoids, in particular PGD_2, from guinea-pig trachea and human bronchus [153, 158]. Although they do not appear to contribute significantly to the contractile response it is possible that they are involved in ET-induced effects other than bronchospasm.

4.1.1.2. Pulmonary vascular smooth muscle. The ETs also potently contract human isolated pulmonary artery and vein. There is conflicting evidence as to whether they are more potent bronchoconstrictor or vascoconstrictor agents [131, 133, 134]. The response to ET-1 in human pulmonary artery appears to be mediated predominantly, if not solely, *via* ET_A-receptor activation (Figure 3) [104]. Thus, ET-1-induced responses were potently antagonized by BQ-123, whereas S6c, the ET_B-selective agonist, was without effect on the level of tone on this tissue. However, ET_B receptors have been proposed to mediate contraction in porcine pulmonary vein [115, 116] and artery [115] and rabbit pulmonary artery [117]. In guinea-pig pulmonary artery evidence was provided that ET-1 and ET-2, on the one hand, and ET-3, on the other, produce contraction *via* distinct receptors [162]. ET-1 potently contracted rat pulmonary artery ($EC_{50} = 1.3$ nM) and pulmonary vein ($EC_{50} = 0.6$ nM) [163]. In the same study in perfused lungs, ET-1 increased microvascular pressure and produced oedema which was proposd to be due to venoconstriction.

In rat isolated lungs under low levels of vascular tone, ET-1 elicits an increase in pulmonary artery pressure [164] whereas under conditions of elevated tone ET-1 may produce pulmonary vasodilatation [165, 166]. In rat isolated lungs the pulmonary vasoconstrictor response to ET-1 is not modulated by cyclooxygenase inhibitors, but is increased by inhibitors of the formation of EDRF/NO, e.g., N^G-monomethyl-L-arginine (L-NMMA) or methylene blue [164, 167]. In contrast, in guinea-pig perfused lungs cyclooxygenase products appear to contribute to ET-1-induced pulmonary vasoconstriction [68]. In rat lungs the vasodilatation response to ET-1 or ET-3 was potentiated by cyclooxygenase inhibiton or a thromboxane receptor antagonist but was not affected by L-NMMA or methylene blue [165, 168]. This suggests that ET-1- or ET-3-induced vasodilatation in this tissue is not due to the release of NO. It was proposed that ET_B receptor activation in rat isolated lungs elicits vasodilatation whereas ET_A receptors mediate vasoconstriction, which is antagonized by BQ-123. However, autoradiography revealed few ET_B receptors in the endothelium [168].

4.1.2. Relaxant activity

4.1.2.1. Airways smooth muscle. ET-1 has been reported to elicit a transient relaxant response in guinea-pig trachea [106, 169, 170]. In studies utilizing isolated tracheal spiral preparations, under either basal or precontracted conditions, ET-1-induced relaxation was epithelium-dependent [106, 170]. However, the relaxation induced by ET-1 in guinea-pig trachea *in situ* was not influenced by removal of the epithelium, or by cyclooxygenase inhibitors [169]. In precontracted guinea-pig trachea the epithelium-dependent relaxation induced by ET-1 was markedly inhibited by methlyene blue, oxyhaemoglobin or L-NMMA. Accordingly, it was proposed that the ET-1-induced relaxation was due to release of NO from epithelial cells [170]. Under basal tone, ET-1-induced relaxation, which preceded the sustained contractile response, was attributed to ET_A receptor activation of epithelial cells, as it was inhibited by the ET_A receptor antagonists, BQ-123 or FR 139317, and was not produced by IRL 1620, the ET_B-selective agonist [106]. ET-1 ($\leq 10^{-9}$ M) also elicits an epithelium-dependent relaxation in rabbit isolated trachea which is decreased, but not abolished, by indomethacin (10 μM) and was associated with enhanced release of PGI_2 and PGE_2 [171]. It was concluded that the ET-1-induced relaxation in this tissue was attributed to the epithelium-derived release of prostanoid and non-prostanoid relaxants.

4.1.2.2. Pulmonary vascular smooth muscle. In various isolated or perfused vascular smooth muscles the ETs and sarafotoxins elicit an endothelium-dependent relaxation which is thought to be due to the release of NO/EDRF [75, 114, 172, 173]. There is little information on the relaxant effects of the ETs in pulmonary blood vessels. In precontracted rat pulmonary arteries ET-1 (0.01-5 nM) did not elicit relaxation [165], whereas in porcine pulmonary artery both ET-1 and ET-3 were potent relaxants *via* a mechanism which was not affected by an ET_A receptor antagonist, BQ-153 [115].

4.1.3. Airway epithelium: The epithelium plays a vital role in the normal physiology of the lungs and an alteration in the structural or functional properties of the epithelium has been proposed to be a significant contributor, if not initiator, of airway inflammation and to underlie the pathophysiology of several lung disorders, notably asthma and cystic fibrosis [174–176]. As indicated previously, the airway epithelium is a rich source of the ETs, and ET receptors have been located there (*vide supra*). In addition, there have been a few reports of studies exploring the effects of the ETs on the airway epithelium.

For example, ET-1 increased the negativity of transepithelial potential difference in ferret trachea [177]. Furthermore, ET-1 (0.1 nM – 1 µM), but not ET-2 or ET-3, stimulated the potential difference and short-circuit current in canine tracheal epithelium. It was proposed that this translated into a selective increase in Cl^- secretion, with no effect on Na^+ absorption, *via* a mechanism which was partially dependent on the release of cyclooxygenase products [178]. Leikauf and co-workers [179] also demonstrated an indomethacin-sensitive, ET-1-induced increase in short-circuit current in canine tracheal epithelium ($EC_{50} = 2.2$ nM). In addition, they reported that ET-1 increased mucosal net ^{36}Cl flux, but had no effect on ^{22}Na flux, stimulated [^3H]-arachidonate release from membrane phospholipids and increased intracellular Ca^{2+} and cAMP accumulation [179]. However, in contrast to Satoh and colleagues [178], they observed that ET-2 ($EC_{50} = 7.2$ nM) and to a lesser extent ET-3 ($EC_{50} = 10.4$ nM), also increased short-circuit current. In addition to stimulating short circuit current and Cl^- secretion, ET-1 potently increased ciliary beat frequency ($EC_{50} = 3$ nM) in canine cultured tracheal epithelium; these effects were reduced by indomethacin [180]. Relatively high concentrations of ET-1 (0.1 µM– 10 µM) stimulated the release of several prostanoids from [^3H]-arachidonic acid-labelled cultured feline tracheal epithelial cells [135].

ET-1 potently ($EC_{50} = 0.68$ nM) enhanced phosphatidylcholine secretion from rat alveolar type II cells; the maximum increase, with 10 nM ET-1, was about 200% of control [136]. The secretory effect of ET-1 was inhibited by high concentrations of nifedipine or nitrendipine, or by incubation in Ca^{2+}-free medium, suggesting it is mediated *via* influx of extracellular Ca^{2+} through voltage-dependent membrane channels. Furthermore, ET-1 enhanced diacylglycerol levels (not blocked by nifedipine) and ET-1-induced secretion was inhibited by purported protein kinase C inhibitors. These data suggest a role for the diacyglycerol/ protein kinase C system in mediating ET-1-induced secretion. It was speculated that ET-1 released from pulmonary micro- and macro-vessel endothelial cells may regulate surfactant secretion [136].

4.1.4. Mucous glands: Immunoreactive ET-1 and ET-1 mRNA was found in human nasal mucosal tissue [181, 182], primarily in the endothelium and venous sinusoids. [^{125}I]-ET-1 binding sites were also detected in submucosal glands, venous sinusoids and small muscular arterioles [182]. In addition to the presence of ET in the vascular endothelium and serous cells, ET-1, and also ET-2, but not ET-3, stimulated serous and mucous cell secretion. However, high concentrations of ET-1 and ET-2 (0.1 μM – 10 μM) were needed to increase lactoferrin and mucous glycoprotein release [182]. There is conflicting information from studies exploring the influence of ET-1 on mucous gland secretion. Thus, ET-1 (1 nM – 1 μM) increased glycoconjugate secretion from feline tracheal isolated glands but decreased secretion from tracheal explants [183]. In ferret trachea, ET-1 was without effect on serous cell secretion, and actually inhibited phenylephrine- and methacholine-induced secretion [177]. ET-1 increased prostanoid production in human cultured nasal mucosa, but again high peptide concentrations were required [≥ 0.1 μM) [184].

4.1.5. Smooth muscle and fibroblast proliferation: ET-1 elicited a small (2.3-fold) mitogenic response in human cultured bronchial smooth muscle cells. This potent effect of ET-1 (EC_{50} = ca. 1 nM), which appeared to be due to activation of ET_A receptors, was not inhibited by salbutamol [105]. ET-1 potently (pD_2 = 9.8) but modestly increased rabbit cultured tracheal smooth muscle cell proliferation [185]. This mitogenic activity was proposed to be mediated *via* a pertussis-toxin-sensitive G-protein-linked mechanism and thromboxane release. In ovine airways smooth muscle cells in culture 24-hr exposure to ET-1 (1 nM – 1 μM) was reported to promote mitogenesis, as assessed by cell number; the maximum effect was about a two-fold increase obtained with 0.1 μM ET-1 [186]. In several systems ET-1 increases the expression of the proto-oncogenes c-*myc* and c-*fos* [187, 188] and in this study ET-1 (0.1 and 1 μM) transiently enhanced, by about 2.5-fold, the expression of c-*fos* mRNA, with the maximum effect occurring at 30 min [186]. It was reported that c-*fos* expression was detected in bronchial biopsies from 8 out of 12 asthmatic patients but not in any of the 10 non-asthmatic specimens [189]. Interestingly, the c-*fos* expression was localized to the airway epithelium rather than the smooth muscle.

ET has been shown to be mitogenic for Swiss 3T3 fibroblasts [190]. This may have relevance to the structural changes observed in asthma, which include increased numbers of fibroblasts which may be associated with increased thickness of the collagen layer in asthmatic airways [191]. ET-1 and also ET-3 induced chemotaxis and replication of fibroblasts obtained from rat pulmonary arteries [192]. The effects on replication were small (maximum of 30% above control), and occurred in much higher concentrations (>10 nM) than those which induced chemotaxis (1 pM–0.1 μM). Nevertheless, it was proposed that this may be relevant to the characteristic vascular remodeling in pulmonary hypertension [192]. In contrast, ET-1 (1–1000 ng/ml) was observed to be weakly mitogenic (measured as [^{3}H]-thymidine incorporation) in porcine and bovine pulmonary artery smooth muscle cells and not thought to contribute markedly to remodeling that is a feature of hypoxic pulmonary hypertension [193]. ET-1 and ET-3 produced concentration-dependent mitogenesis of human pulmonary artery smooth muscle cells *via* an ET_A-receptor-mediated mechanism (*i.e.*, sensitive to BQ-123) [194]. ET-1 (1 nM–3 nM) generally stimulated DNA synthesis and proliferation of pig pulmonary artery smooth muscle cells, although under some conditions there was a paradoxical inhibitory effect [195]. Preliminary evidence was presented for a mitogenic influence of ET-1 in canine lung organ cultures, with a maximum effect observed with 10 nM [196].

4.1.6. Mediator release: ET-1 stimulates thromboxane A_2 release from perfused guinea-pig lung [68] and there is evidence, with the use of cyclooxygenase inhibitors and thromboxane receptor antagonists, that ET-1-induced bronchoconstriction in guinea pigs *in vivo* is mediated in part *via* release of thromboxane [107, 197–199]. However, there is conflicting information on the ability of ET-1 to directly stimulate mediator release from inflammatory cells. For example, stimulation of ET_A receptors detected on murine bone marrow-derived mast cells was without effect on histamine release [25]. Furthermore, ET-1 did not stimulate the release of histamine or the peptidoleukotrienes from guinea-pig trachea [158] or human bronchus [153], although increased release of various prostanoids from bronchus was noted [153, 158]. In contrast, it was reported that ET-1 potently and effectively stimulated histamine release from guinea-pig pulmonary but not peritoneal mast cells, with an $EC_{50} = 0.05$ nM, which was similar to the affinity of these former cells for [^{125}I]-ET-1 (about 0.08 nM) [200]. In rats, intravenous administration of ET-1 increased the levels of 15-HETE in broncho-alveolar lavage (BAL) fluid and the generation of oxygen radicals in BAL cells [201]. In addition, ET-1 (10 nM) stimulated 15-lipoxygenase activity in lung homogenates. High concentrations of ET-1 (≥ 0.1 μM) increased prostanoid production in human cultured nasal mucosa [184]. ET-1 (1 nM–10 nM) stimulated thromboxane and PGD_2, but not histamine, release from cells obtained from BALs in canine airways [202].

4.1.7. Microvascular permeability: Microvascular permeability in rat perfused lungs was enhanced by ET-1 (10 nM) *via* a mechanism which required the presence of both leukocytes and plasma components other than complement [203]. Several other studies in guinea-pig and rat perfused lungs reported that ET-1 induced pulmonary oedema, although there is conflicting information regarding the contribution of prostanoids to the response [163, 164, 204–206]. In rabbit perfused lung, ET-1 produced a potent, concentration-dependent pulmonary vasoconstriction, *via* a mechanism which was proposed to involve extracellular Ca^{2+} influx and activation of protein kinase C [207]. It was hypothesized that in rat and guinea-pig lung ET-1 causes lung oedema largely through a hydrostatic mechanism rather than directly increasing vascular permeability [163, 204]. In blood-perfused rather than physiological salt solution-perfused lungs ET-1 produced little oedema [163, 208].

4.1.8. Inflammatory cell function: ET-1 (1 nM–1 µM) was reported to stimulate human blood monocyte chemotaxis [209], although another study failed to demonstrate ET-1-induced stimulation of human peripheral blood monocyte chemotaxis, adhesion or superoxide production [210]. ET-1 increased arachidonic acid and thromboxane release from guinea-pig alveolar macrophages, with a maximum effect at 1 nM [211], and at a concentration of 1 µM increased superoxide production, intracellular Ca^{2+} levels and protein phosphorylation in human alveolar macrophages [212]. Other reports indicate ET-1-induced release of inflammatory cells products [200–202]. ET-1 potently stimulated, in a time- and concentration-dependent fashion, human monocytes and monocyte-derived macrophages to release TNF-α, IL-1β and IL-6; the effects were less than that observed with LPS [213].

4.1.9. Modulation of neurotransmission: There are few reports on the influence of the ETs on neurotransmission in the peripheral nervous system, although both stimulatory and inhibitory effects have been noted. For example, ET-1 had an inhibitory effect on adrenergic and cholinergic neuroeffector transmission in various tissues including guinea-pig pulmonary artery [214–217]. ET-1 inhibits nicotinic transmission in feline colonic parasympathetic ganglia [218]. This phenomenon appears to be mediated prejunctionally *via* inhibition of neurotransmitter release. In addition, ET-1 stimulates neurotransmission in some of the above tissues by postjunctional mechanisms [214–217].

ET-3 (10 or 100 nM) increased responses elicited by parasympathetic stimulation in rabbit bronchus but not those produced by exogenous administration of acetylcholine [219]. It was proposed that this increase in response (which was 2- and 3-fold with 10 nM and 100 nM ET-3, respectively) was the result of a prejunctional mechanism. In a recent study, Takimoto and co-workers [220] demonstrated that ET receptors, predominantly ET_B, were localized to cell bodies, processes and varicosities of adrenergic and cholinergic intramural autonomic neurons that were observed in primary cultures of guinea-pig tracheal smooth muscle [220]. Stimulation of these "neuronal" receptors elicited a tetrodotoxin (TTX)-sensitive elevation in $[Ca^{2+}]_i$ with subsequent contraction of adjacent smooth muscle cells. This contrasts with the previous reports of a lack of effect of TTX on ET-1-induced contractions in isolated guinea-pig trachea [149, 158]. Unlike cholinergic neurons, it was proposed that adrenergic neurons may have appreciable quantities of ET_A receptors [220]. As indicated previously (see Section 3.4), binding sites for [^{125}I]-ET-1 have been associated with parasympathetic ganglia, paravascular nerves and nerves in the connective tissues in mammalian lung [129, 130, 132, 134].

Research on the influence of the ETs on neurotransmission in the lung has been very limited. It is likely that work in this neglected area will unravel interesting effects of the ETs which, in the author's opinion, may be more relevant than the more widely and more easily studied "direct" effects such as smooth muscle contraction.

4.2. In Vivo Studies

4.2.1. Actions on bronchoconstrictor tone: There are no reports of direct ET administration to human airways *in vivo* although several studies have explored the pulmonary effects of the ETs following challenge to animals. However, in humans, intravenous (i.v.) infusion of ET-1 (4 pmol/kg/min) for 20-min increased pulmonary vascular resistance by 67% [71]. In contrast, it was observed previously that 60-min infusion of ET-1 at a lower rate (0.4 pmol/kg/min) produced no significant alterations in the human pulmonary circulation [70].

Intravenous or aerosol administration of ET-1 to guinea-pigs produces maintained bronchoconstriction which appears to be mediated to a significant extent by an indirect mechanism involving the release of secondary mediators, predominantly thromboxane but also PAF [197–199]. The extent to which secondary mediators contribute to ET-1-induced bronchoconstriction is dependent upon the route of administration [197–199, 205].

The bronchospasm induced by i.v. administration of ET-1 in guinea-pigs was not associated with a change in the number of circulating PMNs or platelets, suggesting that the response was independent of these cells [222]. ET-1-induced bronchospasm was not affected by Ca^{2+}-channel inhibitors, but was potentiated by hexamethonium or propranolol, suggesting a modulatory influence of the autonomic nervous system [198, 199]. It was noted that there was an increase in the responsiveness of aerosol-sensitized and antigen-exposed guinea-pigs to the bronchoconstrictor effects of ET-1, perhaps due to a change in the proteolytic activity in the airway epithelium which normally metabolizes the peptide [59, 223]. ET-1 administration elicits brochospasm in the dog [224], rat [225] and sheep; ET-3 also produced bronchoconstriction in sheep but it is about 400-fold less potent than ET-1 [226].

There is conflicting information on the ability of ET-1 to elicit airways hyperresponsiveness, a hallmark of asthma. For example, aerosolized or i.v. ET-1 failed to induce airways hyperreactivity in guinea-pigs [199, 227, 228]. However, a low concentration of aerosolized ET-1 (10^{-12} M), which was without direct effect on pulmonary function, enhanced the brochospasm produced by aerosol challenge with histamine [229]. Intravenous infusion of ET-1 (2 nmol/day) for 6 days did not change the responsiveness of guinea-pigs to the bronchoconstrictor effects of acetylcholine [230].

4.2.2. Actions on vasomotor tone: In the pulmonary and other regions of the circulation, the ETs can elicit either vasodilatation or vasoconstriction. In fact, following intravenous bolus administration, ET produces a transient systemic vasodilatation preceding a sustained elevation in pressure [7, 8, 74, 164]. The profile of the haemodynamic response is likely to be dependent upon the level of tone in the vessel and its location, in addition to the relative distribution of the ET receptor subtypes. In relation to the latter issue, ET_A receptors predominate in the vascular smooth muscle and produce vasoconstriction, whereas ET_B receptors are distributed on endothelial cells and smooth muscle cells, activation of which may elicit vasodilatation (perhaps via the release of NO/EDRF) or vasoconstriction [14, 73, 173; see Section 3.3.].

In newborn lambs ET-1 elicits pulmonary vasodilatation under conditions when tone is elevated, or during pulmonary hypertension induced by alveolar hypoxia. The vasodilator response was decreased by N-nitro-L-arginine (inhibitor of NO synthase) or glibenclamide, but not by meclofenamic acid, suggesting a role for NO and ATP-sensitive K^+-channels, but not prostaglandins [231]. In cats, i.v. ET-1, ET-2 and ET-3 elicited both pulmonary and systemic vasodilatation; the former but not the latter was decreased by the ATP-sensitive K^+-channel antagonist, glibenclamide [232, 233]. In the rat pulmonary circulation, ET-1 produced both vasoconstriction and vasodilatation; ET-1 was less potent at eliciting vasoconstriction in the pulmonary than in the systemic circulation. The pulmonary vasodilatation may involve ATP-sensitive K^+-channels and appears to be modulated by the release of EDRF [164, 165]. For example, ET-1-induced vasoconstriction in rat isolated lungs was potentiated by L-NMMA, an inhibitor of NO synthase, and methylene blue, an inhibitor of guanylyl cyclase [164,

167]. Intravenous ET-1 was reported to be a potent constrictor of canine airway circulation [234]. BQ-123 (1 μM) markedly inhibited ET-1-induced vasoconstriction in rabbit perfused lungs but did not affect ET-1 induced vasodilatation or hydrostatic oedema, or ET-3-induced vasoconstriction or hydrostatic oedema suggesting that these latter phenomena are mediated *via* non-ET_A receptor-linked mechanisms [235].

Exposure of rats to a hypoxia atmosphere (10% O_2) for 3 weeks abolished, in a reversible manner, the pulmonary vasodilatation in perfused lungs that was normally elicited by ET-1 or ET-3 in tissues from control animals (exposed to room air) [168]. In control tissues, the vasodilatation appears to be due to activation of ATP-sensitive K^+-channels and is not blocked by BQ-123. The mechanism of this effect of chronic hypoxia is unclear although no evidence for changes in ET_A and ET_B receptor binding characteristics from pulmonary vessels or in alterations in the responsiveness of the K^+ channels was obtained [168]. Evidence was presented that in neonatal pigs ET-1 induced pulmonary vasodilatation was mediated *via* a non-ET_A, probably ET_B, receptor subtype and involved the release of NO: there was no effect of indomethacin or glibenclamide [236].

4.2.3. Other in vivo *effects:* Intravenous ET-1 increased microvascular permeability in rats in various tissues, including bronchi, by a mechanism which was sensitive to BQ-123 [237]. However, intravenous administration of ET-1 to guinea-pigs did not increase lung permeability, produce epithelial cell damage or elicit inflammatory cell influx into the alveolar or vascular walls or the bronchial epithelium [199]. Similarly, aerosolized ET-1 did not change the levels of eosinophils in guinea pig lung samples, taken 4 or 24 hours after challenge [228], and infusion of ET-1 (2 nmol/day) *via* the jugular vein to guinea pigs for 6 days did not induce histological changes in the lung or cause infiltration of inflammatory cells [230].

5. Signal Transduction Mechanisms

In the original publication on ET-1 it was hypothesized that it mediated contractions in porcine coronary artery by acting as an endogenous stimulator of voltage-dependent, Ca^{2+} channels [7]. However, subsequent research indicated that in most other vascular smooth tissues and in other systems, ET-induced responses appear to be produced by activation of the phosphatidylinositol pathway resulting in the release of intracellular Ca^{2+}, rather than stimulation of extracellular Ca^{2+} influx [14, 74]. Thus, there are various reports that ET-1 stimulates the formation of inositol phosphates and diacylglyerol, the two second messengers formed from the metabolism of phosphatidylinositol (4, 5) bisphosphate by phospholipase C [238–243]. In addition, ET-1 has been shown directly to produce increases in intracellular levels of Ca^{2+} [14, 74, 243–246], and purported protein kinase C inhibitors attenuate ET-1-induced responses [14, 74, 243, 247, 248].

Consideration of signal transduction mechanisms for the ETs in the lung will focus mainly on the bronchospastic effects of ET-1. However, some information in other cells and systems is given here and elsewhere in the text (*vide supra*). In addition, as indicated previously several of the actions of the ETs are elicited, at least in part, indirectly *via* the release of secondary mediators and these mechanisms and signal transduction pathways are discussed in the appropriate sections.

In the first report on the effects of ET-1 in the pulmonary system, it was indicated that contraction induced by ET-1 in guinea-pig trachea was attenuated by the dihydropyridine, nicardipine, a voltage-dependent Ca^{2+} channel inhibitor [148]; and it was concluded that this supported the original hypothesis that ET-1 is an endogenous Ca^{2+} agonist. However, more extensive studies indicated that ET-1-induced contraction in this tissue is resistant to incubation in Ca^{2+}-free physiological buffer or Ca^{2+} channel inhibitors; this was particularly true for higher concentrations of ET-1 [149, 151, 157, 249]. ET-1 stimulated phosphatidylinositol turnover in guinea-pig trachea ($EC_{50} = 46$ nM) with a comparable potency to that which elicits the contractile response ($EC_{50} = 31$ nM) [151]. Overall, the data indicate that in guinea-pig trachea ET-1-induced contraction is mediated predominantly *via* stimulation of phosphatidylinositol turnover and intracellular Ca^{2+} release, with only a minor contribution of extracellular Ca^{2+} influx through voltage-dependent membrane channels. Preliminary evidence suggests that a similar profile occurs in guinea-pig pulmonary arteries and veins [249]. *In vivo* ET-1-induced bronchoconstriction in guinea pigs is not significantly affected by Ca^{2+} channel inhibitors [198, 199].

ET-1 increased phosphatidylinositol turnover in rat trachea [144] and bovine bronchus [155]. Interestingly, ET-3 did not stimulate the phosphatidylinositol pathway in bovine bronchus, and unlike ET-1, ET-3-induced contractions were markedly attenuated by nifedipine [155]. This would suggest that different mechanisms may mediate responses produced by ET-1 and ET-3 in this tissue. Following a prior comprehensive study, involving autoradiographic, biochemical and functional analysis, it was concluded that both ET_A and ET_B receptors contribute to the ET-1-induced contraction in rat tracheal smooth muscle, but *via* different second messenger pathways: stimulation of phosphatidylinositol turnover and activation of extracellular Ca^{2+} influx (*via* non-L-type membrane channels), respectively [144]. Similarly, in rat lung it was proposed that ET_A receptor activation was linked to stimulation of phosphatidylinositol metabolism and arachidonic acid release, whereas ET_B receptors were only associated with the former [142].

Pretreatment of smooth muscle with purported protein kinase C inhibitors was without effect on contractile responses to ET-1 in guinea-pig trachea [151] or bovine bronchus [155], but attenuated contraction in rabbit trachea [171], suggesting that species differences may exist in the contribution of protein kinase C to ET-1-induced responses. In bovine cultured embryonic tracheal cells ET-1 stimulated cAMP formation in addition to the phosphatidylinositol pathway; the significance of this observation is not known [250]. It has been proposed that there may be a role for Na^+/H^+ exchange in ET-1-induced contraction in guinea-pig airways [251].

Early studies provided evidence that ET-1-induced contraction in human airways is not associated with extracellular Ca^{2+} influx *via* voltage-sensitive channels [131, 150, 252]. However, a more recent preliminary study suggests that a component of the response to ET-1 in human bronchus was mediated by this mechanism [253]. In addition, it was hypothesized that, as proposed for rat trachea [142, 144], different signal transduction pathways mediate ET_A and ET_B receptor activation in human bronchus [253]. In human cultured bronchial smooth muscle a biphasic rise in intracellular Ca^{2+} following addition of ET-1 was observed [254]. It was proposed that ET-1 produces intracellular Ca^{2+} mobilization by stimulation of the phosphatidylinositol pathway *via* activation of a receptor (the molecular weight of the receptor was estimated to be about 70 000 from affinity cross-linking experiments) which is coupled to a pertussis toxin-insensitive G protein. In addition, there is influx of extracellular Ca^{2+} *via* a dihydropyridine-insensitive membrane channel which may contribute to a maintained rise in tone in human bronchial smooth muscle [254].

Thus, as in several other systems (*vide supra*), the evidence in isolated airways is at odds with the initial postulate that ET-1 acts as an endogenous activator of dihydropyridine-sensitive, voltage-dependent Ca^{2+} channels [7]. ET-induced contraction of human and animal isolated airways appears to be predominantly *via* stimulation of phosphatidylinositol turnover and mobilization of intracellular Ca^{2+}, although the relative contribution of these potential signal transduction pathways may depend on the ET receptor subtype and also the species.

In rabbit pulmonary vein, ET-1-induced contractions were unaffected by voltage-dependent Ca^{2+} channel inhibitors or the phospholipase A_2 inhibitor, quinacrine, but markedly attenuated by incubation in Ca^{2+}-free medium or by the purported phospholipase C inhibitor, neomycin [255]. Post-treatment, but not pretreatment, with H-7 a protein kinase C inhibitor, significantly inhibited ET-1-induced contraction in rabbit pulmonary vein. It was concluded that ET-1-induced contraction involves activation of phospholipase C, and protein kinase C, but not phospholipase A_2, and is mediated by release of intracellular Ca^{2+} and influx of extracellular Ca^{2+} *via* non-voltage dependent channels [255]. In rabbit perfused lung, verapamil, cadmium, Ca^{2+}-free buffer or protein kinase C inhibitors significantly attenuated ET-1-induced vaso-constriction, and vasodilators which elevated cAMP were more effective in reversing pulmonary vasoconstriction induced by ET-1 than those agents which increase cGMP [207].

6. Potential Pathophysiological Role

6.1. Background

The preceding sections have highlighted the increasing information from preclinical studies which indicate an array of effects of the ETs in the pulmonary system. As with other proposed important mediators, the critical question to be addressed is what is the significance of these diverse and, in some cases, conflicting laboratory observations in relation to the pathophysiology of lung diseases? It should be reiterated that some of the effects observed required high concentrations of the ETs. In essence, the ETs can only be implicated confidently as significant players in the pathophysiology of pulmonary disorders once the following standard criteria have been fulfilled: 1) pathways for the synthesis, release and metabolism of the ETs must be present in the airways; 2) the ETs must mimic several, if not all, of the features of the disease(s); 3) the levels of the ETs must be elevated in disease states, with a correlation between the amounts of ETs and the severity of the disease; 4) drugs which inhibit the release and/or antagonize the biological effects of the ETs must ameliorate the symptoms of the disease(s) of interest. For asthma and pulmonary hypertension many of the aspects of the criteria outlined in 1) and, to a lesser extent, 2) and 3) have been fulfilled, although some of the information is preliminary and requires confirmation. However, the most important and most difficult criteria, the clinical testing of selective ET receptor antagonists or ECE inhibitors in pulmonary disorders will probably not be completed for several years. Notwithstanding the above unknowns and caveats, the evidence in support of a potential pathophys-iological role for the ETs is summarized below. The emphasis is on asthma and pulmonary hypertension, the two diseases which have been the major focus of research.

6.2. Asthma

The first publication on the effects of ET-1 in the pulmonary system reported that it potently contracted isolated guinea-pig trachea and it

Table 2. Potential association of endothelin with some pathological features of asthma, and the ET receptor subtype involved

Asthma	Endothelin	Receptor subtype
Increased bronchial tone	+++	ET_B
Mucus hypersecretion	+/−	nd
Airway hyperresponsiveness	+/−	nd
Airway smooth muscle hyperplasia	+	ET_A
Epithelial cell damage	−	nd
Bronchial oedema	+/−	ET_A
Inflammatory cell influx	−	nd
Inflammatory cell activation	+/−	nd
Associated with increased ir-ET in the bronchial epithelium and in BAL fluid	+	na

+ = stimulatory effect; − = no effect or inhibitory influence; BAL = bronchoalveolar lavage; nd = not determined; na = not applicable. This table was adapted from reference 152.

was hypothesized that ET-1 may be involved in the pathogenesis in asthma [148, 152]. The potent bronchoconstrictor activity of ET-1 was confirmed subsequently in human bronchus, supporting its potential candidacy as a key mediator in asthma [104, 131–134, 150, 153]. However, it is perhaps rather naive and premature to propose ET, or any other potential mediator, as an important player in the overall aetiology of asthma, based solely on its ability to elicit bronchoconstriction of isolated airways. Thus, asthma is recognized as a chronic inflammatory disorder associated with several cardinal pathologies in the pulmonary system in addition to bronchospasm. These features include airways hyperreactivity, mucus hypersecretion, mucus gland and airways smooth muscle cell hyperplasia, subepithelial fibrosis, inflammatory cell infiltration and activation, increased bronchial microvascular permeability and oedema, and epithelial cell damage and desquamation [256]. As indicated in preceding sections, and summarized in Table 2, there is some, albeit limited, data from preclinical studies on the effects of ETs on several of these characteristic features of asthma other than bronchoconstriction.

Probably the most interesting and relevant observations on a potential role of the ETs in asthma come from preliminary studies in asthmatics which suggested that there is increased synthesis and release of ET in their airways compared with non-asthmatic individuals. For example, in the first study in humans, Nomura and co-workers reported that in the one asthmatic patient examined there was a 6-fold elevation in the levels of ir-ET in BAL fluid, from 0.05 to 0.3 pg/ml, during what was called the "bronchospastic phase" of a status asthmaticus attack [257]. Unfortunately, there was no information on the levels of other mediators, and, perhaps not surprisingly, there have been no further reports on the ET levels in airways during asthmatic attacks. In a subsequent study by Mattoli and co-workers it was reported that the levels of ET in BALs of six asthmatics (0.25 ± 0.05 pg/ml) was about 4-fold higher than in five control subjects (0.06 ± 0.02 pg/ml) or five chronic bronchitics (0.08 ± 0.03 pg/ml) [258]. In contrast, there was no difference between the three groups in the ET amounts in the peripheral venous blood samples. It was hypothesized that the enhanced BAL levles of ET resrepsent increased local production in the bronchial mucosa, and were not a consequence of any alterations in microvascular permeability. Furthermore, ET levels decreased to control values, concomitant with an improvement in lung function, after 15-day treatment with inhaled β-agonists and oral corticosteroids. Another study by the same research group indicated that bronchial epithelial cells obtained by bronchoscopy from six patients who had symptomatic asthma expressed preproendothelin-1 mRNA and released significant quantities of ET-1 into the supernatant (the range was $12-40$ pg/ml/10^6 cells over a 48 hr period) [259]. Epithelial cells from five control, non-asthmatic individuals and five chronic bronchitic patients contained little or no preproET-1 mRNA and did not release ET-1 in appreciable amounts.

The most extensive study on ET-1 in asthmatic subjects was conducted by Springall and co-workers who performed a comparative immunohistochemical analysis between ET-1 expression in endobronchial biopsies from 17 asthmatic patients, covering a broad clinical spectrum, and 11 atopic and non-atopic healthy controls. The most striking observation was that there was markedly increased incidence of the expression of ET in airway epithelium, and also the vascular endothelium, of asthmatics (detected in 11 out of 17) compared with non-asthmatic controls (detected in 1 out of 11) [260]. However, in the asthmatic individuals, there was no correlation between positive staining for ET-1 and parameters such as degree of airflow obstruction, level of bronchial responsiveness, atopy or corticosteroid therapy. It was recently reported that there was an elevation in BAL levels of ET in 5 patients with asthma (2.45 pg/ml) compared with 5 normal subjects or 5 patients with chronic extrapulmonary disease [261]; in 9 out of 10 non-asthmatic individuals the ET levels were below the limit of detection of the RIA (0.8 pg/ml).

Interestingly, in a study measuring ET amounts in patients with nocturnal worsening of asthma, there was a significant decrease rather than an increase in BAL ET-1 levels in the night-time asthmatic group (39.3 pg/mg BAL fluid protein) compared to the day-time asthmatic group (56.8 pg/mg) or night-time control group (77.4 pg/mg); there was no difference in serum ET levels between non-asthmatic and asthmatics and also between day-time and night-time asthmatic groups [262]. There was a significant inverse correlation between the fall in FEV_1, overnight and BAL ET-1 levels. To explain this circadian variation in BAL ET-1 levels, it was hypothesized that ET-1 may act locally on airways smooth muscle and that it may be more tightly bound to the airways smooth muscle and epithelial ET receptors which may be upregulated during overnight exacerbations of asthma. It was concluded that the data suggest that ET-1 may be linked to nocturnal worsening of asthma *via* an effect on smooth muscle tone and/or potentiation of airways inflammation.

These intriguing but preliminary data need to be confirmed and expanded so that the critical question of "cause or effect?" can be addressed satisfactorily. Thus, considerable work, in particular using human isolated tissues, and in the clinical setting with ET receptor antagonists and/or ECE inhibitors remains before a definitive link can be established between the ETs and the pathophysiology of asthma.

6.3. Pulmonary Hypertension

Pulmonary hypertension is a progressively deteriorating condition characterized by an increase in vascular tone and vasoreactivity and enhanced proliferation of smooth muscle cells, resulting in a significant elevation in pulmonary vascular resistance, which leads to right-heart failure and death. There may be increased recruitment of myofibroblasts into the intima and "muscularization" or "remodeling" of pulmonary arteries characterized by hyperplasia and hypertrophy of smooth muscle cells [263–265]. The mechanisms underlying pulmonary hypertension have not been fully clarified.

Interestingly, pulmonary hypertension was the first pulmonary disorder in which evidence in patients was provided to implicate ET in its pathogenesis. Thus, soon after the first publication on ET-1 it was reported that plasma levels of ET were elevated (about 6-fold) in four patients with pulmonary hypertension; controls: 0.26 ± 0.24 pg/ml, n = 14; pulmonary hypertension: 1.52 ± 0.45 pg/ml; $P < 0.001$ [266]. There have been several subsequent reports in children and adults confirming the enhanced plasma ET levels in this disease with, in many cases, a correlation observed between the amounts of ET and the severity of the disease [267–271]. Elevated ir-ET-1 levels were detected in arterial blood samples of newborn infants with persistent pulmonary hypertension (PPHN) (27.6 ± 3.6 pg/ml; n = 24) compared with control infants (15.1 ± 4.1 pg/ml in umbilical cord blood, n = 10) [272]. There was a correlation between the amounts of ET-1 and disease severity and the levels declined with disease resolution in patients who did not need extracorporeal membrane oxygenation. PPHN is a clinical syndrome linked to an array of neonatal cardiopulmonary disorders.

In addition to enhanced plasma levels of ET another study reported that the tissue amounts of ET-1 (as assessed by analyzing for immunoreactivity and mRNA) were markedly increased in patients with various causes of pulmonary hypertension compared with controls

[273]. The ET-1-like immunoreactivity was predominant in endothelial cells of pulmonary arteries with medial thickening and fibrosis, and the increased ET-1 mRNA was detected mainly at sites of ET-1-like immunoreactivity. It was found that a correlation existed between the extent of ir-ET-1 and pulmonary vascular resistance in patients with plexogenic pulmonary arteriopathy, but not in individuals with secondary pulmonary hypertension.

The ETs are potent constrictors of isolated pulmonary blood vessels from various species including humans [131, 132, 134] and *in vivo* studies indicate potent vasoconstrictor properties in the pulmonary vasculature [232, 274]. Further support for a potential significant role of the ETs in pulmonary hypertension may lie in their recognized potent mitogenic properties in human pulmonary artery smooth muscle cells, *via* an ET_A-mediated mechanism [194]. This observation may be relevant to the characteristic vascular remodeling and smooth muscle cell proliferation associated with many forms of pulmonary hypertension [263–265].

The mechanism and stimulus for the increase in release of ET in pulmonary hypertension is not known but may be a consequence of abnormal haemodynamic forces such as high pressure or enhanced blood flow which may stimulate ET release from endothelial cells. For example, shear stress was demonstrated to increase ET production from cultured endothelial cells [275]. In contrast, ET has been shown to exert vasodilator properties in the pulmonary system [274], and it is possible that ET exerts a beneficial spasmolytic effect in pulmonary hypertension.

Thus, in addition to several reports of an elevation in ET levels in pulmonary hypertension, the ETs are able to produce two key features of pulmonary hypertension, namely vasoconstriction and enhanced proliferation of vascular smooth muscle cells. However, it remains to be determined clinically whether ET contributes directly to the pathogenesis of the disorder or whether the elevated levels of ET are merely markers of the disease (for example, an indicator of endothelial cell damage or dysfunction).

Data from animal models and isolated tissues provide support for a role for ET-1 in the genesis and maintenance of pulmonary hypertension. For example, an increase in ET-1 mRNA expression was detected in rats with idiopathic pulmonary hypertension [276]. In addition, ET-1 and ET-3 induced chemotaxis and replication of fibroblasts obtained from rat pulmonary arteries, suggesting a role in vascular remodeling, although, high concentrations of the ETs were needed to produce modest mitogenic effects [192]. Furthermore, ET-1 (1 nM–3 nM) stimulated DNA synthesis and proliferation of pig pulmonary artery smooth muscle cells [195]. However, in view of the weakly mitogenic effect of ET-1 in canine and porcine pulmonary artery smooth muscle cells and lack of effect of hypoxia on ET secretion from porcine endothelial cells, Hassoun and co-workers concluded that ET-1 does not contribute significantly to the remodeling seen in hypoxic pulmonary hypertension [193].

In monocrotaline-induced pulmonary hypertension in rats, the plasma ET-1 level increased progressively, and preceded the development of pulmonary hypertension [277]. The expression of preproET-1 mRNA was elevated in the heart and gradually decreased in the lungs. Infusion of BQ-123, the ET_A receptor antagonist, reduced the progression of pulmonary hypertension and right ventricular hypertrophy and prevented the characteristic pulmonary arterial medial thickening. These data provided relatively convincing evidence that ET-1 plays a role in the cardiopulmonary changes that occur in this rat model of pulmonary hypertension. In contrast, evidence was provided, using BQ-123, that ET-1 does not mediate acute hypoxic pulmonary vasoconstriction in newborn lambs, although ET-1 elicited a rapid increase in pulmonary artery pressure which was abolished by BQ-123 [278]. Hypoxia decreased ET-1 release, measured over 48 hr, from bovine pulmonary microvascular endothelial and pulmonary artery cells. This reduction, which was associated with a decrease in ET-1 mRNA expression, was reversed by re-oxygenation [279]. It was concluded that these data are not in support of a role for enhanced release of ET-1 in regulating the proliferation of pulmonary vascular smooth muscle cells and fibroblasts in the pulmonary circulation, a characteristic of hypoxia.

6.4. Other Pulmonary Disorders

Using immunohistochemical and in situ hybridization techniques ET-like immunoreactivity and mRNA was detected in the majority of surgical specimens of various lung tumours, in particular squamous cell carcinoma and adenocarcinoma. Based on its synthesis and storage in lung tumours it was hypothesized that ET may be involved in their growth and/or differentiation [280]. In humans, expression was highest in the developing lung and lowest in adult lung. Accordingly, it was speculated that ET may be important in promoting or regulating cellular growth in lung development [23].

The plasma ET-1 levels in 13 patients with acute respiratory failure (10.7 ± 5.0 pg/ml) were about 7-fold higher than those in control individuals (1.5 ± 0.5 pg/ml, n = 16). In addition, there was a significant correlation between ET-1 levels and several haemodynamic and respiratory parameters, including mean pulmonary arterial pressure and airway resistance [281].

In lung tissue from patients with cryptogenic fibrosing alveolitis (CFA), a fatal condition of unknown cause which is characterized by inflammation, type II pneumocyte and fibroblastic proliferation, and collagen deposition, there was markedly increased expression of ET-1, most notably in airway epithelium and type II pneumocytes, compared with control tissue or tissues from patients with focal fibrosis [282]. It was hypothesized that ET-1 may be involved in the pathogenesis of CFA and associated pulmonary hypertension, in particular, in the characteristic ultrastructural changes. Furthermore, there was a correlation between the expression of ET-1 and the histological parameters of the disease and it was suggested that ET-1 may be an appropriate marker for disease activity. Another study by the same group detected expression of ET-1 in alveolar epithelial cells of patients with pulmonary fibrosis, but rarely in patients with pulmonary hypertension without fibrosis [273]. Marciniak and co-workers reported that in two patients with cystic fibrosis and one with CFA, ir-proET-1 and ir-proET-3, but not ir-proET-2, was detected in airway epithelium, whereas immunoreactivity for the three isoforms was localized in submucosal glands; unfortunately, no data from analysis of control samples was provided [283].

There was an elevation in the BAL ET levels in 10 patients with idiopathic lung fibrosis (mean = 12.4 pg/ml) and 9 patients with miscellaneous interstitial lung disease (mean = 2.9 pg/ml) compared with 5 normal individuals (< 0.8 pg/ml, limit of detection of RIA) and 5 patients with chronic extrapulmonary disease (1.15 pg/ml) [261].

6.5. Animal Studies

There is some further information from studies in animals – in addition to the material outlined in preceding sections – for a role of ET in pulmonary disorders. In a rat model of lung inflammation induced by intratracheal instillation of Sephadex beads, there was a 3.5 fold increase in lung ET-1 content which was abolished by the glucocorticoid, budesonide [284]. An increase in ET levels in sheep plasma and pulmonary lymph, concomitant with pulmonary vasoconstriction, was detected during endotoxin shock. The increase in circulating ET was thought to be due to endotoxin-induced endothelial cell injury [285]. Furthermore, based on the enhanced plasma levels of ET noted following intravenous administration of ovalbumin in actively and passively sensitized guinea-pigs, it was hypothesized that the ETs may be involved in the pulmonary insufficiency and peripheral circulatory collapse which develops during anaphylactic shock [286]. In a canine model of post-obstructive pulmonary vasculopathy, which is produced by chronic ligation of one pulmonary artery resulting in bronchial collateral vessel proliferation and pulmonary arterial abnormalities, ir-ET-1 was more intense in the endothelium of pulmonary arteries and new bronchial vessels of ligated versus non-ligated lungs. In control lungs, ir-ET-1 was distributed largely in the airway epithelium. Plasma levels of ET-1 measured distally to ligation, in left pulmonary artery, were similar in control and ligated tissues [287]. In a rat model of oleic acid-induced respiratory distress syndrome elevated plasma and BAL levels of the ETs, predominantly, ET-1, were noted which preceded the maximum hypoxia [288]. In rats, acute pulmonary alveolar hypoxia increases plasma and lung levels of ir-ET-1 with a correlation between the severity of hypoxia and the peptide amounts [289].

During acute rejection of lung allografts in dogs ET-1 levels are elevated in bronchial tissue homogenates (4-fold increase) [290] and BALs (2-fold increase) [291]. There was no alteration in the amounts of ANP or histamine in BALs [291]. The elevated levels in BALs in rejected lung allografts were reduced to control levels by immunosuppresive therapy. The cellular source of ET-1 was not identified although it was speculated to be either the vascular endothelium, bronchial epithelium and/or inflammatory cells, such as alveolar macrophages which are increased during rejection [291] and which have been shown to synthesize and release ETs [24]. It was noted that there was no change in plasma ET-1 levels from samples taken during rejection compared to pre-rejection and following immunosuppressive treatment [291]. This highlights the caution one should exercise in trying to relate plasma ET levels to local tissue concentrations.

In an intriguing recent article, it was reported that mice which were deficient in ET-1 ("knockout mice") – generated by disruption in mouse embryonic stem cells by homologous recombination – died of respiratory failure at birth; they also had morphological abnormalities of craniofacial tissues but not in other organs, including lung and trachea [292]. This provides further evidence for a role of ET-1 in normal development [23]. It was proposed that the anoxia-induced death was, in part, due to a dysfunction in the central control of respiration or in the respiratory muscle, and it was suggested that ET-1 may play a key role in neural regulation of the respiratory system [292]. This introduces the intriguing possibility that the ETs may have some beneficial influences in the lung.

7. Conclusions

It is recognized that our understanding of the effects and potential pathophysiological role of the ETs in the pulmonary system is significantly less compared to the cardiovascular system. Nevertheless, since the first publication describing the potent contractile effects in the guinea-pig isolated trachea six years ago [48], considerable progress has been made and a significant amount of information outlining multiple effects of the ETs, in particular ET-1, in different cells in the respiratory tract has been accumulated. Furthermore, although it is too early to realistically propose a role for the ETs in the pathophysiology of pulmonary disorders, the preliminary data, especially in relation to asthma and pulmonary hypertension, are intriguing and worthy of further investigation.

To data, research has focused, to a large extent, on the contractile effects of the ETs. This activity of the ETs is the most facile to study, but for several pulmonary disorders it is unlikely to be the most relevant. Thus, for many lung diseases, including asthma, a chronic inflammatory disorder, it is important that future studies should be directed towards examination of the effects of ET on parameters other than bronchoconstriction e.g., influence on nerves, inflammatory cell function, also the effects of chronic exposure on smooth muscle and fibroblast proliferation and other structural components of the lung. Another critical area of research will be the elucidation and classification of the ET receptor subtypes mediating these effects; the availability of potent and selective receptor antagonists for the various ET receptors will assist greatly in this endeavour. Furthermore, it will be important to assess whether the ETs possess any normal physiological influences (e.g., the proposed role in the neural regulation of the lung [292]), and which receptor subtype(s) mediates these beneficial effects. It is worth reiterating that, in large part, because of the abundance of ET receptors, the lung has been extensively used in biochemical and molecular biological studies, investigating ET receptor subtypes.

As yet it remains to be clarified whether the most appropriate therapeutic agent(s) for the various pulmonary diseases is an ET_A-, ET_B- or perhaps an ET_C-selective antagonist. The required profile will likely depend on the specific disorder. In addition, the probability is that future research will uncover additional ET receptor subtypes. Furthermore, will there be an advantage of an ECE inhibitor over an ET receptor subtype-selective antagonist? In the authors opinion there is the possibility that the former strategy, because of its inherent lack of selectivity compared with subtype-selective receptor antagonists, could have liabilities, as it may attenuate any proposed beneficial effects of the ETs (which may be associated with a specific receptor subtype) as well as their deleterious actions. Preclinical research will be critical to address these and other pivotal questions and to assist the clinicians in the selection of the most appropriate novel therapeutic strategy for individual pulmonary disorders in which the ETs have been implicated.

Thus far, the most convincing evidence and scientific rationale for a pathophysiological role for ET is in pulmonary vascular diseases such as pulmonary hypertension. The unequivocal determination of the pathophysiological role of the ETs in pulmonary disorders awaits the clinical evaluation of potent and selective receptor antagonists for the various ET receptor subtypes and/or ECE inhibitors. It is anticipated that these studies will be conducted in the not-too-distant future, with potent and selective, non-peptide receptors antagonists that are now being identified.

Acknowledgements

The author would like to thank Dotti Lavan and Shirley Wilson for typing this material.

References

1. Furchgott RF, Vanhoutte PM. Endothelium-derived relaxing and contracting factors. FASEB J. 1989; 3: 2007–18.
2. Lüscher TF. Endothelium-derived relaxing and contracting factors: potential role in coronary artery disease. Eur. Heart J. 1989; 10: 847–57.
3. Moncada S, Palmer RMJ, Higgs, EA. Nitric oxide: Physiology, Pathophysiology, and Pharmacology. Pharmacol. Rev. 1991; 43: 109–42.
4. Hickey KA, Rubanyi G, Paul RJ, Highsmith RF. Characterization of a coronary vasoconstrictor produced by cultured endothelial cells. Am. J. Physiol. 1985; 248: C550–6.
5. Gillespie MN, Owasoyo JO, McMurtry IF, O'Brien RF. Sustained coronary vasoconstriction provoked by a peptidergic substance released from endothelial cells in culture. J. Pharmacol. Exp. Ther. 1986; 236: 339–43.
6. O'Brien RF, Robbins RJ, McMurtry IF. Endothelial cells in culture produce a vasoconstrictor substance. J. Cell Physiol. 1987; 132: 263–70.
7. Yanagisawa M, Kurihara H, Kimura S, Tomobe Y, Kobayashi M, Mitsui Y et al. A novel potent vasoconstrictor peptide produced by vascular endothelial cells. Nature 1988; 322: 411–5.
8. Inoue A, Yangisawa M, Kimura S, Kasuya Y, Miyauchi T, Goto K et al. The human endothelin family: three structurally and pharmacologically distinct isopeptides predicted by three separate genes. Proc. Natl. Acad. Sci. U.S.A. 1989; 86: 2863–7.
9. Bloch KD, Friedrich SP, Lee M-E, Eddy RL, Shows TB, Quetermous T. Structural organization and chromosomal assignment of the gene encoding endothelin. J. Biol. Chem. 1989; 264: 10851–7.
10. Bloch KD, Hong CC, Eddy RL, Shows TB, Quertermous T. cDNA cloning and chromosomal assignment of the endothelin 2 gene: Vasoactive intestinal contractor peptide is rat endothelin 2. Genomics 1991; 10: 236–42.
11. Bloch KD, Eddy RL, Shows TB, Quertermous T. cDNA cloning and chromosomal assignment of the gene encoding endothelin 3. J. Biol. Chem. 1989; 264: 18156–61.
12. Yanagisawa M, Masaki T. Endothelin, a novel endothelium-derived peptide. Pharmacological activities, regulation and possible roles in cardiovascular control. Biochem. Pharmacol. 1989; 38: 1877–83.
13. Yanagisawa M, Masaki T. Molecular biology and biochemistry of the endothelins. Trends Pharmacol. Sci. 1989; 10: 374–8.
14. Masaki T, Yanagisawa M, Goto K. Physiology and pharmacology of endothelins. Medicinal Res. Rev. 1992; 12: 391–421.
15. Kitamura K, Tanaka T, Kata J, Eto T, Tanaka K. Regional distribution of immunoreactive endothelin in porcine tissue: abundance in inner medulla of kidney. Biochem. Biophys. Res. Comm. 1989; 161: 348–52.
16. Matsumoto H, Suzuki N, Onda H, Funjo M. Abundance of endothelin-3 in rat intestine, pituitary gland and brain. Biochem. Biophys. Res. Comm. 1989; 164: 74–80.
17. Pernow J, Hemsén A, Lundberg JM. Tissue specific distribution, clearance and vascular effects of endothelin in the pig. Biochem. Biophys. Res. Comm. 1989; 161: 647–53.
18. Yoshimi H, Hirata Y, Fukuda Y, Kawano Y, Emori T, Kuramochi M et al. Regional distribution of immunoreactive endothelin in rats. Peptides 1989; 10: 805–8.

19. Nunez DJR, Brown MJ, Davenport AP, Neylon CB, Schofield JP, Wyse RK. Endothelin-1 mRNA is widely expressed in porcine and human tissues. J. Clin. Invest. 1989; 85: 1537–41.
20. Rozengurt N, Springall DR, Polak JM. Localization of endothelin-like immunoreactivity in airway epithelium of rats and mice. J. Pathol. 1990; 707: 5–8.
21. Rennick RE, Loesch A, Burnstock G. Endothelin, vasopressin, and substance P like immunoreactivity in cultured and intact epithelium from rabbit trachea. Thorax 1992; 47: 1044–9.
22. MacCumber MW, Ross CA, Glaser BM, Snyder SH. Endothelin: visualization of mRNAs by in situ hybridization provides evidence for local action. Proc. Natl. Acad. Sci. USA 1989; 86: 7285–9.
23. Giaid A, Polak JM, Gaitonde V, Hamid QA, Moscoso G, Legon S et al. Distribution of endothelin-like immunoreactivity and mRNA in the developing and adult human lung. Am. J. Respir. Cell Mol. Biol. 1991; 4: 50–8.
24. Ehrenreich H, Anderson RW, Fox CH, Rieckmann P, Hoffman GS, Travis WD et al. Endothelins, peptides with potent vasoactive properties, are produced by human macrophages. J. Exp. Med. 1990; 172: 1741–8.
25. Ehrenreich H, Burd PR, Rottem M, Hültner L, Hylton JB, Garfield M et al. Endothelins belong to the assortment of mast cell-derived and mast cell-bound cytokines. New Biologist 1992; 4: 147–56.
26. Itoh Y, Yanagisawa M, Ohkubo S, Kimura C, Kosaka T, Inoue A et al. Cloning and sequence analysis of cDNA encoding the precursor of a human endothelium-derived vasoconstrictor peptide, endothelin: identity of human and porcine endothelin. FEBS Letts. 1988; 231: 440–4.
27. Emori T, Hirata Y, Ohta K, Shichiri M, Shimokado K, Marumo F. Concomitant secretion of big endothelin and its C-terminal fragment from human and bovine endothelial cells. Biochem. Biophys. Res. Comm. 1989; 162: 217–23.
28. Sawamura T, Kimura S, Shinmi O, Sugita Y, Yanagisawa M, Masaki T. Analysis of endothelin related peptides in culture supernatant of porcine aortic endothelial cells: evidence for biosynthetic pathway of endothelin-1. Biochem. Biophys. Res. Comm. 1989; 162: 1287–94.
29. Miyauchi T, Yanagisawa M, Tomizawa T, Sugishita Y, Suzuki N, Funino M et al. Increased plasma concentration of endothelin-1 and big endothelin-1 in acute myocardial infarction. Lancet 1989; 2: 53–4.
30. Kashiwabara T, Inagaki Y, Ohta H, Iwamatsu A, Nomizu M, Morita A et al. Putative precursors of endothelin have less vasoconstrictor activity in vitro but a potent pressor effect in vivo. FEBS Letts. 1989; 247: 73–6.
31. Kimura S, Kaysuya Y, Sawamura T, Shinmi O, Sugita Y, Yanagisawa M et al. Conversion of big endothelin-1 to 21-residue endothelin-1 is essential for expression of full vasoconstrictor activity: structure-activity relationships of big endothelin-1. J. Cardiovasc. Pharmacol. 1989; 13 (Suppl. 5): S5–S7.
32. Opgenorth TJ, Wu-Wong JR, Shiosaki K. Endothelin-converting enzymes. FASEB J. 1992; 6: 2653–9.
33. Takaoka M, Shiragami K, Fujino K, Miki K, Miyake Y, Yasuda M et al. Phosphoramidon-sensitive endothelin converting enzyme in rat lung. Biochem. Internat. 1991; 25: 697–704.
34. Sawamura, T, Shinmi O, Kishi N, Sugita Y, Yanagisawa M, Goto K et al. Characterization of phosphoramidon-sensitive metalloproteinases with endothelin-converting enzyme activity in porcine lung membrane. Biochim. Biophys. Acta 1993; 1161: 295–302.
35. Takahashi M, Matsushita Y, Iijima Y, Tanzawa K. Purification and characterization of endothelin-converting enzyme from rat lung. J. Biol. Chem. 1993; 268: 21394–8.
36. Gui G, Xu D, Emoto N, Yanagisawa M. Intracellular localization of membane-bound endothelin-converting enzyme from rat lung. J. Cardiovasc. Pharmacol. 1993; 22 (Suppl. 8): S53–6.
37. D'Orléans-Juste P, Télémaque S, Claing A. Different pharmacological profiles of big-endothelin-3 and big-endothelin-1 in vivo and in vitro. Br. J. Pharmacol. 1991; 104: 440–4.
38. Ishikawa S, Tsukada H, Yuasa H, Fukue M, Wei S, Onizuka M et al. Effects of endothelin-1 and conversion of big endothelin-1 in the isolated perfused rabbit lung. J. Appl. Physiol. 1992; 72: 2387–92.

39. Pons F, Touvay C, Lagente V, Mencia-Huerta JM, Braquet P. Involvement of a phosphoramidon-sensitive endopeptidase in the processing of big endothelin-1 in the guinea-pig. Eur. J. Pharmacol. 1992; 217: 65–70.
40. Vemulapalli S, Rivelli M, Chiu PJS, del Prado M, Hey JA. Phosphoramidon abolishes the increases in endothelin-1 release induced by ischemia-hypoxia in isolated perfused guinea pig lungs. J. Pharmacol. Exp. Therap. 1992; 262: 1062–9.
41. Wu-Wong JR, Budzik GP, Devine EM, Opgenorth TJ. Characterization of endothelin converting enzyme in rat lung. Biochem. Biophys. Res. Comm. 1990; 171: 1291–6.
42. Watanabe T, Yokosawa H. The generation of big-endothelin (1–22) (endothelin-valine) from big-endothelin in the soluble fraction of porcine lung. Biochem. Internat. 1992; 27: 1–8.
43. Wypij DM, Nichols JS, Novak PJ, Stacy DL, Berman J, Wiseman JS. Role of mast cell chymase in the extracellular processing of big-endothelin-1 to endothelin-1 in the perfused rat lung. Biochem. Pharmacol. 1992; 43: 845–53.
44. Schwartz J-C, Costentin J, Lecomte J-M. Pharmacology of enkephalinase inhibitors. Trends Pharmacol. Sci. 1985; 6: 472–6.
45. Matsumura Y, Fujita K, Takaoka M, Morimoto S. Big endothelin-3-induced hypertension and its inhibition by phosphoramidon in anaesthetized rats. Eur. J. Pharmacol. 1993; 230: 89–93.
46. Télémaque S, Gratton J-P, Claing A, D'Orleans-Juste P. Pharmacologic evidence for the specificity of the phosphoramidon-sensitive endothelin-converting enzyme for big endothelin-1. J. Cardiovasc. Pharmacol. 1993; 22 (Suppl. 8): S85–9.
47. Black PM, Ghatei MA, Takahashi K, Bretherton-Watt D, Krausz T, Dollery CT et al. Formation of endothelin by cultured airway epithelial cells. FEBS Letts., 1989; 225: 129–32.
48. Mattoli S, Mezzetti M, Riva G, Allegra L, Fasoli A. Specific binding of endothelin on human bronchial smooth muscle cells in culture and secretion of endothelin-like material from bronchial epithelial cells. Am. J. Respir. Cell Mol. Biol. 1990; 3: 145–51.
49. Ninomiya H, Uchida Y, Ishii Y, Nomura A, Kameyama M, Saotome M et al. Endotoxin stimulates endothelin release from cultured epithelial cells of guinea-pig trachea. Eur. J. Pharmacol. 1991; 203: 299–302.
50. Endo T, Uchida Y, Matsumoto H, Suzuki N, Nomura A, Hirata F et al. Regulation of endothelin-1 synthesis in cultured guinea pig airway epithelial cells by various cytokines. Biochem. Biophys. Res. Comm. 1992; 186: 1594–9.
51. Rennick RE, Milner P, Burnstock G. Thrombin stimulates release of endothelin and vasopressin, but not substance P, from isolated rabbit tracheal epithelial cells. Eur. J. Pharmacol. 1993; 230: 367–70.
52. Franco-Cereceda A, Rydh M, Lou Y-P, Dalsgaard C-J, Lundberg JM. Endothelin as a putative sensory neuropeptide in the guinea-pig: different properties in comparison with calcitonin gene-related peptide. Regul. Peptides 1991; 31: 253–65.
53. Hay DWP. Guinea-pig tracheal epithelium and endothelin. Eur. J. Pharmacol. 1989; 171: 241–6.
54. Johnson AR, Asthon J, Schulz WW, Erdös EG. Neutral metalloendopeptidases in human lung tissue and cultured cells. Am. Rev. Respir. Dis. 1985; 132: 564–8.
55. Vijayaraghavan J, Scicli AG, Carretero OA, Slaughter C, Moomaw C, Hersh LB. The hydrolysis of endothelins by neutral endopeptidase 24.11 (enkephalinase). J. Biol. Chem. 1990; 265: 14150–5.
56. Fagny C, Michel A, Léonard I, Berkenboom G, Fontaine J, Deschodt-Lanckman M. In vitro degradation of endothelin-1 by endopeptidase 24.11 (enkephalinase) Peptides 1991; 12: 773–8.
57. Noguchi K, Fukuroda T, Ikeno Y, Hirose H, Tsukada Y, Nishikibe M et al. Local formation and degradation of endothelin-1 in guinea pig airway tissues. Biochem. Biophys. Res. Comm. 1991; 179: 830–5.
58. Di Maria GU, Katayama M, Borson DB, Nadel JA. Neutral endopeptidase modulates endothelin-1-induced airway smooth muscle contraction in guinea-pig trachea. Regul. Peptides 1992; 39: 137–45.
59. Boichot E, Pons F, Lagente V, Touvay C, Mencia-Huerta JM, Braquet P. Phosphoramidon potentiates the endothelin-1-induced bronchopulmonary response in guinea-pigs. Neurochem. Internat. 1991; 18: 477–9.

60. McKay KO, Black JL, Armour CL. Phosphoramidon potentiates the contractile response to endothelin-3, but not endothelin-1 in isolated airway tissue. Br. J. Pharmacol. 1992; 105: 929–32.
61. Candenas M-L, Naline E, Sarria B, Advenier C. Effect of epithelium removal and of enkephalin inhibition on the bronchoconstrictor response to three endothelins of the human isolated bronchus. Eur. J. Pharmacol. 1992; 210: 291–7.
62. Yamaguchi T, Kohrogi H, Kawano O, Ando M, Araki S. Neutral endopeptidase inhibitor potentiates endothelin-1-induced airway smooth muscle contraction. J. Appl. Physiol. 1992; 73: 1108–13.
63. Murphy LJ, Greenhough KJ, Turner AJ. Processing and metabolism of endothelin peptides by porcine lung membranes. J. Cardiovasc. Pharmacol. 1993; 22 (Suppl. 8): S94–7.
64. Fagny C, Michel A, Nortier J, Deschodt-Lanckman M. Enzymatic degradation of endothelin-1 by activated human polymorphonuclear neutrophils. Regul. Peptides 1992; 42: 27–37.
65. Koseki C, Imai M, Hirata Y, Yanagisawa M, Masaki, T. Autoradiographic distribution in rat tissues of binding sites for endothelin: a neuropeptide? Am. J. Physiol. 1989; 256: R858–66.
66. Shiba R, Yanagisawa M, Miyauchi T, Ishii Y, Kimura S, Uchiyama Y et al. Elimination of intravenously injected endothelin-1 from the circulation of the rat. J. Cardiovasc. Pharmacol. 1989; 13 (Suppl. 5): S98–101.
67. Sirviö M-L, Metsärinne K, Saijonmaa O, Fyhrquist F. Tissue distribution and half-life of ^{125}I-endothelin in the rat: importance of pulmonary clearance. Biochem. Biophys. Res. Comm. 1990; 167: 1191–5.
68. De Nucci G, Thomas R, D'Orleans-Juste P, Antunes E, Walder C, Warner TD et al. Pressor effects of circulating endothelin are limited by its removal in the pulmonary circulation and by the release of prostacyclin and endothelium-derived relaxing factor. Proc. Natl. Acad. Sci. USA 1988; 85: 9797–800.
69. Rimar S, Gillis CN. Differential uptake of endothelin-1 by the coronary and pulmonary circulations. J. Appl. Physiol. 1992; 73: 557–62.
70. Wagner OF, Vierhapper H, Gasic S, Nowotny P, Waldhäusl W. Regional effects and clearance of endothelin-1 across pulmonary and splanchnic circulation. Eur. J. Clin. Invest. 1992; 22: 277–82.
71. Weitzberg E, Ahlborg G, Lundberg JM. Differences in vascular effects and removal of endothelin-1 in human lung, brain, and skeletal muscle. Clin. Physiol. 1993; 13: 653–62.
72. Ray SG, McMurray JJ, Morton JJ, Dargie HJ. Circulating endothelin is not extracted by the pulmonary circulation in man. Chest 1992; 102: 1143–4.
73. Sakurai T, Yanagisawa M, Masaki T. Molecular characterization of endothelin receptors. Trends Pharmacol. Sci. 1992; 13: 103–8.
74. Masaki T, Kimura S, Yanagisawa M, Goto K. Molecular and cellular mechanism of endothelin regulation. Implications for vascular function. Circulation 1989; 30: 219–33.
75. Warner TD, de Nucci G, Vane JR. Rat endothelin is a vasodilator in the isolated perfused mesentery of the rat. Eur. J. Pharmacol. 1989; 159: 325–6.
76. Yanagisawa M, Inoue A, Ishikawa T, Kasuya Y, Kimura S, Kumagaye S-I et al. Primary structure, synthesis and biological activity of rat endothelin, an endothelin-derived vasoconstrictor peptide. Proc. Natl. Acad. Sci. 1988; 85: 6964–7.
77. Masuda Y, Miyazaki H, Kondoh M, Watanabe H, Yanagisawa M, Masaki T et al. Two different forms of endothelin receptors in rat lung. FEBS Letts. 1989; 257: 208–10.
78. Takayanagi R, Ohnaka K, Takasaki C, Ohashi M, Nawata H. Multiple subtypes of endothelin receptors in porcine tissues: characterization by ligand binding, affinity labeling and regional distribution. Regul. Peptides 1991; 32: 23–37.
79. Kundu GC, Misono KS. Affinity labeling of endothelin receptors in bovine and rat lung membranes by N^{e9}-azidobenzoyl-^{125}I-endothelin-1. Mol. Cell. Endocrin. 1991; 79: 85–92.
80. Arai H, Hori S, Aramori I, Ohkubo H, Nakanishi S. Cloning and expression of a cDNA encoding and endothelin receptor. Nature Lond. 1990; 348: 730–2.
81. Sakurai T, Yanagisawa M, Takuwa Y, Miyazaki H, Kimura S, Goto K et al. Cloning of a cDNA encoding a non-isopeptide-selective subtype of the endothelin receptor. Nature 1990; 348: 732–5.

82. Sakamoto A, Yanagisawa M, Sakurai T, Takuwa Y, Yanagisawa H, Masaki T. Cloning and functional expression of human cDNA for the ET_B endothelin receptor. Biochem. Biophys. Res. Comm. 1991; 178: 656–63.

83. Martin ER, Brenner BM, Ballermann BJ. Heterogeneity of cell surface endothelin receptors. J. Biol. Chem. 1990; 265: 14044–9.

84. Samson WK, Skala KD, Alexander BD, Huang F-LS. Pituitary site of action of endothelin: selective inhibition of prolactin release in vitro. Biochem. Biophys. Res. Comm. 1990; 169: 737–43.

85. Karne S, Jayawickreme CK, Lerner MR. Cloning and characterization of an endothelin-3 specific receptor (ET_C Receptor) from Xenopus laevis dermal melanophores. J. Biol. Chem. 1993; 268: 19126–33.

86. Adachi M, Yang Y-Y, Furuichi Y, Miyamoto C. Cloning and characterization of cDNA encoding human A-type endothelin receptor. Biochem. Biophys. Res. Comm. 1991; 180: 1265–72.

87. Cyr C, Huebner K, Druck T, Kris R. Cloning and chromosomal localization of a human endothelin ETA receptor. Biochem. Biophys. Res. Comm. 1991; 181: 184–90.

88. Hosoda K, Nakao K, Arai-H, Suga S-i, Ogawa Y, Mukoyama M et al. Cloning and expression of human endothelin-1 receptor cDNA. FEBS Letts. 1991; 287: 23–6.

89. Nakamuta, M, Takayanagi R, Sakai Y, Sakamoto S, Hagiwara H, Mizuno T, et al. Cloning and sequence analysis of a cDNA encoding human non-selective type of endothelin receptor. Biochem. Biophys. Res. Comm. 1991; 177: 34–9.

90. Ogawa Y, Nakao K, Arai H, Nakagawa O, Hosoda K, Suga S-i et al. Molecular cloning of a non-iosopeptide-selective human endothelin receptor. Biochem. Biophys. Res. Comm. 1991; 178: 248–55.

91. Kondoh M, Miyazaki H, Uchiyama Y, Yanagisawa M, Masaki T, Murakami K. Solubilization of two types of endothelin receptors, ET_A and ET_B, from rat lung with retention of binding activity. Biomed. Res. 1991; 12: 417–23.

92. Kozuka M, Ito T, Hirose S, Lodhi KM, Hagiwara H. Purification and characterization of bovine lung endothelin receptor. J. Biol. Chem. 1991; 266: 16892–6.

93. Hagiwara H, Nagasawa T, Lodhi KM, Kozuka M, Ito T, Hirose S. Affinity chromatographic purification of bovine lung endothelin receptor using biotinylated endothelin and avidin-agarose. J. Chromat. 1992; 597: 331–4.

94. Maggi CA, Giuliani S, Patacchini R, Rovero P, Giachetti A, Meli A. The activity of peptides of the endothelin family in various mammalian smooth muscle preparations. Eur. J. Pharmacol. 1989; 174: 23–31.

95. Maggi CA, Giuliani S, Patacchini R, Santicioli P, Rovero P, Giachetti A. The C-terminal hexapeptide, endothelin-(16–21), discriminates between different endothelin receptors. Eur. J. Pharmacol. 1989; 166: 121–2.

96. Maggi CA, Giuliani S, Patacchini R, Santicioli P, Giachetti A, Meli A. Further studies on the response of the guinea-pig isolated bronchus to endothelins and sarafotoxin S6b. Eur. J. Pharmacol. 1990; 176: 1–9.

97. Tschirhart EJ, Drijfhout JW, Pelton JT, Miller RC, Jones CR. Endothelins: functional and autoradiographic studies in guinea pig trachea. J. Pharmacol. Exp. Therap. 1991; 258: 381–7.

98. Cardell LO, Uddman R, Edvinsson L. Evidence for multiple endothelin receptors in the guinea-pig pulmonary artery and trachea. Br. J. Pharmacol. 1992; 105: 376–80.

99. Williams Jr DL, Jones KL, Pettibone DJ, Lis EV, Clineschmidt BV Sarafotoxin S6c: an agonist which distinguishes between endothelin receptor subtypes. Biochem. Biophys. Res. Comm. 1991; 175: 556–61.

100. Ihara M, Noguchi K, Saeki T, Fukuroda T, Tsuchida S, Kimura S et al. Biological profiles of highly potent novel endothelin antagonists selective for the ET_A receptor. Life Sci. 1991; 50: 247–55.

101. Sogabe K, Nirei H, Shoubo M, Nomoto A, Henmi K, Notsu Y et al. A novel endothelin receptor antagonist: studies with FR 139317. Jap. J. Pharmacol. 1992; 58: 105P.

102. Hay, DWP. Pharmacological evidence for distinct endothelin receptors in guinea-pig bronchus and aorta. Br. J. Pharmacol. 1992; 106: 759–61.

103. Cardell LO, Uddman R, Edvinsson L. A novel ET_A-receptor antagonist, FR 139317, inhibits endothelin-induced contractions of guinea-pig pulmonary arteries, but not trachea. Br. J. Pharmacol. 1993; 108: 448–52.

104. Hay DWP, Luttmann MA, Hubbard WC, Undem BJ. Endothelin receptor subtypes in human and guinea-pig pulmonary tissues. Br. J. Pharmacol. 1993; 110: 1175–83.
105. Tomlinson PR, Wilson JW, Stewart AG. Inhibition by salbutamol of the proliferation of human airway smooth cells grown in culture. Br. J. Pharmacol. 1994; 111: 641–7.
106. Battistini B, Warner TD, Fournier A, Vane, JR. Characterization of ET_B receptors mediating contractions induced by endothelin-1 or IRL 1620 in guinea-pig isolated airways: effects of BQ-123, FR139317 or PD 145065. Br. J. Pharmacol. 1994; 111: 1009–16.
107. Noguchi K, Noguchi Y, Hirose H, Niskikibe M, Ihara M, Ishikawa K et al. Role of endothelin ET_B receptors in bronchoconstrictor and vasoconstrictor responses in guinea-pigs. Eur. J. Pharmacol. 1993; 233: 47–51.
108. D'Orléans-Juste P, Télémaque S, Claing A, Ihara M, Yano M. Human big-endothelin-1 and endothelin-1 release prostacyclin via the activation of ET_1 receptors in the rat perfused lung. Br. J. Pharmacol. 1992; 105: 773–5.
109. Fujimoto M, Mihara S-i, Nakajima S, Ueda M, Nakamura M, Sakurai K-s. A novel non-peptide endothelin antagonist isolated from bayberry, *Myrica cerifera*. FEBS Letts. 1992; 305: 41–4.
110. Ihara M, Saeki T, Fukuroda T, Kimura S, Ozaki S, Patel AC et al. A novel radio-ligand [^{125}I]BQ-3020 selective for endothelin (ET_B) receptors. Life Sci. 1992; 51: PL47–52.
111. Clozel M, Breu V, Burri K, Cassal J-M, Fischli W, Gray GA, Hirth G, Löffler B-M, Müller M, Neidhart W, Ramuz H. Pathophysiological role of endothelin revealed by the first orally active endothelin receptor antagonist. Nature 1993; 365: 759–61.
112. Ohlstein EH, Nambi P, Douglas SA, Edwards RM, Gellai M, Lago A et al. SB 209670, a rationally designed potent nonpeptide endothelin receptor antagonist. Proc. Natl. Acad. Sci. 1994; 91: 8052–56.
113. Shetty SS, Okada T, Webb RL, DelGrande D, Lappe RW. Functionally distinct endothelin B receptors in vascular endothelium and smooth muscle. Biochem. Biophys. Res. Commun. 1993; 191: 459–64.
114. Karaki H, Sudjarwo SA, Hori M, Sakata K, Urade Y, Takai M et al. ET_B receptor antagonist, IRL 1038, selectively inhibits the endothelin-induced endothelium-dependent vascular relaxation. Eur. J. Pharmacol. 1993; 231: 371–4.
115. Fukuroda T, Nishikibe M, Ohta Y, Ihara M, Yano M, Ishikawa K et al. Analysis of responses to endothelins in isolated porcine blood vessels by using a novel endothelin antagonist, BQ-153. Life Sci. 1991; 50: 107–12.
116. Sudjarwo SA, Hori M, Takai M, Urade Y, Okada T, Karaki H. A novel subtype of endothelin B receptor mediating contraction in swine pulmonary vein. Life Sci. 1993; 53: 431–7.
117. White DG, Cannon TR, Garratt H, Mundin JW, Sumner MJ, Watts IS. Endothelin ET_A and ET_B receptors mediate vascular smooth-muscle contraction. J. Cardiovasc. Pharmacol. 1993; 22 (Suppl. 8): S144–8.
118. Sokolovsky M, Ambar I, Galron R. A novel subtype of endothelin receptors. J. Biol. Chem. 1992; 267: 20551–4.
119. Lippton HL, Hauth TA, Cohen GA, Hyman AL. Functional evidence for different endothelin receptors in the lung. J. Appl. Physiol. 1993; 75: 38–48.
120. Saeki T, Ihara M, Fukuroda T, Yamagiwa M, Yano M. [Ala^{1,3,11,15}]Endothelin-1 analogs with ET_B agonistic activity. Biochem. Biophys. Res. Comm. 1991; 179: 286–92.
121. Takai M, Umemura I, Yamasaki K, Watakabe T, Fujitani Y, Oda K et al. A potent and specific agonist, Suc-[Glu^9, Ala^{11,15}]-Endothelin-1 (8–21), IRL 1620, for the ET_B receptor. Biochem. Biophys. Res. Comm. 1992; 184: 953–9.
122. Urade Y, Fujitani Y, Oda K, Watakabe T, Umemura I, Takai M et al. An endothelin B receptor-selective antagonist: IRL 1038, [Cys^{11}-Cys^{15}]-endothelin-1 (11–21). FEBS Letts. 1992; 311: 12–6.
123. Mugrage B, Moliterni J, Robinson L, Webb RL, Shetty SS, Lipson KE et al. CGS 27830, a potent nonpeptide endothelin receptor antagonist. Bioorganic Med. Chem. Letts. 1993; 3: 2099–104.
124. Mihara S-i, Fujimoto M. The endothelin ET_A receptor-specific effect of 50-235, a nonpeptide endothelin antagonist. Eur. J. Pharmacol. 1993; 246: 33–8.
125. Harrison VJ, Randriantsoa A, Schoeffter P. Heterogeneity of endothelin-sarafoxin receptors mediating contraction of pig artery. Br. J. Pharmacol. 1992; 105: 511–3.

126. Eglezos A, Cucchi P, Patacchini R, Quartara L, Maggi CA, Mizrahi J. Differential effects of BQ-123 against endothelin-1 and endothelin-3 on the rat vas deferens: evidence for an atypical endothelin receptor. Br. J. Pharmacol. 1993; 109: 736–8.
127. Schoeffter P, Randriantsoa A. Differences between endothelin receptors mediating contraction of guinea-pig aorta and pig coronary artery. Eur. J. Pharmacol. 1993; 249: 199–206.
128. Warner TD, Allcock GH, Mickley EJ, Corder R, Vane JR. Comparative studies with the endothelin receptor antagonists BQ-123 and PD 142893 indicate at least three endothelin receptors. J. Cardiovasc. Pharmacol. 1993; 22 (Suppl. 8): S117–20.
129. Power RF, Wharton J, Zhao Y, Bloom SR, Polak JM. Autoradiographic localization of endothelin-1 binding sites in the cardiovascular and respiratory systems. J. Cardiovas. Pharmacol. 1989; 13 (Suppl. 5): S50–6.
130. Turner NC, Power RF, Polak JM, Bloom SR, Dollery CT. Endothelin-induced contractions of tracheal smooth muscle and identification of specific endothelin binding sites in the trachea of the rat. Br. J. Pharmacol. 1989; 98: 361–6.
131. Hemsén A, Franco-Cereceda A, Matran R, Rudehill A, Lundberg JM. Occurrence, specific binding sites and functional effects of endothelin in human cardiopulmonary tissue. Eur. J. Pharmacol. 1990; 191: 319–28.
132. Henry PJ, Rigby PJ, Self GJ, Preuss JM, Goldie RG. Relationship between endothelin-1 binding site densities and constrictor activities in human and animal airway smooth muscle. Br. J. Pharmacol. 1990; 100: 786–92.
133. Brink C, Gillard V, Roubert P, Mencia-Huerta JM, Chabrier PE, Braquet P et al. Effects and specific binding sites of endothelin in human lung preparations. Pulmon. Pharmacol. 1991; 4: 54–9.
134. McKay KO, Black JL, Diment LM, Armour CL. Functional and autoradiographic studies of endothelin-1 and endothelin-2 in human bronchi, pulmonary arteries, and airway parasympathetic ganglia. J. Cardiovasc. Pharm. 1991; 17 (Suppl. 7): S206–9.
135. Wu T, Rieves RD, Larivee P, Logun C, Lawrence MG, Shelhamer JH. Production of eicosanoids in response to endothelin-1 and identification of specific endothelin-1 binding sites in airway epithelial cells. Am. J. Respir. Cell Mol. Biol. 1993; 8: 282–90.
136. Sen M, Grunstein MM, Chander A. Stimulation of lung surfactant secretion by endothelin-1 from rat alveolar type II cells. Am. J. Physiol. 1994; 266: L255–62.
137. Durham SK, Goller NL, Lynch JS, Fisher SM, Rose PM. Endothelin receptor B expression in the rat and rabbit lung as determined by in situ hybridization using nonisotropic probes. J. Cardiovasc. Pharmacol. 1993; 22 (Suppl. 8): S1–3.
138. Bolger GT, Liard F, Krogsrud R, Thibeault D, Jaramillo J. Tissue specificity of endothelin binding sites. J. Cardiovasc. Pharmacol. 1990; 16: 367–75.
139. Hagiwara H, Nagasawa T, Yamamoto T, Lodhi KM, Ito T, Takemura N et al. Immunochemical characterization and localization of endothelin ET_B receptor. Am. J. Physiol. 1993; 264: R777–83.
140. Goldie RG, Henry PJ, Self GJ, Knott PG, Luttmann M, Hay DWP. Endothelin receptor subtype distribution, density and function in human isolated asthmatic and non-diseased bronchus. Am. J. Respir. Crit. Care. Med., 1994; 149: A472.
141. Nakamichi K, Ihara M, Kobayashi M, Saeki T, Ishikawa K, Yano M. Different distribution of endothelin receptor subtypes in pulmonary tissues revealed by the novel selective ligands BQ-123 and [Ala^{1,3,11,15}]ET-1. Biochem. Biophys. Res. Comm. 1992; 182: 144–50.
142. Cioffi CL, Neale RF Jr, Jackson RH, Sills MA. Characterization of rat lung endothelin receptor subtypes which are coupled to phosphoinositide hydrolysis. J. Pharmacol Exp. Ther. 1992; 262: 611–8.
143. Kobayashi M, Ihara M, Sato N, Saeki T, Ozaki S, Ikemoto F et al. A novel ligand, [^{125}I]BQ-3020, reveals the localization of endothelin ET_B receptors. Eur. J. Pharmacol. 1993; 234: 95–100.
144. Henry PJ. Endothelin-1 (ET-1)-induced contraction in rat isolated trachea: involvement of ET_A and ET_B receptors and multiple signal transduction systems. Br. J. Pharmacol. 1993; 110: 435–41.
145. Noguchi K, Ishikawa K, Yano M, Ahmed A, Cortes A, Hallmon J et al. An endothelin $(ET)_A$ receptor antagonist, BQ-123, blocks ET-1 induced bronchoconstriction and tracheal smooth muscle (TSM) contraction in allergic sheep. Am. Rev. Respir. Dis. 1992; 145: A858.

146. Goldie RG, Grayson PS, Henry PJ. Endothelin-1 (ET-1)-induced contraction of ovine tracheal smooth muscle is mediated via ET_A receptors. Am. Rev. Respir. Dis. 1993; 147: A182.

147. Inui T, James AF, Fujitani Y, Takimoto M, Okada T, Yamamura T et al. ET_A and ET_B receptors on single smooth muscle cells cooperate in mediating guinea pig tracheal contraction. Am. J. Physiol. 1994; 266: L113–24.

148. Uchida Y, Ninomiya H, Saotome M, Nomura A, Ohtsuka M, Yanagisawa M et al. Endothelin, a novel vasoconstrictor peptide, as potent bronchoconstrictor. Eur. J. Pharmacol. 1988; 154: 227–8.

149. Maggi CA, Patacchini S, Meli A. Potent contractile effect of endothelin in isolated guinea-pig airways. Eur. J. Pharmacol. 1989; 160: 179–82.

150. Advenier C, Sarria B, Naline E, Puybasset L, Lagente V. Contractile activity of three endothelins (ET-1, ET-2 and ET-3) on the human isolated bronchus. Br. J. Pharmacol. 1990; 100: 168–72.

151. Hay DWP. Mechanism of endothelin-induced contraction in guinea-pig trachea: comparison with rat aorta. Br. J. Pharmacol. 1990; 100: 383–92.

152. Hay DWP, Henry PJ, Goldie RG. Endothelin and the respiratory system. Trends Pharmacol. Sci. 1993; 14: 29–32.

153. Hay DWP, Hubbard WC, Undem BJ. Endothelin-induced contraction and mediator release in human bronchus. Br. J. Pharmacol. 1993; 110: 392–8.

154. Lee H-K, Leikauf GD, Sperelakis N. Electromechanical effects of endothelin on ferret bronchial and tracheal smooth muscle. J. Appl. Physiol. 1990; 68: 417–20.

155. Nally JE, McCall R, Young LC, Wakelam MJO, Thomson NC, McCrath JC. Mechanical and biochemical responses to endothelin-1 and endothelin-3 in bovine bronchial smooth muscle. Br. J. Pharmacol. 1994; 111: 1163–9.

156. Filep JG, Battistini B, Sirois P. Endothelin induces thromboxane release and contraction of isolated guinea-pig airways. Life Sci. 1990; 47: 1845–50.

157. Sarriá B, Naline E, Morcillo E, Cortijo J, Esplugues J, Advenier C. Calcium dependence of the contraction produced by endothelin (ET-1) in isolated guinea-pig trachea. Eur. J. Pharmacol. 1990; 187: 445–53.

158. Hay DWP, Hubbard WC, Undem BJ. Relative contributions of direct and indirect mechanisms mediating endothelin-induced contraction of guinea-pig trachea. Br. J. Pharmacol. 1993; 110: 955–62.

159. Battistini B, Sirois P, Braquet P, Filep JG. Endothelin-induced constriction of guinea-pig airways: role of platelet-activating factor. Eur. J. Pharmacol. 1990; 186: 307–10.

160. Filep JG, Battistini B, Sirois P. Pharmacological modulation of endothelin-induced contraction of guinea-pig isolated airways and thrmoboxane release. Br. J. Pharmacol. 1991; 103: 1633–40.

161. Ninomiya H, Uchida Y, Saotome M, Nomura A, Ohse H, Matsumoto H et al. Endothelins constrict guinea pig tracheas by multiple mechanisms. J. Pharmacol. Exp. Therap. 1992; 262: 570–6.

162. Cardell LO, Uddman R, Edvinsson L. Two functional endothelin receptors in guinea-pig pulmonary arteries. Neurochem. Internat. 1991; 18: 571–4.

163. Rodman DM, Stelzner TJ, Zamora MR, Bonvallet ST, Oka M, Sato K et al. Endothelin-1 increases the pulmonary microvascular pressure and causes pulmonary edema in salt solution but not blood-perfused rat lungs. J. Cardiovasc. Pharmacol. 1992; 20: 658–63.

164. Raffestin B, Adnot S, Eddahibi S, Macquin-Mavier I, Braquet P, Chabrier PE. Pulmonary vascular response to endothelin in rat. J. Appl. Physiol. 1991; 70: 567–74.

165. Hasunuma K, Rodman DM, O'Brien RF, McMurtry IF. Endothelin 1 causes pulmonary vasodilation in rats. Am. J. Physiol. 1990; 259: H48–54.

166. Eddahibi S, Adnot S, Carville C, Blouquit Y, Raffestin B. L-arginine restores endothelium-dependent relaxation in the pulmonary circulation of chronically hypoxic rats. Am. J. Physiol. 1992; 263: L194–200.

167. Adnot S, Raffestin B, Eddahibi S, Braquet P, Chabrier PE. Loss of endothelium-dependent relaxant activity in the pulmonary circulation of rats exposed to chronic hypoxia. J. Clin. Invest. 1991; 87: 155–62.

168. Eddahibi S, Springall D, Mannan M, Carville C, Chabrier P-E, Levame M et al. Dilator effect of endothelins in pulmonary circulation: changes associated with chronic hypoxia. Am. J. Physiol. 1993; 265: L571–80.

169. White SR, Hathaway DP, Umans JG, Tallet J, Abrahams C, Leff AR. Epithelial modulation of airway smooth muscle response to endothelin-1. Am. Rev. Respir. Dis. 1991; 144: 373-8.
170. Filep JG, Battistini B, Sirois P. Induction by endothelin-1 of epithelium-dependent relaxation of guinea-pig trachea *in vitro*: role for nitric oxide. Br. J. Pharmacol. 1993; 109: 637-44.
171. Grunstein MM, Chuang ST, Schramm CM, Pawlowski NA. Role of endothelin 1 in regulating rabbit airway contractility. Am. J. Physiol. 1991; 260: L75-82.
172. Cocks TM, Broughton A, Dib M, Sudhir K, Angus JA. Endothelin is blood vessel selective: studies on a variety of human and dog vessels in vitro and on regional blood flow in the conscious rabbit. Clin. Exp. Pharmacol. Physiol. 1989; 16: 243-6.
173. Clozel M, Gray GA, Breu V, Löffler B-M, Osterwalder R. The endothelin ET_B receptor mediates both vasodilation and vasoconstriction *in vivo*. Biochem. Biophys. Res. Comm. 1992; 186: 867-73.
174. Johnson DE. Pulmonary neuroendocrine cells. In: Farmer SG, Hay DWP, editors. The Airway Epithelium. Physiology, Pathophysiology, and Pharmacology. New York: Marcel Dekker, Inc, 1991: 335-97.
175. O'Byrne PM. The airway epithelium and asthma. In: Farmer SG and Hay DWP, editors. The Airway Epithelium. Physiology, Pathphysiology, and Pharmacology. New York: Marcel Dekker, Inc, 1991: 171-86.
176. Stutts MJ, Knowles MR, Chinet T, Boucher RC. Abnormal ion transport in cystic fibrosis airway epithelium. In: Farmer SG and Hay DWP, editors. The Airway Epithelium. Physiology, Pathphysiology, and Pharmacology. New York: Marcel Dekker, Inc., 1991: 301-34.
177. Webber SE, Yurdakos E, Woods AJ, Widdicombe JG. Effects of endothelin-1 on tracheal submucosal gland secretion and epithelial function in the ferret. Chest 1992; 101: 63S-67S.
178. Satoh M, Shimura S, Ishihara H, Nagaki M, Sasaki H, Takishima T. Endothelin-1 stimulates chloride secretion across canine tracheal epithelium. Respiration 1992; 59: 145-50.
179. Plews PI, Abdel-Malek ZA, Doupnik CA, Leikauf GD. Endothelin stimulates chloride secretion across canine tracheal epithelium. Am. J. Physiol. 1991; 261: L188-L194.
180. Tamaoki J, Kanemura T, Sakai N, Isono K, Kobayashi K, Takizawa T. Endothelin stimulates ciliary beat frequency and chloride secretion in canine cultured tracheal epithelium. Am. J. Respir. Cell Mol. Biol. 1991; 4: 426-31.
181. Casasco A, Benazzo M, Casasco M, Cornaglia AI, Springall DR, Calligaro A et al. Occurrence, distribution and possible role of the regulatory peptide endothelin in the nasal mucosa. Cell Tissue Res. 1993; 274: 241-7.
182. Mullol J, Chowdhury BA, White MV, Ohkubo K, Rieves RD, Baraniuk J et al. Endothelin in human nasal mucosa. Am. J. Respir. Cell Mol. Biol. 1993; 8: 393-402.
183. Shimura S, Ishihara H, Satoh M, Masuda T, Nagaki N, Sasaki H et al. Endothelin regulation of mucus glycoprotein secretion from feline tracheal submucosal glands. Am. J. Physiol. 1992; 262: L208-13.
184. Wu T, Mullol J, Rieves RD, Logun C, Hausfield J, Kaliner MA et al. Endothelin-1 stimulates eicosanoid production in cultured human nasal mucosa. Am. J. Respir. Cell Mol. Biol. 1992; 6: 168-74.
185. Noveral JP, Rosenberg SM, Anbar RA, Pawlowski NA, Grunstein MM. Role of endothelin-1 in regulating proliferation of cultured rabbit airway smooth muscle cells. Am. J. Physiol. 1992; 263: L317-24.
186. Glassberg MK, Ergul A, Wanner A, Puett D. Endothelin-1 promotes mitogenesis in airway smooth muscle cells. Am. J. Respir. Cell Mol. Biol. 1994; 10: 316-21.
187. Komuro I, Kurihara H, Sugiyama T, Takaku F, Yazaki Y. Endothelin stimulates c-*fos* and c-*myc* expression and proliferation of vascular smooth muscle cells. FEBS Letts. 1988; 238: 249-52.
188. Bobik A, Grooms A, Millar JA, Mitchell A, Grinpukel S. Growth factor activity of endothelin on vascular smooth muscle. Am. J. Physiol. 1990; 258: C408-15.
189. Demoly P, Basset-Seguin N, Chanez P, Campbell AM, Gauthier-Rouvière C, Godard P et al. c-*fos* proto-oncogene expression in bronchial biopsies of asthmatics. Am. J. Respir. Cell. Biol. 1992; 7: 128-33.
190. Takuwa Y, Kasuya Y, Takuwa N, Kudo M, Yanagisawa M, Goto K et al. Endothelin receptor is coupled to phospholipase C via a pertussis toxin-insensitive guanine-nucleotide

binding regulatory protein in vascular smooth muscle cells. J. Clin. Invest. 1990; 85: 653–8.

191. Brewster CEP, Howarth PH, Djukanovic R, Wilson J, Holgate ST, Roche WR. Myofibroblasts and subepithelial fibrosis in bronchial asthma. Am. J. Respir. Cell Mol. Biol. 1990; 3: 507–11.

192. Peacock AJ, Dawes KE, Shock A, Gray AJ, Reeves JT, Laurent GJ. Endothelin-1 and endothelin-3 induce chemotaxis and replication of pulmonary artery fibroblasts. Am. J. Respir. Cell Mol. Biol. 1992; 7: 492–9.

193. Hassoun PM, Thappa V, Landman MJ, Fanburg BL. Endothelin 1: mitogenic activity on pulmonary artery smooth muscle cells and release from hypoxic endothelial cells. Proc. Soc. Exp. Biol. Med. 1992; 199: 165–70.

194. Zamora MA, Dempsey EC, Walchak SJ, Stelzner TJ. BQ123, an ET_A receptor antagonist, inhibits endothelin-1-mediated proliferation in human pulmonary artery smooth muscle cells. Am. J. Respir. Cell Mol. Biol. 1993; 9: 429–33.

195. Janakidevi K, Fisher MA, Del Vecchio PJ, Tiruppathi C, Figge J, Malik AB. Endothelin-1 stimulates DNA synthesis and proliferation of pulmonary artery smooth muscle cells. Am. J. Physiol. 1992; 263: C1295–303.

196. Ricagna F, Miller VM, McGregor CGA. Mitogenic activity of endothelin-1 in canine lung organ cultures. Clin. Res. 1993; 41: 149A.

197. Payne AN, Whittle BJR. Potent cyclo-oxygenase-mediated bronchoconstrictor effects of endothelin in the guinea-pig vivo. Eur. J. Pharmacol. 1988; 158: 303–4.

198. Lagente V, Chabrier PE, Mencia-Huerta JM, Braquet P. Pharmacological modulation of the bronchopulmonary action of the vasoactive peptide, endothelin, administered by aerosol in the guinea-pig. Biochem. Biophys. Res. Comm. 1989; 158: 625–32.

199. Macquin-Mavier I, Levame M, Istin N, Harf A. Mechanisms of endothelin-mediated bronchoconstriction in the guinea pig. J. Pharmacol. Exp. Therap. 1989; 250: 740–5.

200. Uchida Y, Ninomiya H, Sakamoto T, Lee JY, Endo T, Nomura A et al. ET-1 released histamine from guinea pig pulmonary but not peritoneal mast cells. Biochem. Biophys. Res. Comm. 1992; 189: 1196–201.

201. Nagase T, Fukuchi Y, Jo C, Teramoto S, Uejima Y, Ishida K et al. Endothelin-1 stimulates arachidonate 15-lipoxygenase activity and oxygen radical formation in the rat distal lung. Biochem. Biophys. Res. Comm. 1990; 168: 485–9.

202. Ninomiya H, Yu XY, Hasegawa S, Spannhake EW. Endothelin-1 induces stimulation of prostaglandin synthesis in cells obtained from canine airways by bronchoalveolar lavage. Prostaglandins 1992; 43: 401–11.

203. Helset E, Kjæva J, Hauge A. Endothelin-1-induced increases in microvascular permeability in isolated, perfused rat lungs requires leukocytes and plasma. Circul. Shock 1993; 39: 15–20.

204. Horgan MJ, Pinheiro JMB, Malik AB. Mechanism of endothelin-1-induced pulmonary vasoconstriction. Circul. Res. 1991; 69: 157–64.

205. Pons F, Touvay C, Lagente V, Mencia-Huerta JM, Braquet P. Comparison of the effects of intra-arterial and aerosol administration of endothelin-1 (ET-1) in the guinea-pig isolated lung. Br. J. Pharmacol. 1991; 102: 791–6.

206. Ercan ZS, Kiling M, Yazar Ö, Korkusuz P, Türker RK. Endothelin-1-induced oedema in rat and guinea-pig isolated perfused lungs. Arch. int. Pharmacodyn. 1993; 323: 74–84.

207. Mann J, Farrukh IS, Michael JR. Mechanisms by which endothelin 1 induces pulmonary vasoconstriction in the rabbit. J. Appl. Physiol. 1991; 71: 410–6.

208. Barnard JW, Barman SA, Adkins WK, Longenecker GL, Taylor AE. Sustained effects of endothelin-1 on rabbit, dog and rat pulmonary circulations. Am. J. Physiol. 1991; 261: H479–86.

209. Achmad TH, Rao GS. Chemotaxis of human blood monocytes toward endothelin-1 and the influence of calcium channel blockers. Biochem. Biophys. Res. Comm. 1992; 189: 994–1000.

210. Bath PMW, Mayston SA, Martin JF. Endothelin and PDGF do not stimulate peripheral blood monocyte chemotaxis, adhesion to endothelium, and superoxide production. Exper. Cell Res. 1990; 187: 339–42.

211. Millul V, Lagente V, Gillardeaux O, Boichot E, Dugas B, Mencia-Huerta K-M et al. Activation of guinea pig alveolar macrophages by endothelin-1. J. Cardiovasc. Pharmacol. 17, (Suppl. 7) 1991; S233–5.

212. Haller H, Schaberg T, Lindschau C, Lode H, Distler A. Endothelin increases $[Ca^{2+}]_i$, protein phosphorylation, and $O_2^{\bullet}$ production in human alveolar macrophages. Am. J. Physiol. 1991; 261: L478–84.

213. Helset E, Sildnes T, Seljelid R, Konoski ZS. Endothelin-1 stimulates human monocytes in vitro to release TNF-α, IL-1ß and IL-6. Mediators Inflamm. 1993; 2: 417–22.

214. Tabuchi Y, Nakamaru M, Rakugi H, Nagano M, Mikami H, Ogihara T. Endothelin inhibits presynaptic adrenergic neurotransmission in rat mesenteric artery. Biochem. Biophys. Res. Commun. 1989; 161: 803–8.

215. Wiklund NP, Öhlén A, Cederqvist B. Adrenergic neuromodulation by endothelin in guinea pig pulmonary artery. Neurosci. Letts. 1989; 101: 269–73.

216. Wiklund NP, Wiklund CU, Öhlén A, Gustafsson LE. Cholinergic neuromodulation by endothelin in guinea pig ileum. Neurosci. Letts. 1989; 101: 342–46.

217. Wiklund NP, Öhlén A, Wiklund CU, Hedqvist P, Gustafsson LE. Endothelin modulation of neuroeffector transmission in rat and guinea pig vas deferens. Eur. J. Pharmacol. 1990; 185: 25–33.

218. Nishimura T, Kier J, Akasu T. Endothelin causes prolonged inhibition of nicotinic transmission in feline colinic parasympathetic ganglia. Am. J. Physiol. 1991; 261: G628–33.

219. McKay KO, Armour CL, Black BL. Endothelin-3 increases transmission in the rabbit pulmonary parasympathetic nervous system. J. Cardiovasc. Pharmacol. 1993; 22 (Suppl. 8): S181–84.

220. Takimoto M, Inui T, Okada T, Urade Y. Contraction of smooth muscle by activation of endothelin receptors on autonomic neurons. FEBS Letts. 1993; 324: 27–82.

221. Lueddeckens G, Becker K, Rappold R, Förster W. Influence of aminophylline and ketotifen in comparison to the lipoxygenase inhibitors NDGA and esculetin and the PAF antagonists WEB 2107 and BN 52021 on endothelin-1 induced vaso- and bronchoconstriction. Prost. Leuk. Essential Fatty Acids 1991; 44: 155–8.

222. Touvay C, Vilain B, Pons F, Chabrier P-E, Mencia-Huerta JM, Braquet P. Bronchopulmonary and vascular effect of endothelin in the guinea pig. Eur. J. Pharmacol. 1990; 176: 23–33.

223. Boichot E, Lagente V, Mencia-Huerta JM, Braquet P. Effect of phosphoramidon and indomethacin on the endothelin-1 (ET-1) induced bronchopulmonary response in aerosol sensitized guinea pigs. J. Vasc. Med. Biol. 1990; 2: 206.

224. Uchida Y, Hamada M, Kameyama M, Ohse H, Nomura A, Hasegawa S et al. ET-1 induced bronchoconstriction in the early phase but not late phase of anesthetized dogs is inhibited by indomethacin and ICI 198615. Biochem. Biophys. Res. Comm. 1992; 183: 1197–202.

225. Matsuse T, Fukuchi Y, Suruda T, Nagase T, Ouchi Y, Orimo H. Effect of endothelin-1 on pulmonary resistance in rats. J. Appl. Physiol. 1990; 68: 2391–3.

226. Abraham WM, Ahmed A, Cortes A, Spinella MJ, Malike AB, Anderson TT. A specific endothelin-1 antagonist blocks inhaled endothelin-1-induced bronchoconstriction in sheep. J. Appl. Physiol. 1993; 74: 2537–42.

227. Lagente V, Boichot E, Mencia-Huerta J, Braquet P. Failure of aerosolized endothelin (ET-1) to induced bronchial hyperreactivity in the guinea pig. Fundam. Clin. Pharmacol. 1990; 4: 275–80.

228. Boichot E, Carré C, Lagente V, Pons F, Mencia-Huerta JM, Braquet P. Endothelin-1 (ET-1) and bronchial hyperresponsiveness in the guinea-pig. J. Cardiovasc. Pharmacol. 1991; 17 (Suppl. 7): S329–31.

229. Kanazawa H, Kurihara N, Kirata K, Fujiwara H, Matsushita H, Takeda T. Low concentration endothelin-1 enhanced histamine-mediated bronchial contractions of guinea pigs in vivo. Biochem. Biophys. Res. Comm. 1992; 187: 717–21.

230. Pons F, Boichot E, Lagente V, Touvay C, Mencia-Huerta JM, Braquet P. Role of endothelin in pulmonary function. Pulmon. Pharmacol. 1992; 5: 213–9.

231. Wong J, Vanderford PA, Fineman JR, Chang R, Soifer SJ. Endothelin-1 produces pulmonary vasodilation in the intact newborn lamb. Am. J. Physiol. 1993; 265: H1318–25.

232. Minkes RK, Bellan JA, Saroyan RM, Kerstein MD, Coy DH, Murphy WA et al. Analysis of cardiovascular and pulmonary responses to endothelin-1 and endothelin-3 in the anesthetized cat. J. Pharmacol. Exp. Therap. 1990; 253: 1118–25.

233. Lippton HL, Cohen GA, McMurtry IF, Hyman AI. Pulmonary vasodilation to endothelin isopeptides in vivo is mediated by potassium channel activation. J. Appl. Physiol. 1991; 70: 947–52.

234. Barman SA, Ardell JL, Taylor AE. Effect of endothelin-1 on canine airway blood flow. J. Cardiovasc. Pharm. 1993; 22 (Suppl. 8): S274–7.

235. Bonvallet ST, Oka M, Yano M, Zamora MR, McMurtry IF, Stelzner TJ. BQ123, an ET_A receptor antagonist, attenuates endothelin-1-induced vasoconstriction in rat pulmonary circulation. J. Cardiovasc. Pharmacol. 1993; 22: 39–43.

236. Pinheiro JMB, Malik AB. Mechanisms of endothelin-1-induced pulmonary vasodilation in neonatal pigs. J. Physiol. 1993; 469: 739–52.

237. Filep JG, Sirois MG, Földes-Filep É, Rousseau A, Plante GE, Fournier A et al. Enhancement by endothelin-1 of microvascular permeability via the activation of ET_A receptors. Br. J. Pharmacol. 1993; 109: 880–6.

238. Berridge MJ. Inositol trisphosphate and diacylglycerol: Two interacting second messengers. Ann. Rev. Biochem. 1987; 56: 159–93.

239. Resink T, Scott-Burden T, Bühler FR. Endothelin stimulates phospholipase C in cultured vascular smooth muscle cells. Biochem. Biophys. Res. Comm. 1988; 157: 1360–8.

240. Van Renterghem C, Vigne P, Barhanin J, Schmid-Alliana A, Frelin C, Lazdunski M. Molecular mechanism of action of the vasoconstrictor peptide endothelin. Biochem. Biophys. Res. Comm. 1988; 157: 977–85.

241. Griendling KK, Tsuda T, Alexander RW. Endothelin stimulates diacylgylcerol accumulation and activates protein kinase C in cultured vascular smooth muscle cells. J. Biol. Chem. 1989; 264: 8237–40.

242. Muldoon LL, Rodland KD, Forsythe ML, Magun BE. Stimulation of phosphatidylinositol hydrolysis, diacylglycerol release, and gene expression in response to endothelin, a potent new agonist for fibroblasts and smooth muscle cells. J. Biol. Chem. 1989; 264: 8529–36.

243. Ohlstein EH, Horohonich S, Hay DWP. Cellular mechanisms of endothelin in rabbit aorta. J. Pharmacol. Exp. Ther. 1989; 250: 548–55.

244. Marsden PA, Danthuluri NR, Brenner BM, Ballermann BJ, Brock TA. Endothelin action of vascular smooth muscle involves inositol and calcium mobilization. Biochem. Biophys. Res. Comm. 1989; 158: 86–93.

245. Xuan, Y-T, Whorton AR, Watkins WD. Inhibition by nicardipine on endothelin-mediated inositol phosphate formation and Ca^{2+} mobilization in smooth muscle cell. Biochem. Biophys. Res. Comm. 1989; 160: 758–64.

246. Takuwa N, Takuwa Y, Yanagisawa M, Yamashita K, Masaki T. A novel vasoactive peptide endothelin stimulates mitogenesis through inositol lipid turnover in Swiss 3T3 fibroblasts. J. Biol. Chem. 1989; 264: 7856–61.

247. Sugiura M, Inagami T, Hare GMT, Johns JA. Endothelin action: inhibition by a protein kinase C inhibitor and involvement of phosphoinositols. Biochem. Biophys. Res. Comm. 1989; 158: 170–6.

248. Pitkänen M, Mäntymaa P, Ruskoaho H. Staurosporine, a protein kinase C inhibitor, inhibits atrial natriuretic peptide secretion induced by sarafotoxin, endothelin and phorbol ester. Eur. J. Pharmacol. 1991; 195: 307–15.

249. Cardell LO, Uddman R, Edvinsson L. Analysis of endothelin-1-induced contractions of guinea-pig trachea, pulmonary veins and different types of pulmonary arteries. Acta Physiol. Scand. 1990; 139: 130–11.

250. Oda K, Fujitani Y, Watakabe T, Inui T, Okada T, Urade Y et al. Endothelin stimulates both cAMP formation and phosphatidylinositol hydrolysis in cultured embryonic bovine tracheal cells. FEBS Letts. 1992; 299: 187–91.

251. Battistini B, Filep JG, Cragoe EJ Jr., Fournier A, Sirois P. A role of Na^+/H^+ exchange in contraction of guinea-pig airways by endothelin-1 in vitro. Biochem. Biophys. Res. Comm. 1991; 175: 583–8.

252. McKay, KO, Black JL, Armour CL. The mechanism of action of endothelin in human lung. Br. J. Pharmacol. 1991; 102: 422–8.

253. Hay DWP, Luttmann MA, Goldie RG. Calcium (Ca^{2+}) translocation mechanisms mediating endothelin-1 (ET-1)- and sarafotoxin S6c (S6c)-induced contractions in isolated human bronchus. Am. J. Respir. Crit. Care Med. 1994; 149: A1083.

254. Mattoli S, Soloperto M, Mezzetti M, Fasoli A. Mechanisms of calcium mobilization and phosphoinositide hydrolysis in human bronchial smooth muscle cells by endothelin 1. Am. J. Respir. Cell Mol. Biol. 1991; 5: 424–30.
255. Steffan M, Russell JA. Signal transduction in endothelin-induced contraction of rabbit pulmonary vein. Pulmonary Pharmacol. 1990; 3: 1–7.
256. Kay AB. Asthma and inflammation. J. Allergy Clin. Pharmacol. 1991; 87: 893–910.
257. Nomura A, Uchida Y, Kameyama M, Saotome M, Oki K, Hasegawa S. Endothelin and bronchial asthma. Lancet 1989; 2 (8665): 747–8.
258. Mattoli S, Soloperto M, Marini M, Fasoli A. Levels of endothelin in the bronchoalveolar lavage fluid of patients with symptomatic asthma and reversible airflow obstruction. J. Allergy Clin. Immunol. 1991; 88: 376–84.
259. Vittori E, Marini M, Fasoli A, De Franchis R, Mattoli S. Increased expression of endothelin in bronchial epithelial cells of asthmatic patients and effect of coritcosteroids. Am. Rev. Respir. Dis. 1992; 146: 1320–5.
260. Springall DR, Howarth PH, Counihan H, Djukanovic R, Holgate ST, Polak JM. Endothelin immunoreactivity of airway epithelium in asthmatic patients. Lancet 1991; 337: 697–701.
261. Sofia M, Mormile M, Farane S, Alifano M, Zofra S, Romano L et al. Increased endothelin-like immunoreactive material on bronchoalveolar lavage fluid from patients with bronchial asthma and patients with interstitial lung disease. Respiration 1993; 60: 89–95.
262. Kraft M, Beam WR, Wenzel SE, Zamora MR, O'Brien RF, Martin RJ. Blood and bronchoalveolar lavage endothelin-1 levels in nocturnal asthma. Am J. Respir. Crit. Care Med. 1994; 149: 947–52.
263. Reid, L. The pulmonary circulation: remodeling in growth and disease. Am. Rev. Respir. Dis. 1979; 119: 531–47.
264. Wagenvoort CA. Grading of pulmonary vascular lesions – a reappraisal. Histopathol. 1981; 5: 595–8.
265. Heath D, Smith P, Gosney J, Mulcahy D, Fox K, Yacoub M et al. The pathology of the early and late stages of primary pulmonary hyptertension. Br. Heart J. 1987; 58: 204–13.
266. Cernacek P, Steward DJ. Immunoreactive endothelin in human plasma: marked elevations in patients in cardiogenic shock. Biochem, Biophys. Res. Comm. 1989; 161: 562–7.
267. Stewart DJ, Levy RD, Cernacek P, Langleben D. Increased plasma endothelin-1 in pulmonary hypertension: marker or mediator of disease? Ann. Intern. Med. 1991; 114: 464–9.
268. Yoshibayashi M, Nishioka K, Nakao K, Saito Y, Matsumura M, Ueda T et al. Plasma endothelin concentrations in patients with pulmonary hypertension associated with congenital heart defects. Circulation 1991; 84: 2280–5.
269. Allen SW, Chatfield BA, Koppenhafer SA, Schaffer MS, Wolfe RR, Abman SH. Circulating immunoreactive endothelin-1 in children with pulmonary hypertension. Am. Rev. Respir. Dis. 1993; 148: 519–22.
270. Cacoub P, Dorent R, Maistre G, Nataf P, Carayon A, Piette JC et al. Endothelin-1 in primary pulmonary hypertension and the Eisenmenger Syndrome. Am. J. Cardiol. 1993; 71: 448–50.
271. Chang H, Wu G-J, Wang S-M, Hung C-R. Plasma endothelin levels and surgically correctable pulmonary hypertension. Ann. Thorac. Surg. 1993; 55: 450–8.
272. Rosenberg AA, Kennaugh J, Koppenhafer SL, Loomis M, Chatfield BA, Abman SH. Elevated immunoreactive endothelin-1 levels in newborn infants with persistent pulmonary hypertension. J. Pediatr. 1993; 123: 109–14.
273. Giaid A, Yanagisawa M, Langleben D, Michel RP, Levy R, Shennib H et al. Expression of endothelin-1 in the lungs of patients with pulmonary hypertension. New Eng. J. Med. 1993; 328: 1732–9.
274. Lippton HL, Hauth TA, Summer WR, Hyman AL. Endothelin produces pulmonary vasoconstriction and systemic vasodilation. J. Appl. Physiol. 1989; 66: 1008–12.
275. Yoshizumi M, Kurihara H, Sugiyama T, Takaku F, Yanagisawa M, Masaki T et al. Hemodynamic shear stress stimulates endothelin production by cultured endothelial cells. Biochem. Biophys. Res. Comm. 1989; 161: 859–64.
276. Stelzner TJ, O'Brien RF, Yanagisawa M, Sakurai T, Sato K, Webb S et al. Increased lung endothelin-1 production in rats with idiopathic pulmonary hypertension. Am. J. Physiol. 1992; 262: L614–20.

277. Miyauchi T, Yorikane R, Sakai S, Sakurai T, Okada M, Nishikibe M et al. Contribution of endogenous endothelin-1 to the progression of cardiopulmonary alterations in rats with monocrotaline-induced pulmonary hypertension. Circul. Res. 1993; 73: 887–97.
278. Wong J, Vanderford PA, Winters JW, Chang R. Soifer SJ, Fineman JR. Endothelin-1 does not mediate acute hypoxic pulmonary vasoconstriction in the intact newborn lamb. J. Cardiovasc. Pharmacol. 1993; 22 (Suppl. 8): S262–66.
279. Wiebke JL, Montrose-Rafizadeh C, Zeitlin PL, Guggino WB. Effect of hypoxia on endothelin-1 production by pulmonary vascular endothelial cells. Biochim. Biophys. Acta 1992; 1134: 105–11.
280. Giaid A, Hamid QA, Springall DR, Yanagisawa M, Shinmi O, Sawamura T et al. Detection of endothelin immunoreactivity and mRNA in pulmonary tumours. J. Pathol. 1990; 162: 15–22.
281. Mitaka C, Hirata Y, Nagura T, Tsunoda Y, Amaha K. Circulating endothelin-1 concentrations in acute respiratory failure. Chest 1993; 104: 476–80.
282. Giaid A, Michel RP, Stewart DJ, Sheppard M, Corrin B, Hamid Q. Expression of endothelin-1 in lungs of patients with cryptogenic fibrosing alveolitis. Lancet 1993; 341: 1550–4.
283. Marciniak SJ, Plumpton C, Barker PJ, Huskisson NS, Davenport AP. Localization of immunoreactive endothelin and proendothelin in the human lung. Pulmon. Pharmacol. 1992; 5: 175–82.
284. Andersson SE, Zackrisson C, Hemsén A, Lundberg JM. Regulation of lung endothelin content by the glucocorticosteroid budesonide. Biochem. Biophys. Res. Comm. 1992; 188: 1116–21.
285. Morel DR, Lacroix JS, Hemsen A, Steinig DA, Pittet J-F, Lundberg JM. Increased plasma and pulmonary lymph levels of endothelin during endotoxin shock. Eur. J. Pharmacol. 1989; 167: 427–8.
286. Filep JG, Télémaque S, Battistini B, Sirois P, D'Orléans-Juste P. Increased plasma levels of endothelin during anaphylactic shock in the guinea-pig. Eur. J. Pharmacol. 1993; 239: 231–6.
287. Giaid A, Stewart DJ, Michel RP. Endothelin-1-like immunoreactivity in postobstructive pulmonary vasculopathy. J. Vasc. Res. 1993; 30: 333–8.
288. Simmet T, Pritz S, Thelen KI, Peskar BA. Release of endothelin in the oleic acid-induced respiratory distress syndrome in rats. Eur. J. Pharmacol. 1992; 211: 319–22.
289. Shirakami G, Nakao K, Saito Y, Magariibuchi T, Jougasaki M, Mukoyama M et al. Acute pulmonary alveolar hypoxia increases lung and plasma endothelin-1 levels in conscious rats. Life Sci. 1991; 48: 969–76.
290. McLarty AJ, Miller VM, Tazelaar HD, McGregor CGA. Bronchial contractions in transplanted lungs: influence of denervation, acute rejection, and the bronchial epithelium. J. Thorac. Cardiovasc. Surg. 1993; 106: 797.
291. Scherstén H, Aarnio P, Burnett Jr JC, McGregor CGA, Miller VM. Endothelin-1 in bronchoalveolar lavage during rejection of allotransplanted lungs. Transplant. 1994; 57: 159–61.
292. Kurihara Y, Kurihara H, Suzuki H. Kodama T, Maemura K, Nagai R et al. Elevated blood pressure and craniofacial abnormalities in mice deficient in endothelin-1. Nature 1994; 368: 703–10.

Airways Smooth Muscle: Peptide Receptors, Ion Channels and Signal Transduction
ed. by D. Raeburn and M. A. Giembycz

CHAPTER 2
Bradykinin

Stephen G. Farmer

Pulmonary Pharmacology Section, Zeneca Pharmaceuticals Group, Wilmington, Delaware, USA

1   Introduction
2   Bradykinin Receptors
2.1 B$_1$ Receptors
2.2 B$_2$ Receptors
2.3 B$_2$ Receptor Antagonists
2.4 B$_3$ Receptors – Evidence for Novel BK Receptors in Airways Smooth Muscle
3   Pharmacology of Bradykinin in Isolated Airways Smooth Muscle Preparations
3.1 Human Bronchi
3.2 Guinea-Pig Trachea
3.3 Ferret Trachea
4   Bradykinin Receptor Signal Transduction in Airways Smooth Muscle Cells
5   Conclusion: Bradykinin Receptors in Airways Smooth Muscle – Why?
    References

1. Introduction

Bradykinin (BK) is a nonapeptide (Arg-Pro-Pro-Gly-Phe-Ser-Pro-Phe-Arg) that is released during trauma and inflammation. The events involved in the biosynthesis and release of BK and related kinins such as kallidin (Lys-BK, KD) have been reviewed [1–4], and will not be detailed further at present. BK exerts numerous effects, via cell-surface receptors, on virtually every mammalian organ, tissue and cell type [4–7]. In addition, kinins have been implicated as mediators of pain and inflammation in many conditions including pancreatitis, burns, and arthritic diseases [3, 8–11]. BK may also have an important role in the regulation of blood pressure in normal individuals, as well as in hypertension and septic shock [3, 12–15]. Of some relevance here, BK may also contribute to inflammatory diseases of the upper [16, 17] and lower airways [18, 19].

Although BK is a potent bronchoconstrictor in asthmatic subjects [20–22], it exhibits minimal mechanical effects in most published studies utilizing human isolated airways smooth muscle [20]. Nevertheless, BK causes contraction (or, depending upon the circumstances, relaxation) of airways isolated from some species. The present article discusses mechanical and biochemical effects of BK and related kinins on human and animal isolated airways smooth muscle tissues and cells, and the pharmacology of BK receptors therein.

2. Bradykinin Receptors

Kinin receptors were originally classified according to the relative potencies of agonists in isolated vascular smooth muscle, and divided into B_1 and B_2 subtypes [23]. B_2 receptors in rabbit jugular vein or dog carotid artery respond to kinin receptor agonists with the following order of potency:

$$KD > [Tyr(Me)^8]\text{-}BK > BK > desArg^{10}\text{-}KD > desArg^9\text{-}BK$$

B_1 receptors, in rabbit aorta or mesenteric vein, exhibit almost the opposite order of sensitivity:

$$desArg^{10}\text{-}KD > desArg^9\text{-}BK > KD > BK > [Tyr(Me)^8]\text{-}BK$$

Conventionally, B_2 receptors exhibit much higher affinity for KD or BK than for the carboxypeptidase N metabolites, desArg^{10}-KD or desArg^9-BK, whereas B_1 receptors are more sensitive to the latter peptides. Accordingly, the relative potencies of desArg^9-BK and BK are utilized by many investigators to classify a particular response as being mediated, respectively, by B_1 or B_2 receptors [23].

2.1. B_1 Receptors

B_1 receptors are best described in rabbit vascular smooth muscle [8, 24], although they have also been demonstrated in some cells maintained in tissue culture. In human lung fibroblasts, for example, B_1 receptors mediate collagen formation and cell proliferation [25]. Similarly, some types of mouse macrophages express B_1 receptors [26]. However, there is no evidence for the expression of B_1 receptors in airways smooth muscle from any species yet studied, including humans [21, 27, 28], ferrets [29, 30] or guinea pigs [31–33]. B_1 receptors, therefore, will not be considered further in this chapter.

2.2. B_2 Receptors

Traditionally, a lack of effect of desArg^9-BK in a tissue that nevertheless responded to BK resulted, by default, in the classification of the receptor as B_2 [8, 23] B_2 receptors are expressed in most, if not all smooth muscle tissues, where they mediate contraction and/or relaxation (reviewed in [4, 7, 18, 34]). Briefly, kinin-induced responses such as contraction of guinea-pig intestine [35–37], contraction and relaxation of rat duodenum [38–40], and contraction of rat uterus [37, 41, 42] are mediated by B_2 receptors. In addition, endothelium-dependent relaxation of arterial smooth muscle, from several species including man, is mediated by B_2 receptors [43, 44]. Intestinal and airway anion secretion induced by kinins are also mediated by B_2 receptors [36, 45].

2.3. B_2 Receptor Antagonists

Substitution with DPhe for LPro in position 7 of the BK sequence yields [DPhe^7]-BK (NPC 361), and this was the first described B_2 receptor antagonist [46]. Numerous analogues of NPC 361 have been synthesized, and their pharmacology is reviewed elsewhere [5, 47, 48]. Many of the [DPhe^7]-substituted analogues of BK have relatively weak affinity for B_2 receptors [5, 48, 49], and they often elicit contractile responses of isolated smooth muscle [32, 37, 50]. Indeed, some analogues of [DPhe^7]-BK are partial agonists in stimulating increased intracellular calcium ion concentration ($[Ca^{2+}]_i$) in Chinese hamster ovary (CHO) cells transfected with human B_2 receptor DNA [51], as well as in eliciting efflux of radiolabelled Ca^{2+} ($^{45}Ca^{2+}$) in human lung fibroblasts [52].

Recently B_2 receptor antagonists which contain unnatural amino acids were described [53–56]. Peptides such as DArg[Hyp^3, Thi^5, DTic^7, Oic^8]-BK (HOE 140), DArg-[Hyp^3, Thi^5, DTic^7, Tic^8]-BK (NPC 16731), DArg-[Hyp^3, DHypS(transphenyl)^7, Oic^8]-BK (NPC 17761) are several orders of magnitude more potent B_2 receptor antagonists than the [DPhe^7]-substituted analogues both *in vitro* [53, 54, 56, 57] and *in vivo* [55, 56, 58, 59].

2.4. B_3 Receptors – Evidence for Novel BK Receptors in Airways Smooth Muscle

BK-induced contraction of guinea-pig trachea is resistant to inhibition by several B_2 receptor antagonist analogues of NPC 361 [32, 60, 61]. DesArg^9-BK was inactive, and BK-induced contraction was unaltered by desArg^9-[Leu^8]-BK, a B_1 receptor antagonist. In ligand binding studies in lung parenchyma or tracheal smooth muscle, neither desArg^9-[Leu^8]-BK nor DArg-[Hyp^3, DPhe^7]-BK (NPC 567) displaced tracheal binding of [^{3}H]-BK and, in lung membranes, the B_2 receptor antagonist displaced only 60% of bound ligand [32]. Thus, B_2 receptor antagonists had only minor effects against BK-induced contractions of guinea-pig trachea, and did not significantly displace BK binding. We proposed that the spasmogenic activity of BK in this tissue may be mediated by a novel receptor, designated B_3 [32]. Likewise, membranes prepared from sheep trachealis contain receptors from which NPC 567 displaced only 28% of [^{3}H]-BK binding [32]. That desArg^9-[Leu^8]-BK also had no effect suggests that sheep trachealis, like that of guinea pigs, may express B$_3$ receptors.

Although our binding data in guinea-pig lungs have been verified [27], in that a B$_1$ receptor antagonist had no effect, and a B$_2$ antagonist displaced only 70% of BK binding, other studies have failed to reproduce our original findings. For example, a detailed examination of BK binding sites in guinea-pig lungs indicated the presence of two BK binding sites, with high and low affinity. B$_1$ receptor ligands had no effect on [^{3}H]-BK binding, whereas several B$_2$ receptor antagonists caused 100% displacement [62]. Although these data are at variance with other studies in guinea-pig lungs [27, 32], the reasons are unknown, and may reflect dissimilar experimental conditions, or differences in the health status of the animals. In addition, the relative physiological significance, if any, of high and low affinity binding sites in guinea-pig lungs is unknown.

Field and colleagues [61], essentially using the same experimental protocols as us [32], reported that several analogoues of NPC 361 caused complete inhibition of BK binding to guinea-pig tracheal smooth muscle membranes. Similarly, in membranes prepared from guinea-pig trachealis, Trifilieff et al. [63] reported that NPC 361 and NPC 567 each completely displaced BK binding. These contrasting data derived from ligand binding experiments in guinea-pig trachea [32, 57, 61, 63] are puzzling and, as yet, unexplained.

Nevertheless, there are other indications of a novel BK receptor in guinea-pig tracheal smooth muscle cells maintained in tissue culture. In these cells, BK stimulated prostaglandin synthesis and efflux of ^{45}Ca^{2+} [64]. NPC 567 inhibited BK-induced prostaglandin synthesis, indicating the presence of B$_2$ receptors in the cultured cells. The efflux of ^{45}Ca^{2+}, however, was resistant to the B$_2$ receptor antagonist. Moreover desArg^9-BK had no activity [64]. These studies, accordingly, illustrated that ^{45}Ca^{2+} efflux in trachealis may be mediated by B$_3$ receptors.

Also in guinea-pig tracheal smooth muscle cells, Pyne and Pyne [33] reported that BK-induced phospholipase D (PLD) activity, in contrast to phospholipase C (PLC), was only very weakly inhibited by NPC 567. DesArg^9-BK had no apparent agonist effects, and the authors proposed that BK-induced activation of PLC is mediated by B$_2$ receptors, whereas BK-induced stimulation of PLD is mediated by B$_3$ receptors. Studies on airways smooth muscle signal transduction responses, elicited by BK, are discussed in further detail in Section 4 of this chapter.

Until recently, known BK receptor antagonists were all peptides. The discovery of WIN 64338 (phosphonium, [[4-[[2-[[bis(cyclo-hexylamino)methylene]amino]-3-(2-naphthalenyl) 1-oxopropyl]amino]-phenyl]-methyl]-tributyl, chloride, monohydrochloride) provides the first *nonpeptide* B_2 receptor antagonist [65]. In a preliminary study WIN 64338 was reported to competitively antagonize BK-induced contractions of guinea-pig ileum with a pA_2 value of 8.20 [66].

We have compared the effects of WIN 64338 on BK-induced contractions of guinea-pig ileum with those in guinea-pig and ferret trachea. Although it was a selective and potent B_2 receptor antagonist in ileum (pA_2 7.97), WIN 64338 was completely without effect upon BK-induced contractions of tracheal smooth muscle from either species [67]. These data provide convincing evidence that the BK receptor subtype expressed in ferret and guinea-pig tracheal smooth muscle is dissimilar from the classical B_2 receptors in guinea-pig ileum. The current lack of an antagonist that specifically blocks tracheal contractions in response to kinins, however, renders the existence of B_3 receptors moot.

3. Pharmacology of Bradykinin in Isolated Airways Smooth Muscle Preparations

3.1. Human Bronchi

As noted in the Introduction, BK has minimal direct mechanical effects on isolated airways smooth muscle preparations from most species, including man [18]. Over thirty years ago, Bhoola *et al.* [68] first noted that BK, at concentrations as high as (667 µg/ml, or approx. 5×10^{-4} M), failed to contract bronchial smooth muscle preparations isolated from cancer patients. Similar data have since been reported by others [20] although Simonsson *et al.* [69] noted that, in segmental bronchi of 3–5 mm diameter, BK caused "weak contraction" in two out of four preparations from patients without obstructive lung disease. "In contrast, all 7 tested preparations from patients with obstructive lung disease (asthma, bronchitis, emphysema) showed a stronger sensitivity to bradykinin" [69].

More recently, it was reported that BK elicits contractions in smaller preparations of human bronchus than utilized in the earlier studies [28]. In segments of bronchi with diameters of 0.5–1 mm, isolated from nonasthmatic patients undergoing surgery for lung cancer, BK was reasonably potent (EC_{50} 2×10^{-8} M) in eliciting contractions from these tissues. It was, however, not a particularly efficacious agonist as the maximum response was less than 40% of that to acetylcholine. In agreement with earlier studies, Advenier's group [28] noted that they could elicit no response to BK in larger bronchi (3–5 mm diameter). In the smaller diameter tissues, [Hyp³, Tyr(Me)⁸]-BK, a B_2 receptor agonist exhibited similar potency and efficacy to BK, whereas [Sar¹, DPhe⁸, desArg⁹]-BK, a B_1 receptor-selective agonist, caused only a small contraction at a relatively high concentration (1 µM). The potent B_2 (and, possibly, B_3) receptor antagonist, HOE 140, inhibited BK-induced contractions in an apparently noncompetitive manner, whereas desArg⁹-[Leu⁸]-BK was without effect [28]. Thus, B_2 receptors may mediate BK-induced contractions of human isolated bronchi.

In these tissues, indomethacin or capsaicin abolished BK-induced contractions indicating the involvement, respectively, of cyclooxygenase metabolites of arachidonic acid metabolism, and neuropeptides released from sensory nerve endings within the tissues [28]. However, neither a neurokinin (NK)₁ receptor antagonist nor a NK_2 receptor antagonist had any effect on BK-induced contractions. The effects of a combination of both neurokinin antagonists was not examined, and it is possible that the effect of capsaicin on BK-induced contractions reflected some nonspecific action of the alkaloid, although its effects upon other agonists were not examined.

3.2. Guinea-Pig Trachea

The qualitative nature of responses of guinea-pig trachea to BK is highly variable, and depends on the basal tone as well as the integrity of the tracheal epithelium. In early studies, in which tone was raised with histamine, B_2 receptor agonists, but not desArg⁹-BK, caused concentration-dependent relaxation of guinea-pig trachea [31, 70]. That the responses were abolished by inhibitors of cyclooxygenase [31, 70, 71] indicates that BK-induced relaxation is mediated by endogenous prostaglandins. Moreover, there is a recent report to the effect that BK-induced epithelium-dependent relaxation of this tissue involves not only prostaglandin (PG)E_2 but also epithelium-derived nitric oxide (NO) [72].

In guinea-pig trachea under basal tone, BK causes weak relaxation which is converted to contraction by removing the epithelium [32, 73, 74]. As noted above, epithelium-dependent relaxation, in response to BK, is mediated by epithelium-derived PGE_2 [72, 73]. Similarly, BK stimulates the release of sufficient quantities of PGE_2 from dog tracheal epithelial cells, maintained in tissue culture, to cause relaxation of dog tracheal smooth muscle [75].

BK-induced contractions of guinea-pig epithelium-denuded trachea under basal tone are also inhibited by indomethacin [32, 73], suggesting they, too, are mediated largely by protaglandins. Furthermore, in guinea-pig tracheal smooth muscle cells in culture BK, but not desArg⁹-BK, caused concentration-dependent release of PGE_2 and the stable prostacyclin metabolite, 6-keto-$PGF_{1\alpha}$ [64]. BK-induced prostaglandin release was abolished by indomethacin. Even in the presence of indomethacin though, higher concentrations of BK (0.1 µM), elicited contraction [32]. BK, therefore, causes prostaglandin-dependent and

-independent contraction of this tissue. As with other bioactive peptides, guinea-pig tracheal responses to BK are subject to modulation by peptidase enzymes such as neutral endopeptidase 24.11 (NEP) [74] and endopeptidase 24.15 [76].

3.3. Ferret Trachea

BK is a potent excitatory agonist in ferret trachea [29, 77, 78]. BK-induced contractions are not modified by drugs such as atropine, tetrodotoxin, indomethacin, propranolol, phentolamine, ketanserin or chlorpheniramine [30] indicating that BK exerts a direct effect on ferret tracheal smooth muscle BK receptors. The B_2 receptor agonists, BK, KD and $[Tyr(Me)^8]$-BK, were equipotent in causing contraction, whereas desArg^9-BK was very weak and inefficacious [30, 78]. Moreover, peptide B_2 receptor antagonists, but not a B_1 receptor antagonist, inhibited kinin-induced responses in this tissue [30, 78]. These data, then, suggested that ferret tracheal smooth muscle contractions are mediated by B_2 receptors. However, as noted earlier, WIN 64338, a nonpeptide B_2 receptor antagonist, was without effect upon BK-induced contractions in ferret trachea [67], raising the possibility that tracheal smooth muscle from this species, as in guinea pigs, may express putative B_3 receptors.

Also like guinea-pig trachea, responses of ferret trachea to kinins are protentiated by inhibitors of NEP and angiotensin-converting enzyme (ACE) [29]. This study also demonstrated that homogenates prepared from ferret trachea contain both of these peptidases [29]. Recently, we found that many preparations of ferret trachea responded to a cocktail of captopril and thiorphan, inhibitors of ACE and NEP, respectively, by exhibiting strong, well-maintained contractions [30]. These responses were inhibited completely by B_2 receptor antagonists but not a B_1 receptor antagonist. Thus, ferret trachea appears capable of releasing and degrading kinins *in vitro*. Inhibition of kinin catabolism may allow the endogenous kinin to accumulate in concentrations sufficient to elicit tracheal contraction [30].

4. Bradykinin Receptor Signal Transduction in Airways Smooth Muscle Cells

Activation of B_2 receptors, in virtually all tissues, leads to stimulation of phosphatidylinositol (PI)-specific PLC, resulting in the formation of inositol phosphates and diacyglycerol, and subsequent elevations in $[Ca^{2+}]_i$ (see [7, 34, 49, 79, 80] for reviews). Airways smooth muscle appears to be no exception to this [33, 64, 80–82]. Thus, in bovine [81,

82] and guinea-pig [33] tracheal smooth muscle cells maintained in tissue culture, BK receptor activation leads to activation of PLC and elevated $[Ca^{2+}]_i$. In addition, in human tracheal smooth muscle, BK-induced increases in $[Ca^{2+}]_i$ appear to be mediated by Ca^{2+} influx via receptor-operated [83] and voltage-gated Ca^{2+} channels [84].

As noted earlier, it has been demonstrated recently that BK stimulates PLD activity in addition to PLC in guinea-pig tracheal smooth muscle cells [85]. Activation of PLD is abrogated by staurosporine, a protein kinase C (PKC) inhibitor, or following down-regulation of PKC with phorbolesters [85]. Moreover, BK-induced stimulation of PLD is markedly attenuated by removal of extracellular Ca^{2+}. It was suggested that BK receptor activation stimulates PKC which, in turn, increases PLD activity [85]. However, if PKC is activated by 1,2-diacylglycerol (DAG), then this lipid mediator must be generated independently of phosphatidylinositol 4,5-biphosphate (PIP$_2$) hydrolysis by PLC. That is, stimulation of PKC and PLD occur in the absence of PLC activation [33, 85]. Furthermore, these effects are not secondary to generation of prostanoids, as BK-induced PLC activity is unaffected by indomethacin [85].

In an examination of the receptors underlying BK receptor signal transduction in guinea-pig tracheal smooth muscle cells, Pyne and Pyne [33] reported that B$_2$ receptor antagonists abolished BK-induced PLC activity, while having only minimal actions against PLD. The authors proposed that inositol (1,4,5)trisphosphate (IP$_3$) formation, as a result of PLC stimulation, occurs via B$_2$ receptors. B$_3$ receptor effects, on the other hand, may be transduced by PKC which, in turn, activates PLD [33]. Interestingly, whereas BK-induced IP$_3$ synthesis in bovine [81] and guinea-pig [33] tracheal smooth muscle is transient and may be subject to desensitization, PLD activity was sustained, perhaps reflecting its mediation of $^{45}Ca^{2+}$ efflux [64] and the sustained phase of BK-induced tracheal contraction [33]. Recently, BK was shown to enhance adenylyl cyclase activity and intracellular cyclic AMP accumulation, secondarily to PLD activation, in guinea-pig tracheal smooth muscle cells. This, too, was unaffected by a B$_2$ receptor antagonist [86].

In most tissues, eicosanoid synthesis is also enhanced by B$_2$ receptor activation, potentially by release of arachidonic acid from the diacylglycerol produced by inositol lipid hydrolysis [5, 49]. In guinea-pig tracheal smooth muscle cells, as reported earlier in this review, BK stimulates the biosynthesis of relatively large quantities of PGE$_2$ and PGI$_2$, and this was inhibited by indomethacin or a B$_2$ receptor antagonist [64]. We have also found, in human bronchial smooth muscle cells, that BK releases significant quantities of PGE$_2$ and PGI$_2$ (R.M. Burch and S.G. Farmer, unpublished observations).

5. Conclusion: Bradykinin Receptors in Airways Smooth Muscle – Why?

The physiological role of airways smooth muscle BK receptors is unclear, but probably does not involve mediation of contraction to any significant extent. A discussed, inhaled BK is a potent bronchoconstrictor *in vivo* in asthmatic subjects and, yet, it has little or only feeble effects on human isolated bronchial smooth muscle. Furthermore, the pharmacological mechanisms underlying bronchoconstriction in response to BK in humans differs from the pharmacology of BK-induced contractions of isolated airways. For example, cyclooxygenase inhibitors are without effect on BK-induced bronchonconstriction *in vivo* [20, 87], while BK-induced contraction of human bronchi *in vitro* is abolished by indomethacin [28]. Similarly, although antagonists of NK_1 or NK_2 receptors have no effect on BK-induced contraction of human isolated bronchi [28], FK 224, an NK_1/NK_2 receptor antagonist inhibits BK-induced bronchoconstriction *in vivo* [88]. Also, although muscarinic cholinoceptor antagonists markedly attenuate the effect of BK on human airway resistance [20, 22], atropine was absolutely without effect on BK-induced contractions of human isolated bronchi [28]. In summary, in humans, BK-induced bronchoconstriction *in vivo* appears to result largely indirectly via activation of airway nerves but independently of cyclooxygenase products of arachidonic acid metabolism (reviewed in [22]). Thus, direct effects of BK on airways smooth muscle evidently have little or no importance *in vivo*.

As in man, the pharmacology of BK *in vivo* and *in vitro* in guinea-pig airways differs. While inhibitors of cyclooxygenase clearly abrogate BK-induced contraction of guinea-pig isolated trachea [32, 73], their effects in intact animals varies considerably depending upon the route of BK administration [89–93]. Nevertheless, indomethacin was without effect increases lung resistance elicited by BK instilled into the trachea of anesthetized guinea pigs [91]. Also, in preliminary studies, we found that the bronchoconstrictor response following inhalation of aerosolized BK in conscious guinea pigs was unaffected by indomethacin [94]. Moreover, while capsaicin densensitization has a small inhibitory effect on BK-induced guinea-pig tracheal contraction [95], BK-induced bronchoconstriction *in vivo* is abolished by capsaicin [91, 94].

What then is the function of airways smooth muscle BK receptors? BK stimulates mitogenesis and proliferation of other cell types including various fibroblast lines [25, 96, 97] and, in kidney, mesangial cells [98, 99]. In contrast, there is evidence that BK inhibits proliferation of vascular smooth muscle cells [100, 101]. As discussed above, the signal transduction events coupled to airways smooth muscle BK receptors involve activation of PLC and PLD, IP_3 and DAG production, elevations in $[Ca^{2+}]_i$, and stimulation of PKC. Each of these mechanisms,

which often interact with each other, has been implicated in cell proliferation [97–99, 102–106]. Also, PLD may have pivotal role in mitogenesis [107]. Moreover, in several types of cells, including smooth muscle,
BK receptors may be coupled to one or more tyrosine kinases [106,
108–111]. Several colony-stimulating factors and growth factors activate receptors, activation of which results in tyrosine phosphorylation
and cell proliferation [106–108, 112–114]. Thus, it is feasible that BK
receptors may have an important role in the regulation of airways
smooth muscle proliferation and this is clearly an area that is ripe for
future research.

References

1. Proud D, Kaplan AP. Kinin formation: mechanisms and role in inflammatory disorders. Ann Rev Immunol 1988; 6: 49–84.
2. Clements JA. The glandular kallikrein family of enzymes: tissue-specific expression and hormonal regulation. Endocrine Rev 1989; 10: 393–419.
3. Margolius HS. Tissue kallikreins and kinins: regulation and roles in hypertensive and diabetic diseases. Ann Rev Pharmacol Toxicol 1989; 29: 343–64.
4. Bhoola KD, Figueroa CD, Worthy K. Bioregulation of kinins: Kallikreins, kininogens, and kininases. Pharmacol Rev 1992; 44: 1–80.
5. Burch RM, Farmer SG, Steranka LR. Bradykinin receptor antagonists. Med Res Rev 1990, 10: 237–69.
6. Farmer SG, Burch RM. The pharmacology of bradykinin receptors. In: Burch RM, ed. Bradykinin antagonists. Basic and clinical research. New York: Marcel Dekker, Inc, 1991: 1–31.
7. Hall JM. Bradykinin receptors: pharmacological properties and biological roles. Pharmacol Ther 1992; 56: 131–90.
8. Marceau F, Lussier A, Regoli D, Giroud JP. Pharmacology of kinins: their relevance to tissue injury and inflammation. Gen Pharmacol 1983; 14: 209–29.
9. Kozin K, Cochrane CG. The contact activation system of plasma: biochemistry and pathophysiology. In: Gallin JI, Goldstein IM, Snyderman R, eds. Inflammation: basic principles and clinical correlates. New York: Raven Press, Ltd, 1988: 101–120.
10. Steranka LR, Burch RM. Bradykinin antagonists in pain and inflammation. In: Burch RM, ed. Bradykinin antagonists. Basic and clinical research. New York: Marcel Dekker, Inc, 1991: 191–211.
11. Dray A, Perkins M. Bradykinin and inflammatory pain. Trends Neurosci 1993; 16: 99–104.
12. Gavras H, Gavras I. Effects of bradykinin antagonists on the cardiovascular system. In: Burch RM, ed. Bradykinin antagonists. Basic and clinical research. New York: Marcel Dekker Inc, 1991: 171–189.
13. Bao G, Gohlke, Unger T. Role of bradkykinin in chronic antihypertensive actions of ramipril in different hypertension models. J. Cardiovasc Pharmacol 1992; 20 Suppl. 9: S96–9.
14. Busse R, Fleming I, Hecker M. Endothelium-derived bradykinin: Implications for angiotensin-converting enzyme-inhibitor therapy. J. Cardiovasc Pharmacol 1993; 22 Suppl. 5: S31–6.
15. Madeddu P, Glorioso N, Varoni MV, Demontis MP, Fattaccio MC, Anania V. Cardiovascular effects of brain kinin receptor blockade in spontaneously hypertensive rats. Hypertension 1994; 23 Suppl.: I189–92.
16. Pongracic JA, Churchill L, Proud D. Kinins in rhinitis. In: Burch RM, ed. Bradykinin antagonists. Basic and clinical research. New York: Marcel Dekker, Inc, 1991: 237–259.
17. Rajakulasingam K, Polosa R, Church MK, Holgate ST, Howarth PH. Kinins and rhinitis. Clin Exp Allergy 1992; 22: 734–40.

18. Farmer SG. Role of kinins in airway diseases. Immunopharmacology 1991; 22: 1–20.
19. Barnes PJ. Bradykinin and asthma. Thorax 1992; 47: 979–83.
20. Fuller RW, Dixon CMS, Cuss FMC, Barnes PJ. Bradykinin-induced bronchoconstriction in humans: mode of action. Am Rev Respir Dis 1987; 135: 176–80.
21. Polosa R, Holgate ST. Comparative airway responses to inhaled bradykinin, kallidin, and [des-Arg^9]bradykinin in normal and asthmatic subjects. Am Rev Respir Dis 1990; 142: 1367–71.
22. Farmer SG. The pharmacology of bradykinin in human airways. In: Page CP, Metzger WJ, eds. Drugs and the lung. New York: Raven Press, 1994: 449–465.
23. Regoli D, Barabé J. Pharmacology of bradykinin and related kinins. Pharmacol Rev 1980; 32: 1–46.
24. Marceau F, Regoli D. Kinin receptors of the B1 type and their antagonists. In: Burch RM, ed. Bradykinin antagonists. Basic and clinical research. New York: Marcel Dekker, Inc, 1991: 33–49.
25. Goldstein RH, Wall M. Activation of protein formation and cell division by bradykinin and Des-Arg^9-bradykinin. J. Biol Chem 1984; 259: 9263–8.
26. Tiffany CW, Burch RM. Bradykinin stimulates tumor necrosis factor and interleukin 1 release from macrophages. FEBS Lett 1989; 247: 189–92.
27. Mak JCW, Barnes PJ. Autoradiographic visualization of bradykinin receptors in human and guinea pig lung. Eur J Pharmacol 1991; 194: 37–43.
28. Molimard M, Martin CAE, Naline E, Hirsch A, Advenier C. Contractile effects of bradykinin on the isolated human small bronchus. Am Rev Respir Dis 1994; 149: 123–7.
29. Dusser DJ, Nadel JA, Sekizawa K, Graf PD, Borson DB. Neutral endopeptidase and angiotensin converting enzyme inhibitors potentiate kinin-induced contraction of ferret trachea. J. Pharmacol Exp Ther 1988; 244: 531–6.
30. Farmer SG, Broom T, DeSiato MA. Effects of bradykinin receptor agonists, and captopril and thiorphan in ferret isolated trachea: evidence for bradykinin generation in vitro. Eur J Pharmacol 1994; 259: 309–313.
31. Mizrahi J, Couture R, Caranikas S, Regoli D. Pharmacological effects of peptides on tracheal smooth muscle. Pharmacology 1982; 25: 39–50.
32. Farmer SG, Burch RM, Meeker SN, Wilkins DE. Evidence for a pulmonary bradykinin B$_3$ receptor. Mol Pharmacol 1989; 36: 1–8.
33. Pyne S, Pyne NJ. Differential effects of B$_2$ receptor antagonists upon bradykinin-stimulated phospholipase C and D in guinea-pig cultured tracheal smooth muscle. Br J Pharmacol 1993; 110: 477–81.
34. Farmer SG, Burch RM. Airway bradykinin receptors. Ann NY Acad Sci 1991; 629: 237–49.
35. Manning DC, Vavrek R, Stewart JM, Snyder SH. Two bradykinin binding sites with picomolar affinities. J Pharmacol Exp Ther 1986; 237: 504–12.
36. Kachur JF, Allbee W, Danho W, Gaginella TS. Bradykinin receptors: functional similarities in guinea-pig gut muscle and mucosa. Regul Pept 1987; 17: 63–70.
37. Farmer SG, Burch RM, Dehaas CJ, Togo J, Steranka LR. [Arg^1-DPhe^7]-substituted analogs of bradykinin inhibit vasopressin- and bradykinin-induced contractions of uterine smooth muscle. J. Pharmacol Exp Ther 1989; 248: 677–81.
38. Boschcov P, Paiva ACM, Paiva TB, Shimuta SI. Further evidence for the existence of two receptor sites for bradykinin responsible for the diphasic effect in the rat isolated duodenum. Br J Pharmacol 1984; 83: 591–600.
39. Paiva ACM, Paiva TB, Pereira CC, Shimuta SI. Selectivity of bradykinin analogues for receptors mediating contraction and relaxation of the rat duodenum. Br J Pharmacol 1989; 98: 206–10.
40. Hall JM, Morton IKM. Bradykinin B$_2$ receptor evoked K$^+$ permeability increase mediates relaxation in the rat duodenum. Eur J Pharmacol 1991; 193: 231–8.
41. Liebmann C, Offermanns S, Spicher K, Hinsch K-D, Schnittler M, Morgat JL, Reissmann S, Schultz G, Rosenthal W. A high affinity bradykinin receptor in membranes from rat myometrium is coupled to pertussis toxin-sensitive G-proteins of the G$_i$ family. Biochem Biophys Res Commun 1990; 167: 910–7.
42. Snell PH, Phillips E, Burgess CA, Snell CR, Webb M. Characterization of bradykinin receptors solubilized from rate uterus and NG108-15 cells. Biochem Pharmacol 1990; 39: 1921–8.

43. Whalley ET, Amure YO, Lye RH. Analysis of the mechanism of action of bradykinin on human basilar artery in vitro. Naunyn-Schmiedeberg's Arch Pharmacol 1987; 335: 433–7.
44. Schini VB, Boulanger C, Regoli D, Vanhoutte PM. Bradykinin stimulates the production of cyclic GMP *via* activation of B_2 kinin receptors in cultured porcine aortic endothelial cells. J. Pharmacol Exp Ther 1990; 252: 581–5.
45. Gaginella TS, Kachur JF. Kinins as mediators of intestinal secretion. Am J Physiol 1989; 256: G1–G15.
46. Vavrek RJ, Stewart JM. Competitive antagonists of bradykinin. Peptides 1985; 6: 161–4.
47. Steranka LR, Farmer SG, Burch RM. Antagonists of B_2 bradykinin receptors. FASEB J 1989; 3: 2019–25.
48. Stewart JM, Vavrek RJ. Chemistry of peptide B_2 bradykinin antagonists. In: Burch RM, ed. Bradykinin antagonists. Basic and clinical research. New York: Marcel Dekker Inc, 1991: 51–96.
49. Farmer SG, Burch RM. Biochemical and molecular pharmacology of kinin receptors. Ann Rev Pharmacol Toxicol 1992; 32: 511–36.
50. Llona I, Vavrek R, Stewart J, Huidobro-Toro JP. Identification of pre- and postsynaptic bradykinin receptor sites in the vas deferens: evidence for different structural prerequisites. J. Pharmacol Exp Ther 1987; 241: 608–14.
51. Eggerickx D, Raspe E, Bertrand D, Vassart G, Parmentier M. Molecular cloning, functional expression and pharmacological characterization of a human bradykinin B_2 receptor gene. Biochem Biophys Res Commun 1992; 187: 1306–13.
52. Sawutz DG, Faunce DM, Houck WT, Haycock D. Characterization of bradykinin B_2 receptors on human IMR-90 lung fibroblasts: Stimulation of $^{45}Ca^{2+}$ efflux by D-Phe^7 substituted bradykinin analogues. Eur J Pharmacol Mol Pharmacol 1992; 227: 309–15.
53. Hock FJ, Wirth K, Albus U, Linz W, Gerhards HJ, Wiemer G, St. Henke G, Breipohl G, König W, Knolle J, Schölkens BA. Hoe 140 a new potent and long acting bradykini-antagonist: *in vitro* studies. Br J. Pharmacol 1991; 102: 769–73.
54. Kyle DJ, Martin JA, Farmer SG, Burch RM. Design and conformational analysis of several highly potent bradykinin receptor antagonists. J. Med Chem 1991; 34: 1230–3.
55. Wirth K, Hock FJ, Albus U, Linz W, Alpermann HG, Anagnostopolous H, St. Hencke G, Breipohl G, König, W. Knolle, J. Schölkens BA. Hoe 140 a new potent and long acting bradykinin-antagonist: *in vivo* studies. Br J Pharmacol 1991; 102: 774–7.
56. Kyle DJ, Burch RM. A survey of bradykinin receptors and their antagonists. Curr Opin Invest Drugs 1993; 2: 5–20.
57. Farmer SG, Burch RM, Kyle DJ, Martin JA, Meeker SN, Togo J. DArg[Hyp^3-Thi^5-DTic^7-Tic^8]-bradykinin, a potent antagonist of smooth muscle BK_2 receptors and BK_3 receptors. Br J Pharmacol 1991; 102: 785–7.
58. Martorana PA, Kettenbach B, Breipohl G, Linz W, Schölkens BA. Reduction of infarct size by local angiotensin-coverting enzyme inhibition by a bradykinin antagonist. Eur J Pharmacol 1990; 182: 395–6.
59. Bao G, Qadri F, Stauss B, Stauss H, Gohlke P, Unger T. HOE 140, a new highly potent and long-acting bradykinin antagonist in conscious rats. Eur J Pharmacol 1991; 200: 179–82.
60. Perkins MN, Burgess GM, Campbell EA, Hallett A, Murphy RJ, Naeem S, Patel IA, Rueff A, Dray A. HOE140: a novel bradykinin analogue that is a potent antagonist at both B_2 and B_3 receptors *in vitro*. Br J Pharmacol 1991; 102: 171P.
61. Field JL, Hall JM, Morton IKM. Putative novel bradykinin B_3 receptors in the smooth muscle of the guinea-pig taenia caeci and trachea. Agents Actions 1992; 38, Suppl. I: 540–5.
62. Trifilieff A, Haddad E-B, Landry Y, Gies J-P. Evidence for two high affinity bradykinin binding sites in the guinea-pig lung. Eur J Pharmacol 1991; 207: 129–34.
63. Trifilieff A, Da Silva A, Landry Y, Gies J-P. Effect of Hoe 140, a new B_2 noncompetitive antagonist, on guinea pig tracheal bradykinin receptors. J Pharmacol Exp Ther 1992; 263: 1377–82.
64. Farmer SG, Ensor JE, Burch RM. Evidence that cultured airway smooth muscle cells contain bradykinin B_2 and B_3 receptors. Am J Respir Cell Miol 1991; 4: 273–7.
65. Salvino JM, Seoane PR, Douty BD, Awad MMA, Dolle RE, Houck WT, Faunce DM, Sawutz DG. Design of potent non-peptide competitive antagonists of the human bradykinin B_2 receptor. J Med Chem 1993; 36: 2583–4.

66. Sawutz DG, Salvino JM, Dolle RE, Casiano F, Ward SJ, Houck WT, Faunce DM, Douty B, Awad M, Soeane P. Discovery and *in vitro* biological activity of the nonpeptide bradykinin receptor antagonist WIN 64338. Pharmacologist 1993; 35: 142.
67. Farmer SG, De Siato MA. Effects of a novel nonpeptide bradykinin B_2 receptor antagonist, on intestinal and airway smooth muscle. Further evidence for the tracheal B_3 receptor. Br J Pharmacol 1994; 112: 461–464.
68. Bhoola KD, Collier HOJ, Schachter M, Shorley PG. Action of some peptides on bronchial muscle. Br J Pharmacol 1962; 19: 190–7.
69. Simonsson BG, Skoogh B-E, Bergh NP, Andersson R, Svedmyr N. *In vivo* and *in vitro* effect of bradykinin on bronchial motor tone in normal subjects and patients with airways obstruction. Respiration 1973; 30: 378–88.
70. Mizrahi, J. D'Orléans-Juste P, Caranikas S, Regoli D. Effects of peptides and amines on isolated guinea pig tracheae as influenced by inhibitors of the metabolism of arachidonic acid. Pharmacology 1982; 25: 320–6.
71. Rhaleb N-E, Dion S, D'Orléans-Juste P, Drapeau G, Regoli D, Browne RG. Bradykinin antagonism: Differentiation between peptide antagonists and antiinflammatory agents. Eur J Pharmacol 1988; 151: 275–9.
72. Schempler V, Calixto JB. Nitric oxide pathway-mediated relaxant effect of bradykinin in the guinea-pig isolated trachea. Br J Pharmacol 1994; 111: 83–8.
73. Bramley AM, Samhoun MN, Piper PJ. The role of the epithelium in modulating the responses of guinea-pig trachea induced by bradykinin *in vitro*. Br J Pharmacol 1990; 99: 762–6.
74. Frossard N, Stretton CD, Barnes PJ. Modulation of bradykinin responses in airway smooth muscle by epithelial enzymes. Agents Actions 1990; Suppl. 31: 204–9.
75. Barnett K. Jacoby DB, Nadel JA, Lazarus SC. The effects of epithelial cell supernatant on contractions of isolated canine tracheal smooth muscle. Am Rev Respir Dis 1988; 138: 780–3.
76. Da Silva A, Dhuy J, Waeldelé F, Bertrand C, Landry Y. Endopeptidase 24.15 modulates bradykinin-induced contraction in guinea-pig trachea. Eur J Pharmacol 1992; 212: 97–9.
77. Kyle H. Widdicombe JG. The effects of peptides and mediators on mucus secretion and smooth muscle tone in the ferret trachea. Agents Actions 1987; 22: 86–90.
78. Farmer SG, DeSiato MA, Broom T. Effects of kinin receptor agonists and antagonists in ferret isolated tracheal smooth muscle. Evidence that bradykinin-induced contractions are mediated by B_2 receptors. Br J Pharmacol 1992; 107: 31P.
79. Burch RM, Kyle DJ. Recent developments in the understanding of bradykinin receptors. Life Sci 1992; 50: 829–38.
80. Pyne NJ, Pyne S. Cellular signal pathways in tracheal smooth muscle. Cell Signal 1993; 5: 401–9.
81. Marsh KA, Hill SJ. Bradykinin B_2 receptor-mediated phosphoinositide hydrolysis in bovine cultured tracheal smooth muscle cells. Br J Pharmacol 1992; 107: 433–7.
82. Marsh KA, Hill SJ. Characteristics of the bradykinin-induced changes in intracellular calcium ion concentration of single bovine tracheal smooth muscle cells. Br J Pharmacol 1993; 110: 29–35.
83. Murray RK, Kotlikoff MI. Receptor-activated calcium influx in human airway smooth muscle cells. J Physiol 1991; 435: 123–44.
84. Tomasic M, Boyle JP, Worley JF, III, Kotlikoff MI. Contractile agonists activate voltage-dependent calcium channels in airway smooth muscle cells. Am J Physiol 1992; 263: C106–13.
85. Pyne S, Pyne NJ. Bradykinin stimulates phospholipase D in primary cultures of guinea-pig tracheal smooth muscle. Biochem Pharmacol 1993; 45: 593–606.
86. Stevens PA, Pyne S, Grady M, Pyne NJ. Bradykinin-dependent activation of adenylate cyclase activity and cyclic AMP accumulation in tracheal smooth muscle occurs via protein kinase C-dependent and -independent pathways. Biochem J 1994; 297: 233–9.
87. Polosa R, Phillips GD, Lai CKW, Holgate ST. Contribution of histamine and prostanoids to bronchoconstriction provoked by inhaled bradykinin. Allergy 1990; 45: 174–82.
88. Ichinose M, Nakajima N, Takahashi T, Yamauchi H, Inoue H, Takishima T. Protection against bradykinin-induced bronchonconstriction in asthmatic patients by neurokinin receptor antagonist. Lancet 1992; 340: 1248–51.

89. Greenberg R, Osman GH, O'Keefe EH, Antonaccio MJ. The effects of captopril (SQ 14,225) on bradykinin-induced bronchoconstriction in the anesthetized guinea pig. Eur J Pharmacol 1979; 57: 287–94.
90. Rossoni G, Omini C, Uigano AT, Mandelli V, Follo GC, Berti F. Bronchoconstriction by histamine and bradykinin in guinea pigs: relationship to thromboxane A_2 generation and the effect of aspirin. Prostaglandins 1986; 20: 547–57.
91. Ichinose M, Belvisi MG, Barnes PJ. Bradykinin-induced bronchoconstriction in guinea pig in vivo: role of neutral mechanisms. J Pharmacol Exp Ther 1990; 253: 594–9.
92. Arakawa H, Kawikova I, Löfdahl C-G, Lötvall J. Bradykinin-induced airway responses in guinea pig: Effects of inhibition of cyclooxygenase and thromboxane synthetase. Eur J Pharmacol 1992; 229: 131–6.
93. Kawikova I, Arakawa H, Löfdahl C-G, Skoogh B-E, Lötvall J. Bradykinin-induced airflow obstruction and airway plasma exudation: Effects of drugs that inhibit acetylcholine, thromboxane A_2 or leukotrienes. Br J Pharmacol 1993; 110: 657–64.
94. Farmer SG, DeSiato MA. The pharmacology of aerosol bradykinin-induced increases in specific airway resistance in conscious and anesthetized guinea pigs. Am Rev Respir Dis 1993; 147: A813.
95. Inoue H, Kato H, Takata S, Aizawa H, Ikeda T. Excitatory role of axon reflex in bradykinin-induced contraction of guinea pig tracheal smooth muscle. Am Rev Respir Dis 1993; 146: 1548–52.
96. Marceau F, Tremblay B. Mitogenic effect of bradykinin and of des-Arg^9-bradykinin on cultured fibroblasts. Life Sci 1986; 39: 2351–8.
97. Issandou M, Rozengurt E. Bradykinin transiently activates protein kinase C in Swiss 3T3 cells. J Biol Chem 1990; 265: 11 890–6.
98. Issandou M, Darbon J-M. DesArg^9 bradykinin modulates DNA synthesis, phospholipase C, and protein kinase C in cultured mesangial cells. Distinction from effects of bradykinin. J Biol Chem 1991; 266: 21 037–43.
99. Bascands J-L, Pecher C, Rouaud S, Emond C, Tack JL, Bastie MJ, Burch RM, Regoli D, Girolami J-P. Evidence for existence of two distinct bradykinin receptors on rat mesangial cells. Am J Physiol 1993; 264: F548–56.
100. Farhy RD, Ho K-L, Carretero OA, Scicli AG. Kinins mediate the antiproliferative effect of ramipril in rat carotid artery. Biochem Biophys Res Commun 1992; 182: 283–8.
101. Farhy RD, Carretero OA, Ho K-L, Scicli AG. Role of kinins and nitric oxide in the effects of angiotensin converting enzyme inhibitors on neointima formation. Circ Res 1993; 72: 1202–10.
102. Olsen R, Santone K, Medler D, Oakes SG, Abraham R, Powis G. An increase in intracellular free Ca^{2+} associated with serum-free growth stimulation of Swiss 3T3 fibroblasts by epidermal growth factor in the presence of bradykinin. J Biol Chem 1988; 263: 18 030–5.
103. Ruggerio M, Srivasava SK, Fleming TP, Ron D, Eva A. NIH 3T3 fibroblasts transformed by the dbl oncogene show altered expression of bradykinin receptors: effect on inositol lipid turnover. Oncogene 1989; 4: 767–71.
104. Etscheid BG, Albert KA, Villereal ML, Palfrey HC. Transduction of the bradykinin response in human fibroblasts. Prolonged elevation of diacyglycerol level and its correlation with protein kinase C activation. Cell Regul 1991; 2: 229–39.
105. Godin C, Smith AD, Riley PA. Bradykinin stimulates DNA synthesis in competent Balb/c 3T3 cells and enhances inositol phosphate formation induced by platelet-derived growth factor. Biochem Pharmacol 1991; 42: 117–22.
106. McAllister BS, Leeb-Lundberg LMF, Javors MA, Olson MS. Bradykinin receptors and signal transduction pathways in human fibroblasts. Integral role for intracellular calcium. Arch Biochem Biophys 1993; 304: 294–301.
107. Boarder MR. A role for phospholipase D in control of mitogenesis. Trends Pharmacol Sci 1994; 15: 57–62.
108. Lee-Lundberg LMF, Song X-H. Bradykinin and bombesin rapidly stimulate tyrosine phosphorylation of a 120-kDa group of proteins in Swiss 3T3 cells. J Biol Chem 1991; 266: 7746–9.
109. Ahn NG, Robbins, DJ, Haycock JW, Seger R, Cobb MH, Krebs EG. Identification of an activator of the microtubule-associated protein 2 kinases ERK1 and ERK2 and PC12 cells stimulated with nerve growth factor or bradykinin. J Neurochem 1992; 59: 147–56.

110. Jong Y-JI, Dalemar LR, Wilhelm B, Baenziger NL. Human bradykinin B_2 receptors isolated by receptor-specific monoclonal antibodies are tyrosine phosphorylated. Proc Natl Acad Sci 1993; 90: 10 994–8.
111. Lee K-M, Toscas K, Villereal ML. Inhibition of bradykinin- and thapsigargin-induced Ca^{2+} entry by tyrosine kinase inhibitors. J Biol Chem 1993; 268: 9945–8.
112. Miyajima AI, Kitamura T, Harada N, Yokota T, Arai K. Cytokine receptors and signal transduction. Ann Rev Immunol 1992; 10: 295–331.
113. Sakamaki K, Miyajima I, Kitamura T, Miyajima A. Critical cytoplasmic domains of the common β subunit of the human GM-CSF, IL-3 and IL-5 receptors for growth signal transduction and tyrosine phosphorylation. EMBO J 1992; 11: 3541–9.
114. Sato N, Sakamaki K, Terada N, Arai K, Miyajima A. Signal transduction by the high-affinity GM-CSF receptor. Two distinct cytoplasmic regions of the common β subunit responsible for different signaling. EMBO J 1993; 12: 4181–9.

Airways Smooth Muscle: Peptide Receptors, Ion Channels and Signal Transduction
ed. by D. Raeburn and M. A. Giembycz

CHAPTER 3
Tachykinins and Calcitonin Gene-Related Peptide

Carlo Alberto Maggi

Department of Pharmacology, A. Menarini Pharmaceuticals, Florence, Italy

1   Tachykinins and Calcitonin Gene-Related Peptide
2   Tachykinin Receptors, Signal Transduction and Airway Effects
2.1 Tachykinin Receptor Types and Subtypes
2.2 Pharmacology of NK_1 and NK_2 Receptors
2.3 Signal Transduction by Tachykinin Receptors
2.4 Airway Effects of Tachykinins
3   CGRP Receptors, Signal Transduction and Airway Effects
3.1 CGRP Receptors
3.2 Signal Transduction by CGRP Receptors
3.3 Airway Effects of CGRP
4   Conclusions
    References

1. Tachykinins and Calcitonin Gene-Related Peptide

The study of airway sensory innervation has been a major topic in respiratory physiology because of the importance of afferent neural input for regulating lung function and activating reflex bronchoconstriction and cough. In this context, capsaicin, the natural pungent ingredient of red peppers, has been extensively used by respiratory physiologists as a chemical probe to activate reflexes arising from the airways.

In the early 1980s, the work by Szolcsanyi and Bartho' [1, 2] and by Lundberg and Saria [3, 4] identified the existence of an "efferent" component in the airway sensory innervation exerted by a particular subset of primary afferent neurones which are characterized by their sensitivity to the actions of capsaicin [5–7 for reviews]. These early investigations [1–4] showed that the capsaicin-sensitive primary afferent neurones release mediators in the airways to produce powerful biological effects such as bronchoconstriction and plasma protein extravasation, which are of pathophysiological relevance for airway function. The constellation of "efferent" responses produced by sensory nerves activation is commonly referred to as "neurogenic inflammation".

It is now well accepted that the specific action of capsaicin on a subpopulation of neuropeptide-containing primary afferent neurones involves the activation of a membrane receptor which recognizes capsaicin and other capsaicin-like agents (the "vanilloid" receptor) [8]. Occupation of the vanilloid receptor leads to the opening of a peculiar type of receptor-operated cation channel [9]. The consequent influx of sodium and calcium ions into the primary afferent neurones initiates afferent impulses and concomitantly produces the secretion of sensory neuropeptides, tachykinins and calcitonin gene-related peptide (CGRP), from the peripheral endings of the afferent neurone. The ability of the capsaicin-sensitive primary afferent neurones (CSPANs) to release neurotransmitters from both their central and peripheral endings enables them to exert a dual, sensory and efferent or local effector, function [5, 6, 10, 11]. Accordingly, tachykinins and CGRP meet the various criteria required to accept the neurotransmitter status of a given mediator in the peripheral nervous system [11].

The study of the biological significance of tachykinins and CGRP in the peripheral regulation of airway function almost totally overlaps with the investigation of the efferent function of CSPANs. This is because the bulk of the tachykinins and CGRP in mammalian airways is stored in and released from the peripheral endings of CSPANs, the most notable and possibly unique exception being represented by the expression of CGRP-like immunoreactivity by neuroepithelial bodies of the airway epithelium [12–15].

For quite a long time, the investigation on the biological roles of sensory neuropeptides (tachykinins and CGRP) on airway function has almost exclusively relied on a comparison of the acute effects of capsaicin, which is assumed to act by releasing tachykinins and CGRP, and the responses produced by the exogenous administration of sensory neuropeptides. In recent years the recognition of the existence of different receptors mediating the effects of sensory neuropeptides in the airways and the availability of receptor-selective agonists and antagonists has enabled a more precise characterization of the relative contribution of tachykinins and CGRP to the various components of neurogenic inflammation. The aim of this chapter is to review the current knowledge of tachykinin and CGRP receptor types and subtypes and their role in producing biological effects in the mammalian airways.

2. Tachykinin Receptors, Signal Transduction and Airway Effects

2.1. Tachykinin Receptor Types and Subtypes

In mammals, three distinct tachykinins, substance P (SP), neurokinin (NK) A and neurokinin (NK) B have gained the status of neurotransmitters either in the peripheral or central nervous system [16, 17 for

Table 1. Amino acid sequences of natural tachykinins. The C-terminal sequence common to the peptides of the tachykinin family is underlined

Substance P	H-Arg-Pro-Lys-Pro-Gln-Gln-Phe-Phe-Gly-Leu-Met-NH$_2$
Neurokinin A	H-His-Lys-Thr-Asp-Ser-Phe-Val-Gly-Leu-Met-NH$_2$
Neurokinin B	H-Asp-Met-His-Asp-Phe-Val-Gly-Leu-Met-NH$_2$

reviews]. The CSPANs innervating mammalian airways express the preprotachykinin I gene which encodes the synthesis of SP and NKA but not that of NKB. Accordingly, SP and NKA have been shown to be present in the airways of various species while no evidence for the presence of NKB at this level has been presented [12, 18]. The application of depolarizing stimuli such as capsaicin itself, high potassium media or electrical stimulation of vagus nerve produces the co-release of SP and NKA in the airways [18, 19]. CGRP is co-released from CSPANs in the airways as well [20]. Furthermore, peptidases which inactivate the released tachykinins have been repeatedly demonstrated in the respiratory tract [19, 21] Among these, special emphasis has been given to endopeptidase 24.11 (enkephalinase, neutral endopeptidase) because of a possible relevance of changes in expression of this enzyme to the pathophysiology of asthma and bronchial hyperreactivity [21].

With few, although notable exceptions (mast cells degranulation, see below), the biological actions of tachykinins are encoded by their common, C-terminal sequence (Table 1). Three main receptor types have been identified by functional and radioligand binding techniques, which have been subsequently isolated and cloned [17, 22, 23]: the three receptors are termed NK$_1$, NK$_2$ and NK$_3$. Among natural tachykinins with established neurotransmitter status, SP has the highest affinity for NK$_1$ receptor, NKA for NK$_2$ receptor and NKB for NK$_3$ receptor, respectively. However, providing that high concentrations of the agonist are used, the three natural tachykinins are equieffective in stimulating the three receptor types. This raises the possibility of an extensive cross-talking between different tachykinins released from sensory nerves in the airways and different receptors expressed on target cells. Because of their extensive cross-talking with various receptors, natural tachykinins are poor instruments to characterize precisely the presence and functional roles of different tachykinin receptors. To obviate this problem, a number of synthetic "receptor-selective" agonists have been developed which possess remarkable affinity for only one of the three tachykinin receptor types [17]. In the mammalian (including human) airways, only SP and NKA are present [13, 14, 24] and, in agreement with this notion, expression of NK$_1$ and NK$_2$ predominates in various target cells of the airways (Table 2). For this reason only the pharmacology of NK$_1$ and NK$_2$ receptors will be considered.

Table 2. Tachykinin receptors expressed on target cells in mammalian airways

Target cells	Species	Tachykinin receptors
Smooth muscle	Guinea-pig, Rabbit	NK_1 NK_2
	Hamster, Human	NK_2
Postganglionic nerves	Guinea-pig	NK_1
	Rabbit	NK_1 NK_2
Tracheobronchial ganglia	Guinea-pig	NK_1 NK_3
Pulmonary vessels		
smooth muscle	Rabbit	NK_2 (vasoconstriction)
endothelium	Guinea-pig, Rabbit	NK_1 (vasodilatation)
Airway epithelium	Rat, Mouse	NK_1 (bronchodilatation)
Postcapillary venules	Rat, Guinea-pig	NK_1
Seromucous glands	Guinea-pig, Ferret	NK_1
Alveolar macrophages	Guinea-pig	NK_2

2.2. Pharmacology of NK_1 and NK_2 Receptors

The pharmacology of NK_1 and NK_2 receptors has developed, since the mid 1980s by the discovery of a number of synthetic, peptide analogues which act as selective agonists or antagonists at these receptors [17]. More recently, a number of nonpeptide antagonists selective for NK_1 or NK_2 receptors have been developed [25–27]. The availability of these powerful tools has permitted a clear characterization of the actions mediated by NK_1 and NK_2 receptors in the respiratory system (Table 2) and provides the basis to speculate about a possible therapeutic profile of NK_1 and NK_2 receptor antagonists in respiratory medicine [13, 14].

The introduction of selective antagonists to block NK_1 and NK_2 receptors has also revealed an unforeseen pharmacological heterogeneity, implying the existence of NK_1 or NK_2 receptor subtypes and/or species variants of these receptor types [17, 28].

For the NK_1 receptor, it has been recognized that the affinity of certain competitive antagonists is markedly species-dependent: the non peptide antagonist $(\pm)$CP 96 345 is about 100 times more potent at the NK_1 receptor expressed in the guinea-pig, rabbit and human species than in rat or mouse species, while the nonpeptide antagonist RP 67 580 is more potent in rat and mouse species than in guinea-pig, rabbit or human species [29–31]. The structural basis for this species-related difference in antagonist affinities has been resolved at the molecular level by showing the existence of species-related variations of aminoacids at discrete positions in the transmembrane segments of the NK$_1$ receptor protein: in particular, exchanging the aminoacids at positions 116 and 290 of the human NK$_1$ receptor protein with the corresponding aminoacids present in the rat NK$_1$ receptor protein, provided a mutant receptor where the affinity of CP 96 345 and RP 67 580 is comparable to that found in the wild type, rat NK$_1$ receptor; the reverse occurs when exchanging aminoacids at the same position of the NK$_1$ receptor with aminoacids present in the human NK$_1$ receptor [32]. These elegant experiments by Fong and coworkers [32] demonstrate the existence of true species variants of the NK$_1$ receptor which should not be considered as NK$_1$ receptor subtypes.

In addition to species-related variants, the existence of true NK$_1$ receptor subtypes (intra-species variants) has been proposed on the basis of the unusual pharmacological profile of certain NK$_1$ receptor agonists (such as septide), including their poor affinity at classical NK$_1$ receptor in binding experiments as opposed to high potency and sensitivity to act as agonists in bioassay conditions [33–35]. The existence of a septide-sensitive receptor, which could be an NK$_1$ receptor subtype could thus be proposed [33–35]. However, this hypothesis is not yet supported from molecular biology data because only one form of functional NK$_1$ receptor protein has been isolated from the various species examined thus far [22, 23].

For the NK$_2$ receptor, the situation is quite similar to that described above for NK$_1$ receptor. Even in this case the introduction of receptor selective antagonists has disclosed a pharmacological heterogeneity [17, 36]. On the basis of various pharmacological criteria the existence of NK$_2$ receptor subtypes, tentatively termed NK$_{2A}$ and NK$_{2B}$ has been forwarded [17]. Although some examples of intraspecies heterogeneity, suggesting the existence of true receptor subtypes, has been presented [37, 38], the situation here is analogous to that described above for the NK$_1$ receptor, i.e. only one form of functional NK$_2$ receptor protein has been isolated from the various species examined thus far [22, 23]. The possibility that the heterogeneity of the NK$_2$ receptor disclosed following the introduction of receptor selective antagonists is totally or partially species-dependent appears very likely. Indeed, when looking at NK$_2$ receptor expressed on smooth muscle cells a clear species-dependency in the pattern of antagonist affinities is evident [17]. If the heterogeneity of NK$_2$ receptor were entirely species-related, then it should be possible to identify species-related variants at crucial aminoacid position(s) of the NK$_2$ receptor protein responsible for the varying affinities of competitive antagonists, as it has been shown to occur for the NK$_1$ receptor.

Summarizing this section, functional, binding and molecular biology data support the existence of species variants of the NK$_1$ receptor, while the existence of NK$_1$ receptor subtypes (classical vs. septide-sensitive) remains a matter of investigation. For the NK_2 receptor the situation is less well defined: a remarkable heterogeneity is evident from functional and binding data but it is not yet possible to firmly ascribe this heterogeneity to species-related variants of the NK_2 and/or to receptor subtypes.

2.3. Signal Transduction by Tachykinin Receptors

The three main types of tachykinin receptors belong to the superfamily of G-protein-coupled, rhodopsin-like, receptors with seven putative transmembrane spanning hydrophobic segments [17]. The main signal transduction system coupled to the three tachykinin receptors is stimulation of phospholipase C (PLC) and activation of phosphoinositide (PI) turnover: the consequent elevation of intracellular calcium is a plausible mechanism for signalling of the various biological responses produced by tachykinin agonists in various target cells [39 for review]. Experiments involving the transfection of cloned tachykinin receptors into cell expression systems have documented the ability of NK_1, NK_2 and NK_3 receptors to stimulate PLC and produce PI accumulation [40]. In these experiments, also a stimulation of adenylyl cyclase and elevation of cyclic AMP levels following the activation of tachykinin receptors has been reported [40]. The stimulation of PLC and adenylyl cyclase was observed in both intact cells and in membrane preparations, suggesting a direct linkage, through G-proteins, to the effector systems [40]. Interestingly, the concentrations of tachykinin agonists required to activate adenylyl cyclase were about 1 order of magnitude higher than those required to activate PLC [40]. Takeda et al. [41] likewise showed that transfected NK_1 and NK_2 receptors both stimulate PI turnover and cyclic AMP accumulation and also showed that the cyclic AMP signal produced by NK_2 receptor is greater and more prolonged than that activated by the NK_1 receptor. At present, however, no clear evidence for elevation of cyclic AMP levels has been reported for the wild form of tachykinin receptors spontaneously expressed in mammalian cells: it remains therefore unclear whether the elevation of cyclic AMP observed in cells transfected with the receptors may involve a coupling to G proteins which do not spontaneously occur in intact cells. Indeed, elevation of intracellular calcium by NK_2 receptor stimulation has been shown to negatively modulate stimulated cyclic AMP accumulation in cultured cells [42].

Eistetter et al. [43, 44] showed that stimulation of the bovine recombinant NK_2 receptor expressed in two different cell lines produces not only activation of PI turnover and calcium mobilization but also cyclic AMP accumulation and [^{3}H]arachidonic acid mobilization: they suggested that adenylyl cyclase stimulation and cyclic AMP accumulation in response to tachykinin receptor stimulation is a kind of autocrine response produced by prostanoids generated by tachykinins.

Another effector response which has been described in response to tachykinin receptor stimulation is a G-protein-mediated closure of inwardly rectifying K^+ channel [45, 46]. More recently, a G-protein-mediated opening of the maxi-Cl^- channel by NK_1 receptor stimulation in smooth muscle has been reported [47, 48] which could produce depolarization through Cl^- efflux and, by consequence of depolarization, produce the opening of voltage-sensitive Ca^{2+} channels.

Therefore at least 4 distinct effector systems (stimulation of PI turnover, cyclic AMP accumulation possibly via prostanoids production, closure of K^+ channels and opening of Cl^- channel) could be produced by occupation of tachykinin receptors via G-protein(s): interestingly, some of the described actions of tachykinins (PLC stimulation, closure of inwardly rectifying K^+ channels) are resistant to pretreatment with pertussis toxin (PTX) [45, 49] and may be mediated by the PTX-insensitive G proteins (G_p [49] or $G_{q/11}$) [50]. On the other hand, the opening of the maxi Cl^- channel in smooth muscle is mediated through a PTX-sensitive G-protein [48].

Summarizing this section, stimulation of PI turnover and consequent elevation of intracellular calcium is the most accepted and investigated signalling system coupled to stimulation of tachykinin receptors. This second messenger system is certainly adequate for producing many of the biological effects of tachykinins on target cells. However, other signalling systems are operated by tachykinins, although the mechanisms underlying these additional mechanisms are less well investigated. Owing to the multiplicity of the tachykinin system, both in terms of peptides synthesized by neurones and in terms of receptors expressed by terget cells, it appears conceivable that multiple effector systems may be operated by tachykinins in physiological conditions. For instance, several neuronal cell types, including primary afferent neurones, synthesize and release both SP and NKA and various target cells express both NK_1 and NK_2 receptors: it would appear that an unreasonable degree of duplication exists if PI turnover were the only signalling mechanism coupled to tachykinin receptor stimulation in physiological conditions. A recent work [50a] showed the functional nonequivalence of structurally homologous domains of the NK_1 and NK_2 receptor for activation of second messenger responses in transfected Chinese hamster ovary (CHO) cells: in particular, the putative third intracellular loop is crucial for interaction of the NK_1 receptor with G-protein(s), while the C-terminal tail is essential to produce second messenger response after NK_2 receptor stimulation. It has been speculated that this functional nonequivalence may underlie a different time-course and different degrees of desensitization observed in response to NK_1/NK_2 receptor stimulation of intact tissues [50a].

2.4. Airway Effects of Tachykinins

Tachykinins produce a variety of biological effects in the airways, mimicking the signs and symptoms of asthma/bronchial hyperreactivity [13, 14]. These effects include changes in bronchomotor tone, vasodilatation, increase in vascular permeability, stimulation of secretions and facilitation of the release of other transmitters (*e.g.* acetylcholine). The bronchomotor response to tachykinins is markedly species-dependent (see below) while the other effects, especially the increase in vascular permeability, seem to be more conserved across species.

In guinea-pig, rabbit, ferret, hamster and human isolated airways (trachea or bronchi) tachykinins produce a powerful bronchoconstriction which equals in magnitude that observed in response to cholinomimetics [13, 14, 51]. In guinea-pig and rabbit isolated airways, both NK_1 and NK_2 receptors mediate contraction to exogenous tachykinins [51, 52], while in hamster and human isolated airways only NK_2 receptors mediate this response; in isolated airways from asthmatics preliminary evidence for NK_1 receptor-mediated bronchoconstriction has been presented [53]. By contrast, in rat or mouse [54, 55] isolated airways tachykinins produced airway relaxation apparently due to the release of airway relaxant factors from the epithelium [13, 14 for review].

Immunoblockade by specific antiserum [56] and receptor selective antagonists [57, 58] have been used to assess the relative contribution of SP and NKA, which are co-released by depolarization in the guinea-pig isolated airways [18, 19] and of NK_1 and NK_2 receptors, both present in guinea-pig airways [51] in mediating the atropine-resistent, nonadrenergic noncholinergic prolonged contractile response which is produced by electrical stimulation of CSPANs [1–3]. The results of these studies indicate that while NK_1 and NK_2 receptor agonists produce a maximal bronchoconstrictor response of comparable magnitude [51], the bronchoconstriction brought about by endogenous tachykinins released from sensory nerves is largely (>80% of total response) mediated by NK_2 receptors [56–58]. This conclusion was based on the following evidence: a) a specific tachykinin antiserum, which blocks the contractile action of NKA but not that of SP, strongly inhibits the contractile response of the guinea-pig isolated airways to capsaicin and the atropine-resistant component of nerve response to electrical field stimulation [56]; b) selective NK_2 receptor antagonists inhibit to a large extent the noncholinergic response of the guinea-pig isolated bronchus to electrical field stimulation while selective NK_1 receptor antagonists are barely effective [57, 58] and, c) the rank order of potency of NK_2 receptor antagonists in inhibiting the noncholinergic response to electrical field stimulation matches their relative potency in antagonizing contractions produced by a selective NK_2 receptor agonist [56].

Quantitatively, the larger part of the bronchoconstrictor response produced by activation of capsaicin-sensitive afferents in guinea-pig isolated airways appears to be mediated by NK_2 receptors [56, 57]. The relative contribution of NK_1 receptors to the overall response can be amplified by the addition of peptidase inhibitors. Pretreatment with peptidases inhibitors has several effects, as follows: a) increased magnitude of the atropine-resistant bronchoconstriction, b) increased potency of the NK_1 receptor-preferring agonist SP, more than that of the NK_2 receptor-preferring agonist NKA, c) decreased effectiveness of NK_2 antagonists in reducing the atropine-resistant bronchoconstriction and d) increased effectiveness of the NK_1 receptor antagonists in reducing the atropine-resistant bronchoconstriction [51, 57]. Therefore, while SP and NKA are co-released from sensory nerves in the guinea-pig airways [18, 19], the bronchoconstrictor response to endogenous tachykinins is largely mediated by NKA via NK_2 receptors because of its greater resistance, as compared to SP, to degradation by peptidases. The results obtained in guinea-pig isolated airways have been substantially confirmed by the results of *in vivo* experiments, in which an atropine-resistant bronchoconstriction was evoked by electrical stimulation of the vagi or by the intravenous administration of capsaicin in anaesthetized guinea-pigs [59–61]. Even under these experimental conditions, NK_2 receptor antagonists proved effective in reducing the bronchoconstrictor response produced by sensory nerves activation demonstrating a major contribution by NKA through NK_2 receptors.

The increase in vascular permeability, detectable as extravasavation of a dye bound to serum proteins, is one of the most striking effects produced by tachykinin receptor agonists in the respiratory system [4, 62]. This response is not only observed in the trachea and main bronchi, but also in the nasal mucosa [62]. A marked regional difference in the intensity of plasma protein extravasation (PPE) to endogenous tachykinins (*e.g.* released by capsaicin) is evident in the airways which parallels the regional difference in density of innervation indirectly assessed by measuring tissue levels of tachykinins [63]. Experiments with receptor selective agonists [64] and antagonists [65–67] have shown that PPE produced by both exogenous and endogenous tachykinins is entirely mediated by activation of NK_1 receptors, at least in the tracheobronchial tree [65–67]. Evidence has been presented to suggest that PPE may be mediated by NK_2 receptor stimulation in guinea-pig lower airways [68].

PPE is thought to occur at the level of postcapillary venules through the formation of gaps between endothelial cells, thus enabling the escape of plasma protein into the interstitial spaces [69]. Surprisingly, the density of tachykininergic innervation is remarkably low at the level of postcapillary venules in rat airways [70]. PPE ascribable to release on endogenous tachykinins in the airways has been documented in response to application of stimuli which, like smoke cigarette or antigen challenge [66, 71] are relevant for airway pathophysiology. From these results, a therapeutic effect on NK_1 receptor antagonists to control airway inflammation has been envisaged.

The NK_1 receptor is also important for inducing mucus secretion in the respiratory tract: thus mucus discharge from both seromucous glands and goblet cells [72–75] induced by tachykinins seems to be exclusively mediated via NK_1 receptors.

Exogenously administered tachykinins produce facilitation of acetylcholine release in the airways [52, 76]. This effect, which may be important in enhancing reflexly-evoked cholinergic drive to airways smooth muscle, seems to occur at two distinct sites in guinea-pig airways: parasympathetic ganglia and postganglionic nerves. Facilitatory tachykinin receptors on postganglionic nerves terminals have been demonstrated by using exogenous agonists in guinea-pig and rabbit airways [52, 76]. In the former species NK_1 receptors predominate, while in rabbit airways both NK_1 and NK_2 receptors facilitate acetylcholine release [52, 76]. Importantly, while the facilitatory action of exogenous agonists on acetylcholine release from postganglionic nerve terminals can be selectively blocked by NK_1 (guinea-pig and rabbit) or NK_2 (rabbit) receptor antagonists, the administration of tachykinin receptor antagonists does not affect the amplitude of cholinergic contractions under resting conditions [52, 76]. This implies that endogenous tachykinins do not modulate acetylcholine release at a postganglionic level to a significant extent. In the guinea-pig trachea, a facilitatory action by NK_1 receptor agonists on cholinergic transmission occurs [76]. Watson et al. [76] showed that the neutral endopeptidase inhibitor, phosphoramidon, which prevents degradation of tachykinins, facilitates ganglionic transmission in response to vagus nerve stimulation; the effect of phosphoramidon is prevented either by capsaicin pretreatment or by administration of an NK_1 receptor antagonist. These findings imply a modulatory action of endogenous tachykinins, via the NK_1 receptor, on cholinergic transmission in airway parasympathetic ganglia [76]. More recently, Myers and Undem [77] showed that both NK_1 and NK_3 receptors mediate tachykinin-induced depolarization of guinea-pig bronchial parasympathetic ganglion neurones: furthermore, after desensitization with an NK_3 receptor selective agonist, the depolarizing response to capsaicin, thought to occur by the release of endogenous tachykinins was prevented. Myers and Undem [77] concluded that NK_3 receptors mediate the depolarizing action of endogenous tachykinins in guinea-pig parasympathetic ganglia. This is to date the only reported effect which would be mediated by NK_3 receptors in mammalian airways.

Finally, Tachykinin receptors are also present on pulmonary blood vessels where they can produce vasoconstriction through a direct action on smooth muscle cells (*e.g.* NK_2 receptors in the rabbit pulmonary artery) and vasorelaxation through the release of vasodilatory mediators from endothelial cells (*e.g.* NK_1 receptors in the rabbit or guinea-pig pulmonary artery) [78]. There is no evidence, however, that endogenous tachykinins mediate pulmonary vasodilatation produced by activation of sensory nerves which appears to be rather mediated by CGRP (see below).

3. CGRP Receptors, Signal Transduction and Airway Effects

3.1. CGRP Receptors

The pharmacology of CGRP and CGRP receptors is underdeveloped [79, 80 for review] as compared to the rapid expansion observed in recent times in the tachykinin field. Notwithstanding, there is clear evidence to suggest the existence of multiple CGRP receptors; since the evidence for CGRP receptor heterogeneity has been provided in different preparations from the same species these multiple receptors would correspond to true receptor subtypes [79, 80]. Until now, however, no CGRP receptor has been isolated and cloned by molecular biology techniques and a structural support to the heterogeneity revealed by functional and radioligand binding studies is lacking. Judged on the basis of changes in the affinity state of CGRP receptors in binding studies upon addition of GTP and analogues, it appears likely that CGRP receptors belong to the superfamily of G-protein coupled, rhodopsin-like, receptors which also include tachykinin receptors [79]. Various studies, reviewed by Poyner [79], have reported the photolabelling of CGRP-binding proteins from various tissues. Taken together, the result of these studies suggest that CGRP receptors are likely to be about 60 kDa glycoproteins with internal disulphide bridges.

A very important tool in CGRP receptor pharmacology is the C-terminal fragment of human αCGRP, hαCGRP(8–37): Chiba *et al.* [81] showed that this CGRP C-terminal fragment binds with relatively high affinity to CGRP receptors in rat liver but does not activate adenylyl cyclase; rather it is able to prevent cyclic AMP accumulation induced by CGRP in a manner consistent with competitive antagonism at CGRP receptors. This finding has been subsequently confirmed and extended by various groups [80, 82–84] and CGRP(8–37) has become the most important tool for assessing the involvement of endogenous CGRP in functional responses produced by activation of the efferent function of sensory nerves. Shorter C-terminal fragments of CGRP also act as CGRP receptor antagonists [85, 86] but their potency becomes progressively lower by shortening of the C-terminal sequence.

Quirion and coworkers [see 80 for review] have provided a number of pharmacological criteria to distinguish two distinct CGRP receptor subtypes by using various modified peptide sequences: the outcome of these studies was the proposal of the existence of two CGRP receptors termed $CGRP_1$ and $CGRP_2$, respectively. Amongst the criteria proposed to distinguish CGRP receptor subtypes, the sensitivity to the blocking action of CGRP(8–37) is the most important: thus the affinity of CGRP(8–37) as an antagonist would be much higher at $CGRP_1$ than $CGRP_2$ receptors. Examples have been provided of organs or systems in which distinct CGRP actions show a remarkable difference in their sensitivity to CGRP(8–37): thus neuronal stimulation and acetylcholine release produced by CGRP in the guinea-pig ileum are totally unaffected by CGRP(8–37) at concentrations which block the direct smooth muscle relaxation produced in this preparation [87]; in rat stomach the reduction in gastric acid secretion produced by CGRP is blocked by CGRP(8–37) while the protective effect of CGRP on acetylsalicilic acid-induced ulcers score was unchanged by CGRP(8–37) [88].

3.2. Signal Transduction by CGRP Receptors

Various types of effector systems mediate the effect of CGRP in different systems. Since CGRP receptors have not yet been isolated and cloned it is unclear whether the various effector systems are activated as a primary and direct consequence of CGRP receptor occupancy by the agonist or may be secondary to other intracellular events. The most widely reported cellular effect observed following stimulation of CGRP receptors is activation of adenylyl cyclase and cyclic AMP accumulation. This effect is likely to underlie several actions of CGRP such as neuronal stimulation, increase in cardiac inotropism and chronotropism and smooth muscle relaxation including vasorelaxation in various blood vessels [79]. The powerful vasodilator action of CGRP has been most investigated in this respect. Many reports have indicated that vasorelaxation by CGRP is endothelium-independent in various blood vessels/ vascular beds from different species; in several instances a positive correlation with cyclic AMP accumulation has been established [89–91]. In the rat aorta, however, CGRP-induced vasodilatation is endothelium-dependent, possibly linked to nitric oxide (NO) release [92–94]. Also the hypotensive effect of CGRP in conscious rats could be partly mediated through NO production [95], while hypotension in anaesthetized rabbits is unaffected by NO synthase inhibition [96]. A recent study [97] proposed a quite complex series of events to account for the endothelium-dependent CGRP-induced relaxation in rat aorta which, paralleled by an increase in both cyclic AMP and cyclic GMP accumulation, is blockable by NO synthase inhibitors: stimulation of CGRP receptors would lead to cyclic AMP increase in endothelial cells and NO generation which, in turn, would diffuse to smooth muscle producing relaxation via cyclic GMP accumulation [97]. $CGRP_1$ and $CGRP_2$ receptors have been implicated in the endothelium-independent (pig coronary artery) and endothelium-dependent (rat aorta) relaxation to CGRP, respectively [98].

A third mechanism has been proposed to account for CGRP-induced vasodilatation, *i.e.* activation of ATP-sensitive K^+ channels (K^+_{ATP}): evidence comes from experiments showing that glibenclamide, a K^+_{ATP} channels inhibitor, prevents CGRP-induced vasodilatation in some vascular beds [99, 100].

3.3. Airway Effects of CGRP

CGRP produces various biological effects in the airways, which include changes in bronchomotor tone, vasodilatation and stimulation of mucus secretion. Among these, the most controversial effect is about a possible influence of CGRP on airways smooth muscle contractility.

The earlier reports of a powerful bronchoconstrictor action of CGRP on guinea-pig and human isolated airways smooth muscle [101,102] have not been confirmed in subsequent studies [103, 104] and, more recently, evidence was presented to suggest that CGRP produces relaxation of airways smooth muscle [105–107] and also antagonism of tachykinin-induced contraction. The issue is further complicated by the proposal that CGRP may instead act as an endogenous bronchoconstrictor released from sensory nerves. Tschirhart *et al.* [108] showed that the capsaicin-induced contraction is bigger in epithelium-intact as compared to epithelium-denuded guinea-pig trachea. The tachykinin antagonist, spantide, abolished the response to capsaicin in epithelium-free preparations while leaving a fraction of the response spared in epithelium intact tracheas. Furthermore, a CGRP antiserum partially reduced the response to capsaicin but only in epithelium intact guinea-pig trachea. Tschirhart *et al.* [108] also showed that rat CGRP produces a tracheal contraction which is larger in epithelium intact than in denuded preparations while human CGRP is ineffective in both cases. They concluded that endogenous CGRP may participate to capsaicin-induced contraction of guinea-pig trachea through the release of contractile mediators from epithelial cells. Facilitation of acetylcholine release by CGRP in rabbit trachea has also been reported [109].

Endogenous CGRP seems the main mediator of the endothelium-independent vasodilatation produced by sensory nerve activation in the guinea-pig main pulmonary artery [78, 110]. In the guinea-pig isolated pulmonary artery, NANC vasorelaxant responses ascribable to sensory neuropeptide release are evoked by electrical nerve stimulation and abolished by capsaicin desensitization. In precontracted vessels, tachykinins produce an endothelium-dependent vasorelaxation while CGRP produces an endothelium-independent vasorelaxation. Although capsaicin or electrical stimulation induces a consistent release of both tachykinins and CGRP from the guinea-pig pulmonary artery, the evoked relaxation is unchanged by mechanical removal of the endothelium implying that, in this vessel, nerve-released tachykinins do not gain access to NK_1 receptors located on endothelial cells [78]. The use of the CGRP receptor antagonist, CGRP(8–37), has been instrumental in demonstrating the involvement of endogenous CGRP in vasorelaxation produced by sensory nerve activation at this level [110].

Similar to tachykinins, CGRP has been reported to be a powerful stimulant of goblet cell secretion in guinea-pig airways and its possible involvement in mucus discharge produced by sensory nerves activation deserves further consideration [75].

4. Conclusions

The basic science data obtained thus far enable us to draw the following conclusions: a) in mammalian airways, the peripheral endings of CSPANs are the major source of tachykinins and CGRP (sensory neuropeptides); b) these sensory neuropeptides are releaased in an efferent fashion from CSPANs in the airways and modulate the activity of a number of target cells (neurogenic inflammation) via the activation of specific receptors; c) the activity of sensory neuropeptides is, in turn, modulated by peptidases which terminate their biological actions; d) a remarkable degree of specialization exists, dictated by receptor expression at postjunctional level, whereby individual transmitters predominate in determining certain biological effects (tachykinins for modulation of bronchomotor tone and vascular permeability, CGRP for pulmonary artery vasodilatation).

Any attempt to understand the significance of this process in airway pathophysiology must start from considering which are the appropriate stimuli to produce release of sensory neuropeptides. The results obtained in several laboratories indicate chemosensitivity as a prominent feature of CSPANs throughout the body, including the airways. A number of chemical agents either exogenous (such as capsaicin itself and other vanilloid receptor agonists, cigarette smoke, toluene diisocyante, ether, acrolein) or endogenous (bradykinin, protons, acetylcholine via nicotinic cholinoceptor, various autacoids and mediators of inflammation) have been shown to be adequate stimuli to trigger a local release of sensory neuropeptides from CSPANs. Accordingly, the hypothesis has been put forward that CSPANs provide a neurogenic support to airway inflammation and are activated in response to inhalation of irritants. This concept has received support from experimental studies demonstrating that either pretreatment with capsaicin, to block sensory neuropeptides release from CSPANs and/or administration of tachykinin receptor antagonists provide a kind of protective effect toward airway responses evoked in animal models of respiratory diseases [62, 65–67, 71, 111–113]. From the results of these latter studies, a therapeutic effect of sensory neuropeptides antagonists in respiratory medicine could be anticipated. On the other hand, it is important to recognize that the major part of studies dealing with the pathophysiological relevance of sensory neuropeptides in airway function have been performed in small rodents, especially rats and guinea-pigs. It appears that in these two species, the relative importance of the efferent function of CSPANs is somewhat overexpressed as compared to other species and, consequently, it would be unwise to overemphasize the results obtained in small rodents with respect to respiratory medicine in humans. Eventually, the clinical testing of selective receptor antagonist will establish the true relevance of tachykinins and CGRP for airway function in humans.

References

1. Szolcsanyi J, Bartho L. Capsaicin-sensitive non-cholinergic excitatory innervation of the guinea-pig tracheobronchial smooth muscle. Neurosci Letters 1982; 34: 247–250.
2. Szolcsanyi J. Tetrodotoxin-resistant noncholinergic neurogenic contraction evoked by capsaicinoids and piperine on the guinea-pig trachea. Neurosci Letters 1983; 42: 83–88.
3. Lundberg JM, Saria A. Bronchial smooth muscle contraction induced by stimulation of capsaicin-sensitive sensory neurons. Acta Physiol Scand 1982; 116: 473–476.
4. Lundberg JM, Saria A. Capsaicin-induced desensitization of airway mucosa to cigarette smoke, mechanical and chemical irritants, Nature 1983; 302: 251–253.
5. Szolcsanyi J. Capsaicin-sensitive chemoceptive neural system with dual sensory-efferent function. In: LA Chahl, J Szolcsanyi and F Lembeck, editors: Antidromic vasodilatation and neurogenic inflammation. Budapest, Hungary: Akademiai Kiado, 1984: 27–55.
6. Maggi CA, Meli A. The sensory-efferent function of capsaicin-sensitive sensory neurons. Gen Pharmacol 1988; 19: 1–43.
7. Maggi CA. Capsaicin and primary afferent neurons: From basic science to human therapy? J. Autonom Nervous System 1991; 33: 1–14.
8. Szallasi A, Blumberg PM. Resininferatoxin and its analogs provide novel insights into the pharmacology of the vanilloid (capsaicin) receptor. Life Sci 1990; 47: 1399–1408.
9. Bevan S, Szolcsanyi J. (1992) Sensory neuron-specific actions of capsaicin: mechanisms and applications. Trends Pharmacol Sci 1992; 11: 330–333.
10. Holzer P. Local effector functions of capsaicin-sensitive sensory nerve endings: involvement of tachykinins, CGRP and other neuropeptides. Neuroscience 1988; 24: 739–68.
11. Maggi CA. Tachykinins and calcitonin gene-related peptide (CGRP) as co-tranasmitters released from peripheral endings of sensory nerves. Progress in Neurobiol. In press.
12. Lundberg JM, Saria A. Polypeptide-containing neurons in airway smooth muscle. Ann Rev Physiol 1987; 49: 557–572.
13. Maggi CA. Tachykinin receptors in the airways and lung: What should we block? Pharmaocl Res 1990; 22: 527–540.
14. Maggi CA. (1993b) Tachykinin receptors and airway pathophysiology. Eur Respir J 1993; 6: 735–742.

15. Cadieaux A, Springall DR, Mulderry PK, Rodrigo J, Ghatei MA, Terenghi O, Blook SR, Polak JM. Occurrence, distribution and ontogeny of CGRP immunoreactivity in the rat lower respiratory tract, effect of capsaicin treatment and surgical denervations. Neurosci. 1986; 19: 605–627.
16. Otsuka M, Yoshioka K. Neurotransmitter function of mammalian tachykinins. Physiol Reviews 1993; 73: 229–308.
17. Maggi CA, Patacchini R, Rovero P, Giachetti A. Tachykinin receptors and tachykinin receptor antagonists. J Autonom Pharmacol 1993; 13: 23–93.
18. Saria A, Martling CR, Yan Z. Theodorsson-Norheim E, Gamse R, Lundberg JM. Release of multiple tachykinins from capsaicin-sensitive sensory nerves in the lung by bradykinin, histamine, DMPP and vagal nerve stimulation. Amer Rev Resp Dis 1988; 137: 1330–1335.
19. Maggi CA, Patacchini R, Perretti F, Meini S, Manzini S, Santicioli P, Del Bianco E, Meli A. The effect of thiorphan and epithelium removal on contractions and tachykinin release produced by activation of the capsaicin-sensitive afferents in the guinea-pig isolated bronchus. Naunyn Schmiedeberg's Arch Pharmacol 1990; 341: 74–79.
20. Martling CR, Saria A, Fischer JA, Hokfelt T, Lundberg JM. CGRP and the lung: Neuronal coexistence with SP release by capsaicin and vasodilatory effects. Regul Peptides 1988; 20: 125–139.
21. Nadel JA. Neutral endopeptidase modulates neurogenic inflammation. Eur Resp J 1991; 4: 745–754.
22. Nakanishi S. Mammalian tachykinin receptors. Annu Rev Neurosci 1991; 14: 123–136.
23. Gerard NP, Bao L, Xiao-Ping H, Gerard C. Molecular aspects of the tachykinin receptors. Regul Peptides 1993; 43: 21–35.
24. Martling CR, Theodorsson-Norheim E, Lundberg JM. Occurrence and effects of multiple tachykinins, SP, NKA and neuropeptide K in human airways. Life Sciences 1987; 40: 1633–1643.
25. Snider RM, Constantine JW, Lowe JA, Longo KP, Lebel WS, Woody HA, Drozda SE, Desai MC, Vinick FJ, Spencer RW, Hess HJ. A potent nonpeptide antagonist of the SP (NK-1) receptor. Science 1991; 251: 435–437.
26. Garret C, Carruette A, Fardin V, Moussaoui S, Peyronel JF, Blanchard JC, Laduron PM. Pharmacological properties of a potent and selective nonpeptide SP antagonist, Proc Natl Acad Sci USA 1991; 88: 10208–10211.
27. Emonds-Alt X, Vilain P, Goulaouic P, Proietto V, Van Broeck D, Advenier C, Naline E, Neliat G, Le Fur G, Breliere JC. A potent and selective nonpeptide antagonist of the neurokinin A (NK-2) receptor. Life Sci-Pharmacol Letters 1992; 50: PL101–106.
28. Maggi CA. Evidence for receptor subtypes/species variants of tachykinin receptors. In: Buck SH, editor: Tachykinin receptors. Humana Press, 1994.
29. Barr AJ, Watson SP. Non-peptide antagonists CP 96,345 and RP 67,580, distinguish species variants in tachykinin NK-1 receptors. Br J Pharmacol 1993; 108: 223–227.
30. Patacchini R, Santicioli P, Astolfi M, Rovero P, Viti G, Maggi CA. Activity of peptide and non-peptide antagonists at peripheral NK-1 tachykinin receptors. Eur J Pharmacol 1992; 215: 93–98.
31. Gitter BD, Waters DC, Bruns RF, Mason NR, Nixon JA, Howbert JJ. Species differences in affinities of nonpeptide antagonists for SP receptors. Eur J Pharmacol 1991; 197: 237–238.
32. Fong TM, Yu H, Strader CD. Molecular basis for the species selectivity of the NK_1 receptor antagonists CP 96,345 and RP 67,580. J Biol Chemistry 1992; 267: 25668–25671.
33. Petitet F, Saffroy M, Torrens Y, Lavielle S, Chassaing G, Loeuillet D, Glowinski J, Beaujouan JC. Possible existence of a new tachykinin receptor subtype in the guinea-pig ileum. Peptides 1992; 13: 383–388.
34. Maggi CA, Patacchini R, Meini S, Giuliani S. Evidence for the presence of a septide-sensitive tachykinin receptor in the circular muscle of the guinea-pig ileum. Eur J Pharmacology 1993; 235: 309–311.
35. Meini S, Patacchini R, Maggi CA. Tachykinin NK-1 receptor subtypes in the rat urinary bladder. Br J Pharmacol 1994; 111: 739–746.
36. Maggi CA, Patacchini R, Giuliani S, Rovero P, Dion S, Regoli D, Giachetti A, Meli A. Competitive antagonists discriminate between NK-2 tachykinin receptor subtypes. Br J Pharmacol 1990; 100: 588–592.

37. Nimmo A, Carstairs JR, Maggi CA, Morrison JFB. Evidence for the co-existence of multiple NK-2 tachykinin receptor subtypes in rat bladder. Neuropeptides 1992; 22: 48.
38. Brunelleschi S, Ceni E, Fantozzi R, Maggi CA. Evidence for tachykinin NK-2B-like receptors in guinea-pig alveolar macrophages. Life Sci-Pharmacology Letters 1992; 51: PL177–181.
39. Guard S, Watson SP. Tachykinin receptor types: Classification and membrane signalling mechanisms. Neurochem Int 1991; 18: 149–165.
40. Nakajima Y, Tsuchida K, Negishi M, Ito S, Nakanishi S. Direct linkage of three tachykinin receptors to stimulation of both phosphatidylinositol hydrolysis and cyclic AMP cascades in transfected chinese hamster ovary cells. J Biol Chemistry 1992; 267: 2437–2442.
41. Takeda Y, Blount P, Sachais BS, Hershey AD, Raddatz R, Krause JE. Ligand binding kinetics of substance P and neurokinin A receptors stably expressed in CHO cells and evidence for differential stimulation of inositol 1,4,5-triphosphate and cAMP second messenger responses. J Neurochem 1992; 59: 740–745.
42. De Bernardi MA, Seki T, Brooker G. Inhibition of cAMP accumulation by intracellular calcium mobilization in C6-2B cells stably transfected with substance K receptor cDNA. Proc Natl Acad Sci USA 1991; 88: 9257–9261.
43. Eistetter HR, Church DJ, Mills A, Godfrey PP, Capponi AM, Brewster R, Schulz MF, Kawashima E, Arkinstall SJ. Recombinant bovine NK-2 receptor stably expressed in chinese hamster ovary cells couples to multiple signal transduction pathways. Cell Regul 1991; 2: 767–779.
44. Eistetter HR, Mills A, Arkinstall SJ. Signal transduction mechanisms of recombinant NK-2 receptor stably expressed in baby hamster kidney cells. J Cell Biochemistry 1993; 52: 84–91.
45. Stansfeld PR, Nakajima Y, Yamaguchi K. Substance P raises neuronal membrane excitability by reducing inward rectification. Nature 1985; 315: 498–501.
46. Nakajima Y, Nakajima S, Inoue M. Pertussis toxin-insensitive G protein mediates substance P-induced inhibition of potassium channels in brain neurons. Proc Natl Acad Sci USA 1988; 85: 3643–3647.
47. Sun XP, Supplisson S, Torres R, Sachs G, Mayer E. Characterization of large conductance chloride channels in colonic smooth muscle. J Physiol (London) 1992; 448: 355–382.
48. Sun XP, Supplisson S, Mayer E. Chloride channels in myoctyes from rabbit colon are regulated by a pertussis toxin-sensitive G protein. Amer J Physiol 1993; 264: G774–G785.
49. Mau SE, Saermark T. Substance P stimulation of polyphosphoinositide hydrolysis in rat anterior pituitary membranes involves a GTP-dependent mechanism. J Endocrinology 1990; 130: 63–70.
50. Watra MM, Schwinn DA, Schreurs J, Blank JL, Kim CM, Benovic JL, Krause JE, Caron MG, Lefkowtiz RJ. The substance P receptor which couples to $G_{q/11}$, is a substrate of β-adrenergic receptor kinase 1 and 2. J Biol Chemistry 1993; 268: 9161–9164.
50a. Blount P, Krause JE. Functional nonequivalence of structurally homologous domains of NK-1 and NK-2 type tachykinin receptors. J Biol Chemistry 1993; 268: 16388–16395.
51. Maggi CA, Patacchini R, Quartara L, Rovero P, Santicioli P. Tachykinin receptors in the guinea-pig isolated bronchi. Eur J Pharmacol 1991; 197: 167–174.
52. Belvisi M, Patacchini R, Barnes PJ, Maggi CA. Facilitatory effects of selective agonists for tachykinin receptors on cholinergic neurotransmission: evidence for species differences. Br J Pharmacol 1994; 111: 103–110.
53. Barnes PJ. Neuropeptides and airways pathophysiology. Neuropeptides 1992; 22: 7.
54. Manzini S. Bronchodilation by tachykinins and capsaicin in the mouse main bronchus. Br J Pharmacol 1992; 105: 968–972.
55. Devillier P, Acker M, Advenier C, Marsac J, Regoli D, Frossard N. Activation of an epithelial neurokinin 1 receptor induces relaxation of rat trachea through release of prostaglandin E_2. J Pharmacol Exp Ther 1992; 263: 767–772.
56. Maggi CA, Patacchini R, Baroldi P, Theodorsson E, Meli A. Immunoblockade by a specific tachykinin antiserum of the noncholinergic contractile responses in the guinea-pig isolated bronchus. J Autonom Pharmacol 1990; 10: 173–179.

57. Maggi CA, Patacchini R, Rovero P, Santicioli P. Tachykinin receptors and noncholiner-gic bronchoconstriction in the guinea-pig isolated bronchi. Am Rev Resp Dis 1991; 144: 363–367.
58. Martin CAE, Naline E, Emonds-Alt X, Advenier C. Influence of CP 96,345 and SR 48,968 on eletrical field stimulation of the isolated guinea-pig main bronchus. Eur J Pharmacol 1992; 224: 137–143.
59. Maggi CA, Guiliani S, Ballati L, Lecci A, Manzini S, Patacchini R, Renzetti AR, Rovero P, Quartara L, Giachetti A. In vivo evidence for tachykininergic transmission using a new NK-2 receptor selective antagonist, MEN 10376. J Pharmacol Exp Ther 1991; 257: 1172–1178.
60. Ballati L, Evangelista S, Maggi CA, Manzini S. Effect of selective tachykinin receptor antagonists on capsaicin- and TK-induced bronchospasm in anaesthetized guinea-pigs. Eur J Pharmacol 1992; 214: 215–221.
61. Lou YP, Lee LY, Satoh H, Lundberg JM. Postjunctional inhibitory effect of the NK-2 receptor antagonist SR 48968 on sensory NANC bronchoconstriction in the guinea-pig. Br J Pharmacol 1993; 109: 765–773.
62. Lundberg JM, Brodin E, Hua XY, Saria A. Vascular permeability changes and smooth muscle contraction in relation to capsaicin-sensitive SP afferents in the guinea-pig. Acta Physiol Scand 1984; 120: 217–227.
63. Manzini S, Conti S, Maggi CA, Abelli L, Somma V, Del Bianco E, Geppetti P. Regional differences in the motor and inflammatory responses to capsaicin in guinea-pig airways. Amer Rev Resp Dis 1989; 140: 936–941.
64. Abelli L, Maggi CA, Rovero P, Del Bianco E, Regoli D, Drapeau G, Giachetti A. Effect of synthetic tachykinin analogues on airway microvascular leakage in rats and guinea-pigs: evidence for the involvement on NK-1 receptors. J Autonom Pharmacol 1991; 11: 267–275.
65. Eglezos A, Giuliani S, Viti G, Maggi CA. Direct evidence that capsaicin-induced plasma protein extravasation is mediated through tachykinin NK-1 receptors. Eur J Pharmacol 1991; 209: 277–279.
66. Delay-Goyet P, Lundberg JM. Cigarette smoke-induced airway oedema is blocked by the NK-1 antagonist, CP 96,345. Eur J Pharmacol 1991; 203: 157–158.
67. Lei YH, Barnes PJ, Rogers DF. Inhibition of neurogenic plasma exudation in guinea-pig airways by CP-96,345 a new non-peptide NK-1 receptor antagonist. Br J Pharmacol 1992; 105: 261–262.
68. Tousignant C, Chan CC, Guevremont D, Brideau C, Hale JJ, MacCoss M, Rodger IW. NK-2 receptor mediate plasma extravasation in guinea-pig lower airways. Br J Pharmacol 1993; 108: 383–386.
69. Majno G, Shea SM, Leventhal M. Endothelial contraction induced by histamine-type mediators. An electron microscopy study. J Cell Biology 1969; 42: 647–670.
70. Baluk P, Nadel JA, McDonald D. SP-immunoreactive sensory axons in the rat respira-tory tract: A quantitative study of their distribution and role in neurogenic inflammation. J Comp Neurol 1992; 319: 586–598.
71. Bertrand C, Geppetti P, Baker J, Yamawaki I, Nadel JA. Role of neurogenic inflamma-tion in antigen induced vascular extravasation in guinea-pig trachea. J Immunology 1993; 150: 1479–1485.
72. Webber SE. Receptors mediatind the effects of substance P and neurokinin A on mucus secretion and smooth muscle tone of the ferret trachea: potentiation by an enkephalinase inhibitor. Br J Pharmacol 1989; 98: 1197–1206.
73. Rogers DF, Aursudkij B, Barnes PJ. Effect of tachykinins on mucus secretion in human bronchi in vitro. Eur J Pharmacol 1989; 174: 283–286.
74. Meini S, Mak JCW, Rohde JAL, Rogers DF. Tachykinin control of ferret airways: Mucus secretion, bronchoconstriction and receptor mapping. Neuropeptides 1993; 24: 81–89.
75. Kuo HP, Rohde JA, Tokuyama K, Barnes PJ, Rogers DF. Capsaicin and sensory neuropeptide stimulation of goblet cell secretion in guinea-pig trachea. J Physiol (Lon-don) 1990; 431: 629–641.
76. Watson N, Maclagan J, Barnes PJ. Endogenous tachykinins facilitate transmission through parasympathetic ganglia in guinea-pig trachea. Br J Pharmacol 1993; 109: 751–759.

77. Myers AC, Undem BJ. Electrophysiological effects of tachykinins and capsaicin on guinea-pig bronchial parasympathetic ganglion neurones. J Physiol (London) 1993; 470: 665–679.

78. Maggi CA, Patacchini R, Perretti F, Tramontana M, Manzini S, Geppetti P, Santicioli P. Sensory nerves, vascular endothelium and neurogenic relaxation of the guinea-pig isolated pulmonary artery. Naunyn Schmiedeberg's Arch Pharmacol 1990; 342: 78–84.

79. Poyner DR. CGRP: multiple actions, multiple receptors. Pharmacol Ther 1993; 56: 23–51.

80. Quirion R, Van Rossum D, Dumont Y, St Pierre S, Fournier A. Characterization of $CGRP_1$ and $CGRP_2$ receptor subtypes. Ann New York Acad Sci 1992; 657: 88–105.

81. Chiba T, Yamaguchi A, Yamatani T, Nakamura A, Morishita T, Inui T, Fukase M, Noda T, Fujita T. CGRP receptor antagonist hCGRP(8–37). Am J Physiol E331–E335.

82. Mimeault M, Fournier A, Dumont Y, St Pierre S, Quirion R. Comparative affinities and antagonistic potencies of various human CGRP fragments on CGRP receptors in brain and periphery. J Pharmacol Exp Ther 1991; 258: 1084–1090.

83. Maggi CA, Chiba T, Giuliani S. Human αCGRP(8–37) as an antagonist of exogenous and endogenous calcitonin gene-related peptide. Eur J Pharmacol 1991; 191: 85–88.

84. Giuliani S, Wimalawansa SJ, Maggi CA. Involvement of multiple receptors in the biological effects of CGRP and amylin in rat and guinea-pig preparations. Br J Pharmacol 1992; 107: 510–514.

85. Dennis TB, Fournier A, Cadieaux A, Pomerlau F, Jolicoeur FB, St Pierre S, Quirion R. Human CGRP(8–37) receptor antagonist revealing CGRP receptor heterogeneity in brain and periphery. J Pharmacol Exp Ther 1990; 254: 123–128.

86. Rovero P, Guiliani S, Maggi CA. CGRP antagonist activity of short C-terminal fragments of human αCGRP, CGRP(23–37) and CGRP(19–37). Peptides 1992; 13: 1025–1027.

87. Bartho L, Koczan G, Maggi CA. Studies on the mechanisms of the contractile action of rat CGRP and capsaicin on the guinea-pig ileum: Effect of hCGRP(8–37) and CGRP tachyphylaxis. 1993; Neuropeptides 25: 325–329.

88. Evangelista S, Tramontana M, Maggi CA. Pharmacological evidence for the involvement of multiple calcitonin gene-related peptide (CGRP) receptors in the antisecretory and antiulcer effect of CGRP in rat stomach. Life Sci Pharmacol Letters 1992; 50: P13–P18.

89. Edvinsson L, Fredholm BB, Hamel E, Jansen I, Verrecchia C, Perivascular peptides relax cerebral arteries concomitant with a rise in cAMP or release of an endothelium-derived relaxant factor. Neurosci Letters 1985; 58: 213–217.

90. Kubota M, Moseley JM, Botera L, Dusting GJ, MacDonald PS, Martin TS. CGRP stimulates cAMP in rat aortic smooth muscle cells. Biochem Biophys Res Comm 1985; 132: 88–94.

91. Grace GC, Dusting GJ, Kemp BE, Martin TJ. Endothelium and the vasodilator action of rat CGRP. Br J Pharmacol 1987; 91: 729–733.

92. Brain SD, Williams TJ, Tippins JR, Morris HR, MacIntyre I. CGRP is a potent vasodilator. Nature 1985; 313: 54–56.

93. Fiscus RR, Zhou HL, Wang X, Han C, Ali S, Joyce CD, Murad F. CGRP-induced cAMP, cGMP and vasorelaxant response in rat thoracic aorta are antagonized by blockers of endothelium-derived relaxing factor (EDRF). Neuropeptides 1991; 20: 133–139.

94. Gray DW, Marshall I. Nitric oxide synthesis inhibitors attenuate CGRP endothelium-dependent vasorelaxation in rat aorta. Eur J Pharmacol 1992; 212: 37–42.

95. Abdelrahman A, Wang YX, Chang SD, Pang CCY. Mechanism of the vasodilator action of CGRP in conscious rats. Br J Pharmacol 1992; 106: 45–48.

96. Andersson SE. Glibenclamide and L-nitro-arginine methyl ester modulate the ocular and hypotensive effects of CGRP. Eur J Pharmacol 1992; 224: 89–91.

97. Gray DW, Marshall I. Human αCGRP stimulates adenylate cyclase and guanylate cyclase and relaxes rat thoracic aorta by releasing nitric oxide. Br J Pharmacol 1992; 107: 691–696.

98. Marshall I. Mechanism of vascular relaxation by the CGRP. New York Acad Sci 1992; 657: 204–215.

99. Nelson MT, Huang Y, Brayden JE, Hescheler J, Standen NB. Arterial dilations in response to CGRP involve activation of K channels. Nature 1990; 344: 770–772.
100. Hood JS, McMahon TJ, Kadowitz PJ. Influence of lemakalim on the pulmonary vascular bed of the cat. Eur J Pharmacol 1991; 202: 101–107.
101. Palmer JBD, Cuss FMC, Mulderry PK, Ghatei MA, Springall DR, Cadieux A, Bloom SR, Polak JM, Barnes PJ. CGRP is localised to human airway nerves and potently constricts human airway smooth muscle. Br J Pharmacol 1987; 91: 95–101.
102. Hamel R, Ford-Hutchison AW. Contractile activity of CGRP on pulmonary tissues. J Pharm Pharmacol 1988; 40: 210–211.
103. Martling CR, Saria A, Fischer JA, Hokfelt T, Lundberg JM. CGRP and the lung: Neuronal coexistence with SP release by capsaicin and vasodilatory effects. Regul Peptides 1988; 20: 125–139.
104. Warner EA, Krell RD, Buckner CK. Pharmacologic studies on the differential influence of inhibitors of neutral endopeptidase on nonadrenergic noncholinergic contractile responses of the guinea-pig isolated hilar bronchus to transmural electrical stimulation and exogenously applied tachykinins. J Pharmacol Exp Ther 1990; 254: 824–830.
105. Gatto C, Lussky RC, Erickson LW, Berg KJ, Wobken JD, Johnson DE. Calcitonin and CGRP block bombesin and SP-induced increases in airway tone. J Appl Physiol 1989; 66: 573–577.
106. Cadieaux A, Lanoue C, Sirois P, Barabe' J. Carbamylcholine and 5-hydroxytryptamine-induced contraction in rat isolated airways: inhibition by CGRP. Br J Pharmacol 1990; 101: 193–199.
107. Lanoue C, Fournier A, St Pierre S, Cadieaux A. Characterization of CGRP receptor sites in rat airways. New York Acad Sci 1992; 657: 441–442.
108. Tschirhart E, Bertrand C, Theodorsson E, Landry Y. Evidence for the involvement of CGRP in the epithelium-dependent contraction of guinea-pig trachea in response to capsaicin. Naunyn Schmiedeberg's Arch Pharmacol 1990; 342: 177–181.
109. Kanemura T, Tamaoki J, Horii S, Sakai N, Kobayashi K, Isono K, Takeuchi S, Takizawa T. CGRP augments parasympathetic contraction of rabbit tracheal smooth muscle in vitro. Agents Actions 1990; 31: 219–224.
110. Butler A, Worton SP, O'Shaughnessy CT, Connor HE. Sensory nerve-mediated relaxation of guinea-pig pulmonary artery: Prejunctional modulation by alpha 2 adrenoceptor agonists but not sumatriptan. Br J Pharmacol 1993; 109: 126–130.
111. Manzini S, Maggi CA, Geppetti P, Bacciarelli C. Capsaicin desensitization protects from antigen induced bronchospasms in conscious guinea-pigs. Eur J Pharmacol 1987; 138: 307–308.
112. Martling CR, Lundberg JM. Capsaicin-sensitive afferents contribute to acute airway edema following tracheal instillation of hydrochloric or gastric juice in the rat. Anesthesiology 1988; 68: 350–356.
113. Solway J, Kao BM, Jordan JE, Gitter B, Rodger IW, Howbert JJ, Alger LE, Necheles J, Leff AR, Garland A. Tachykinin receptor antagonists inhibit hyperpnea-induced bronchoconstriction in guinea-pigs. J Clin Invest 1993; 92: 315–323.

Airways Smooth Muscle: Peptide Receptors, Ion Channels and Signal Transduction
ed. by D. Raeburn and M. A. Giembycz
© 1995 Birkhäuser Verlag Basel/Switzerland

CHAPTER 4
Vasoactive Intestinal Peptide

Sami I. Said

Department of Veterans Affairs, Medical Center at Northport, New York, and University Medical Center, Stony Brook, New York, USA.

1   Historical Background: Discovery of VIP in the Lung and Other Organs
2   Localization, Distribution and Origins in the Lung
3   Molecular Forms and Biosynthesis
4   Co-Existence with Other Neuropeptides and Neurotransmitters
5   Influence on Airway Functions
5.1   Smooth Muscle Tone and Airway Reactivity
5.2   Airway Secretion
5.3   Pulmonary and Bronchial Circulations
5.4   Airway Inflammation
5.5   Airway Smooth Muscle Proliferation
6   Interactions with Other Neurotransmitters
7   Receptors
8   Mechanisms of Action: Role of Second Messengers and Other Transmitters
9   Enzymatic Degradation
10   VIP as a Regulator of Airways Function
10.1   Airway Smooth Muscle Tone and Reactivity
10.1.1   VIP and the NANC Inhibitory System
10.1.2   Nitric Oxide as a Co-Transmitter of NANC Relaxation
10.2   Secretion
10.3   Pulmonary Microvascular Permeability
10.4   Surfactant Synthesis and Secretion
10.5   Airway Smooth Muscle Cell Proliferation
11   VIP in Airway Pathophysiology and Disease
11.1   Airway Hyperreactivity and Inflammation
11.2   Possible Role in Airway Diseases
11.2.1   Cystic Fibrosis
11.2.2   Bronchial Asthma
12   Therapeutic Potential
12.1   Asthma and Airway Inflammation
12.1.1   Clinical Trials in Asthma
12.1.2   VIP Analogues that Resist Enzymatic Degradation
12.2   Acute Lung Injury
12.3   Pulmonary Hypertension
12.4   Lung Cancer
13   Summary and Conclusions; Future Perspectives
         Acknowledgements
         References

1. Historical Background: Discovery of VIP in the Lung and Other Organs

VIP was first isolated from the small intestine, but it had been discovered earlier in the lung, as a vasodilator peptide [1, 2]. Several years later it was identified in the central and peripheral nervous systems [3], and has since been recognized as a neuropeptide with wide distribution, acting as a neuroransmitter or neuromodulator in the lung and practically all other organs and tissues. The discovery and characterization of VIP has spawned an active field of research in the regulation of lung function by peptides. To date, at least 20 other peptides have been demonstrated in the lung, and their functions investigated [4, 5].

2. Localization, Distribution and Origins in the Lung

VIP-containing nerves supply the upper airways, including the nose and the entire tracheobronchial tree. The nerves are distributed to the smooth muscle, around submucosal mucous and serous glands, in the lamina propria and in the walls of pulmonary and systemic blood vessels [6, 7]. These nerves are mainly efferent and often cholinergic [8, 9]. Immunoreactive VIP is strongly present in neuronal cell bodies of airway microganglia, which may provide an intrinsic source of innervation of pulmonary structures with this and other peptides [6, 7]. Supporting their intrinsic nature, VIP-containing nerve fibres that innervate bronchial smooth muscle, glands and bronchial blood vessels, survive culture periods of up to 7 days [10, 11]. VIP-containing nerves are also undiminished in denervated human lungs removed for transplantation [12]. Several lines of evidence suggest that VIP may also be present in some sensory nerves in the lung: VIP is released from the airways, together with substance P, by capsaicin [13]; it is colocalized with substance P in some neurons in cat airways [14]; and is present in some nerve terminals that reach airway epithelial cells [15].

In addition to its neuronal localization, immunoreactive VIP is also present in mast cells of rat lungs, from which it may be released by mast-cell degranulators [16]. Several VIP-derived peptides have been chemically characterized in rat basophilic leukemia cells, which resemble mast cells [17] and, more recently, in mouse mast cells [18]. Of other inflammatory cells, eosinophils have the highest concentrations of VIP, which are also higher than those of all other neuropeptides [19].

Of the VIP-related peptides, immunoreactive helodermin/helospectins have been reported in the lungs of several mammalian species (hamster, mouse, rat, guinea pig), especially in neuroendocrine cells [20]. Pituitary adenylyl cyclase-activating peptide (PACAP)-immunoreactive nerve fibres are present around guinea pig tracheal and bronchial smooth muscle, glands and blood vessels [21, 22]. Simultaneous double-immunostaining with specific antibodies showed that, in the respiratory tract of sheep and ferrets, all PACAP-containing nerve fibres were also immunoreactive to VIP [22].

3. Molecular Forms and Biosynthesis

Several facts about the biosynthesis of VIP have been learned from studies in human neuroblastoma cell lines. In those cells, VIP is synthesized from a precursor molecule (Pro-VIP), which also contains peptide histidine methionine (PHM-27) [23]. The human VIP gene, located on human chromosomal region 6q24 [24], contains seven exons, each encoding a distinct functional domain on the protein precursor or its mRNA [25]. VIP/PHM mRNA synthesis is stimulated by increases in cyclic AMP levels, as well as by increased protein kinase C activity, induced by phorbol esters. Cyclic AMP and phorbol esters act synergistically to stimulate VIP gene transcription, apparently via different sites on the gene [26].

VIP is structurally related to several other peptides. These peptides, which make up a peptide "family", include PHM (in human tissue) or PHI (its counterpart in other mammalian tissues), PHV, secretin, glucagon, helodermin, PACAP, growth hormone-releasing hormone (GHRH), corticotropin-releasing hormone (CRH), sauvagine, urotensin I and gastric inhibitory peptide (or glucose-dependent insulinotropic peptide (GIP)). PACAP exists in two forms, PACAP-38 and the C-terminally truncated PACAP-27. Sequence similarities between these peptides explain certain similarities in their biological actions, but each peptide usually has one or more distinctive actions.

4. Co-Existence with Other Neuropeptides and Neurotransmitters

As mentioned above, VIP is co-synthesized with PHM (in human tissues) or PHI (in other mammalian species). Either of these two peptides therefore is usually colocalized [27] and coreleased with VIP. Differential processing of pre-pro VIP may also result in the production of a third, related peptide, peptide histidine valine (PHV). Colocalization of VIP with numerous other neuropeptides and neurotransmitters has been documented (Table 1) in the lungs and elsewhere [28]. Such neuropeptides include PACAP [22], opioid peptides [29], galanin [30], SP [10, 11, 14, 15], CGRP, and neuropeptide Y [31]. The presence of VIP in cholinergic neurons was first noted in nerves supplying exocrine glands [32]; coexistence with acetylcholine was later confirmed in nerves to the airways [33, Shimosegawa and Said, unpublished observations].

Table 1. Peptides and Neurotransmitters That Coexist with VIP

Other Peptides	Neurotransmitters
PHI (or PHM), PHV, Galanin, Opioid Peptides, Substance P, CGRP, Neuropeptide Y	Acetylcholine, Nitric Oxide, Noradrenaline

Other nonpeptide neurotransmitters that may coexist with VIP in the same neurons are: noradrenaline and nitric oxide (NO). The latter, indentified by the presence of the enzyme of NO synthase, which catalyzes its formation from L-arginine, has been demonstrated to co-exist with VIP in guinea pig [34] and ferret [35] airway ganglia.

These colocalizations facilitate important physiological interactions. For example: a) the innervation of the exocrine glands with acetyl-choline and VIP results in greater secretion and blood flow responses than if either of the two transmitters was present alone [36]; b) VIP modulates cholinergic transmission in respiratory airways [37, 38]; c) the co-presence of VIP and substance P in cholinergic nerves to the airways [14] provides a balance between the bronchconstrictor, pro-inflammatory substance P and the bronchial-relaxant, anti-inflamma-tory VIP [39]; and d) VIP and NO act as co-transmitters of smooth muscle relaxation in blood vessels, respiratory airways and other tissues [40, 41].

5. Influence on Airway Functions

5.1. Smooth Muscle Tone and Airway Reactivity

VIP relaxes airway smooth muscle both *in vitro* and *in vivo*: it relaxes isolated tracheal or bronchial segments from guinea pigs, rabbits, dogs, and humans and prevents or attenuates the constrictor effect of his-tamine, prostaglandin $F_{2\alpha}$, kallikrein, leukotriene D_4, neurokinins A and B, and endothelin [42–50]. The bronchial-relaxant action of VIP is relatively long-lasting and is independent of adrenergic and cholinergic receptors or of cyclooxygenase activity [43, 51–53]. As a relaxant of human bronchi *in vitro*, VIP is almost 100 times more potent than isoprenaline making it the most potent endogenous bronchodilator known [54]. VIP also relaxes strips of human lung parenchyma, but less effectively than it relaxes human bronchi [53]. Inhaled VIP protects against the bronchoconstriction induced by histamine or prostaglandin $F_{2\alpha}$ in dogs [55], and infused VIP reverses 5-HT-induced bronchocon-striction in cats [56].

VIP-induced tracheal relaxation of guinea pig tracheal segments *in vitro* is potentiated and prolonged by the addition of phosphoramidon, a selective inhibitor of neutral endopeptidase (NEP, E.C.34.24.11) [57]. The airway relaxant response of VIP is also enhanced by removal of airway epithelium, probably because of the removal of the peptidases that degrade the peptide [58]. Certain VIP-related peptides, natural and synthetic, that are relatively more resistant to enzymatic digestion, are longer-acting or more potent airway relaxants. An example is heloder-min, orginally isolated from the lizard Gila monster Heloderma, but also present in mammalian tissues. Structurally similar to VIP, helo-dermin has a carboxy-terminal extension that may account for its greater resistance to enzymatic degradation by NEP and other tracheal epithelial cell-derived peptidases [59]. This may also explain its longer-lasting airway relaxation, of isolated tracheal strips [60] or *in vivo* [61]. Related synthetic analogues [62], and other C-terminally extended forms of the peptide [63] are more potent than VIP as airways smooth-muscle relaxants, or longer-acting, or both.

5.2. Airway Secretion

Acting directly on receptors on the surface of epithelial cells, VIP stimulates active chloride ion secretion in canine tracheal epithelium [64]. This effect, elicited under short-circuit conditions, is resistant to a combination of atropine, phentolamine and propranolol, and to tetrodotoxin, indicating its independence of the release of other neural transmitters [64]. VIP also stimulates chloride secretion in the shark rectal gland, probably by a cAMP-mediated mechanism [65], and is a potent secretagogue of the avian salt gland [66, 67]. VIP activates a sustained Cl^- conductance current that is inhibited by Cl^- channel blockers flufenamic acid and niflumic acid and by the inhibitory cAMP isomer, RpcAMP. These findings, together with the similarity of the kinetics of the Cl^- current evoked by VIP to that activiated by cell-per-meant cAMP analogues or by forskolin in these and other secretory cells [68, 69], suggest that VIP-receptor binding activates a cAMP-de-pendent Cl^- channel [70].

The reported effects of VIP on the secretion of sulphated macro-molecules have depended on the species tested. The peptide stimulated the secretion by ferret tracheal explants [71]. This action, like that on Cl^- transport, was not inhibited by adrenergic or cholinergic blockade or by tetrodotoxin, and was mediated by a receptor in the apical membrane of the tracheal epithelium [72]. In human tracheal explants, however, VIP inhibited baseline and methacholine-stimulated release of both glycoconjugates and lysozyme [73].

VIP, at low concentrations ($EC_{50} = 60$ pM), increases ciliary beat activity in rabbit cultured tracheal epithelial cells. This effect, like airway relaxation, is potentiated by phosphoramidon, an inhibitor of NEP [74].

5.3. Pulmonary and Bronchial Circulations

Vasodilatation is the principal action of VIP and the one that guided its discovery and isolation [2]. VIP dilates the vessels supplying the nose, upper airways [75], trachea and bronchi [76], as well as the pulmonary vessels [47, 77]. The pulmonary vasodilator action of VIP is more potent than that of prostacyclin [53], is independent of the endothelium [78] and is exerted on vessels from several species, including the guinea pig, cat, dog, rabbit, cow and from man. VIP also promotes blood flow through the tracheo-bronchial circulation, and may play an important role in regulating airway perfusion [79]. It is approximately equipotent with substance P in decreasing canine tracheal vascular resistance, but is longer-acting [80].

5.4. Airway Inflammation

VIP has now been shown to prevent, or at least reduce or delay, acute inflammatory lung injury evoked by a variety of insults, including pro-oxidants [81]. The related peptide helodermin appears to share this property with VIP. More recently, the anitinflammatory activity of VIP has also been demonstrated in the airways, where the peptide protects against capsaicin-induced airway constriction and oedema ([13], and unpublished observations).

VIP modulates the functions of several types of inflammatory cells: Specifically, it inhibits: 1) mitogen-induced T-lymphocyte proliferation and other aspects of T-lymphocyte function, including the release of certain cytokines, especially interleukin-2 [82–84]; 2) natural killer cell activity [85]; the respiratory burst in human monocytes [86]; 4) phago-cytosis and superoxide production by rat alveolar macrophages [87]; 5) antigen-induced release of histamine (and possibly other mast-cell medi-ators) from guinea pig lung [88]; and rabbit platelet aggregation and 5-HT secretion induced by platelet-activating factor, by a cAMP-depen-dent mechanism [89]. No adenylyl cyclase-linked VIP receptors have been demonstrated on neutrophils [90].

5.5. Airways Smooth Muscle Proliferation

In a recent study, VIP was reported to inhibit the basal proliferation rate of human airways smooth muscle (ASM) cells, by up to 82%. The effect was specific for VIP because equimolar concentrations of the related peptide glucagon were ineffective. Histamine, a mitogen, concen-tration-dependently increased cell counts by up to 66%, but VIP nul-lified this mitogenic effect. Forskolin and 8-bromo-cyclic AMP, separately and especially when combined with VIP, concentration-dependently reduced cell counts and [³H]thymidine incorporation, in direct proportion to the associated increase in cAMP levels. These findings demonstrate that VIP: 1) selectively inhibits the growth and multiplication of human ASM cells in a concentration-dependent manner, and 2) neutralizes the mitogenic effect of histamine on these cells by a cAMP-dependent mechanism [91]. If VIP modulates the proliferation of ASM cells *in vivo*; then its deficiency, as has been reported in asthma [92], may facilitate hyperplasia of these cells due to their unopposed stimulation by histamine and other mitogens. The inhibition of ASM proliferation has important implications for bronchial asthma, where proliferation of these cells contributes significantly to the increased airway resistance [93–96].

6. Interactions with Other Neurotransmitters

As discussed earlier, VIP often co-exists with acetylcholine in cholinergic nerves in the airways, and may therefore function as a co-transmitter in these neurons [8, 29, 33]. The co-release of VIP with acetylcholine is more likely to occur under certain patterns of neural activation, such as high-frequency firing [97]. VIP reduces the contractile effect of acetylcholine on ASM *in vitro* and may therefore modulate cholinergic bronchoconstriction [98]. In the presence of cholinergic nerve activity causing bronchoconstriction, the co-release of VIP from activated cholinergic nerves could dilate nearby blood vessels, thus promoting blood flow to contracting ASM cells, and supplying them with additional nutrients [99].

VIP and constitutive nitric oxide (NO) synthase co-exist in some neurons, e.g., those of the myenteric plexus [100] and tracheal ganglia [35]. As physiological co-transmitters, VIP and NO, normally produced in low concentrations, interact in several ways: 1) Both transmitters are released during NANC relaxation of gastrointestinal and ASM, evoked by electrical field or vagal stimulation [101]. 2) In some tissues, e.g., guinea pig gastric fundus, VIP stimulates NO release and NO promotes VIP release [41], while in other tissues, e.g., guinea pig trachea, VIP-induced relaxation, at least in some studies [102], but not in others [103], appears to be independent of NO release. 3) Released together, either independently or interdependently, VIP and NO synergistically promote smooth muscle relaxation (and other functions) via two separate but complementary pathways, the stimulation of adenylyl cyclase and guanylyl cyclase, respectively [40]. There is also evidence that both VIP and NO can induce smooth muscle relaxation through hyperpolarization of smooth muscle and the activation of ATP-sensitive K^+ channels [104, 105]. Acting through multiple pathways, therefore, the

Table 2. Models of Acute Lung and Airway Injury That May Be Attenuated or Prevented by VIP

Method of Injury *	Preparation
1. HCl Intratracheally	
2. PAF into the PA	
3. Xanthine + xanthine oxidase into the PA	Isolated rat lung
4. Pospholipase C into the PA	
5. Prolonged perfusion *ex vivo*	
6. Paraquat into the PA	Guinea pig isolated lung
7. Capsaicin into the airways	Rat and guinea pig lungs, perfused via airway
8. PAF i.v.	Anaesthetized dogs
9. Cobra venom factor	Anaesthetized rats (*in vivo*)

*Abbrevations: PAF: platelet-activating factor; PA: pulmonary artery.

combined release of NO and VIP could well be an attempt to ensure the most effective smooth muscle relaxation in potentially life-threatening situations, presented by severe constriction of respiratory airways or of blood vessels to vital organs [40].

In the setting of acute tissue injury, however, VIP and NO have opposing roles: VIP prevents or attenuates acute lung injury in a variety of experimental models (Table 2) [106], while excess NO, acting directly or with other free radicals, may be an essential intermediate in the production of lung injury in at least some forms of oxidant stress [107, 108]. Under such conditions, therefore, VIP acts to protect the lung and probably other tissues and organs against inflammatory injury, whereas excess NO may promote that injury.

7. Receptors

Specific receptors for VIP have been identified in membrane preparations of rat and human lung [109, 110] and in human lung tumour cells, including small-cell carcinoma cell lines [111]. The VIP receptor is coupled to the adenylyl cyclase-stimulating protein G_s.

VIP binding sites have been localized by autoradiography [112], and as reflected in tissue cAMP content, which increases as a result of the binding [113]. These receptor sites include the nasal mucosa, submucosal serous and mucous glands, tracheal and bronchial epithelial cells, bronchial smooth muscle and alveolar cells, and systemic and pulmonary blood vessels. Upon binding to its membrane receptors, the VIP-receptor complex may be internalized by endocytosis [114]. In addition, VIP binds to nuclear receptors in HT-29 human colonic adenocarcinima cells lines [115].

VIP receptors in the lung have been characterized by several techniques, including photo-affinity labelling, cross-linking of [125]VIP to its receptor, and detergent solubilization [109, 110, 116, 117]. More recently, the nucleotide sequence and amino acid sequence of the VIP-receptor cDNA was reported [118]. This receptor, isolated from a rat cDNA library, mediated VIP-induced cAMP accumulation and was expressed in brain and other tissues. The cloned receptor, with 459 amino acid residues and a calculated Mr of 52 kDa, contains 7 transmembrane segments, and shows sequence similarity to the secretin, calcitonin and parathyroid hormone receptors [118]. A human homologue of this receptor, now called the VIP-1 receptor, was later cloned from human colon carcinoma cells, and is also present in the lung [119]. A second type of VIP receptor (the VIP-2 receptor, more selective for PACAP) has also been cloned [120] and is present in the CNS.

The full relaxant action of VIP on airways and other smooth muscle requires the entire ptide   sequence [121, 122]. Several VIP fragments, analogue and other related molecules, have been   K   to be specific and competitive antagonists of VIP. These putative antagonists include: VIP (10-28) [123], VIP (1-11) [124], (4-Cl-F-Ph^6, Leu^{17}) VIP [125, 126], (N-Ac-Tyr^1, D-Phe^2) GRF (1-29)-NH$_2$ (growth hormone-releasing factor analogue) [127], and a hybrid peptide made up of a portion of VIP and a portion of neurotensin [128]. Of these, only VIP 1-11 inhibits ^{125}I-VIP binding to guinea pig tracheal epithelial cells and VIP-induced tracheal relaxation while the others have little or no antagonistic activity on the airways (S.I. Said, unpublished observation).

PACAP binds to a receptor in the lung (so-called PACAP-2 receptor), that is probably the same as the VIP-1 receptor, and recognizes VIP, PACAP-27 and PACAP-38 with nearly equal affinities. Another class of receptor, highly specific for PACAP (the PACAP-1 receptor), has been identified in brain and other tissues [129]. This PACAP-preferring receptor has been subdivided into A and B subtypes, which bind with high affinity, respectively, with PACAP-27 and PACAP-38. The effector systems for the PACAP-1 receptor include both adenylyl cyclase and the inositol phosphate cytosolic Ca^{2+} pathways [129]. Helodermin also binds to the VIP receptor, but some small-cell lung cancer cell lines express helodermin-preferring receptors [111].

8. Mechanisms of Action: Role of Second Messengers and Other Transmitters

The actions of VIP are mediated principally by cAMP, but other signal transduction pathways, and other neurotransmitters, may be involved in some instances.

· *Cyclic AMP.* Strong evidence points to adenylyl cyclase stimulation and cAMP production as the dominant mechanism of most actions of VIP. This is true for relaxation of airways, vascular and other smooth muscle [40, 130, 131], stimulation of pancreatic exocrine [132] and intestinal secretion [133], modulation of inflammatory-cell functions [83, 134] and inhibition of ASM cell proliferation [135]. There is a two-way relationship between VIP and cAMP: VIP stimulates cAMP production and higher cAMP levels promote VIP biosynthesis [136].

The increase in cAMP activates cAMP- and cGMP-dependent protein kinases [137], eventually leading to reduction in intracellular (Ca^{2+}), myosin deposphorylation and smooth muscle relaxation [138–141].

Calcium and phosphoinositides. In some cells and tissues, *e.g.*, superior cervical ganglion [142] and adrenal chromaffin cells [143], relatively high (1 µM) concentrations of VIP increase the breakdown of phosphoinositides to inositol phosphates, and the intracellular mobilization of Ca^{2+}. The effect on the adrenal medulla is linked to the induction of catecholamine secretion [143]. VIP, at unusually low concentrations (100 pM), increases intracellular $(Ca^{2+})_i$ in rat cortical astrocytes [Brenneman, personal communication], and, at higher concentrations, has the same effect in cultured rat hippocampal neurons [144]. In the latter preparation, PACAP exerts the same action and with greater potency [144]. The VIP-induced rise in $(Ca^{2+})_i$ in astrocytes is correlated with the production of trophic factors by these cells [145]. Less marked changes in $(Ca^{2+})_i$ due to VIP have been described in prolactin-producing cells of rat anterior pituitary [146]. As mentioned above, however, the inositol phosphate/$(Ca^{2+})_i$ pathway is probably more important in mediating the effect of PACAP than of VIP.

Calmodulin. A VIP-binding protein (Mr18 kDa) that may be calmodulin has been identified in guinea-pig lung membranes [147, 148], but it is unlikely to be the high-affinity VIP receptor in the lung [149].

Role of NO and cGMP. The actions of VIP on some tissues and organs, including gastric and colonic smooth muscle [41, 150–154], internal anal sphincter [122], myenteric plexus ganglia [155] and pinealocytes [156], may be mediated in part by the release of NO from neurons or smooth muscle cells. Partial mediation of VIP-induced tracheal relaxation was reported in one study [157], but not in another, similarly conducted study on guinea pig lungs [102]. Wherever VIP activates NO synthase, it will cause a secondary activation of cytosolic guanylyl cyclase [158] with resultant stimulation of intracellular cGMP formation.

Whether or not the synthesis of NO is stimulated by VIP in the airways, NO is physiologically co-released with VIP and participates in producing many of the same physiological effects [40]. Once thought to have opposing results, especially on mast-cell degranulation and ASM tone in asthma [159], the cAMP and cGMP pathways are now recognized as parallel and complementary means of eliciting smooth muscle relaxation [160, 161].

Chloride channel. As discussed earlier, VIP activates a sustained, cAMP mediated Cl^- conductance, resulting in a membrane depolarization that is mimicked by cAMP-promoting agents [70]. The kinetics of this Cl^- conductance are similar to those of other cAMP-activated Cl^- channels, including the Cystic fibrosis transmembrane conductance regulator (CFTR) [162].

Other mechanisms. Stimulation of Ca^{2+}-activated postassium (K_{Ca}) channels has recently been reported as a novel mechanism of action of β-adrenoceptor agonists on ASM [163]. The same mechanism may be involved in the VIP effects on pancreatic acinar cells [164], but its possible role in mediating VIP actions in the airways has not been investigated.

9. Enzymatic Degradation

VIP is degraded by proteases that are present at or near the airway mucosa, and include mast-cell tryptase and chymase [103, 165, 166], NEP [167] and possibly other enzymes. Tryptase cleaves VIP rapidly at Arg14 and Lys20, two of the five potential cleavage sites (with dibasic residues) in the peptide. Chymase cleaves VIP primarily at a single site, Tyr21 [165]. Human recombinant NEP cleaves amino- and carboxy-terminal oligopeptides from VIP in a time- and enzyme concentration-dependent manner [167]. Both mast cell tryptase and chymase reverse VIP-induced ASM relaxation: tryptase, which cleaves the peptide more avidly, being more potent [168]. Conversely, appropriate protease inhibitors, potentiate and prolong VIP-induced tracheal relaxation. Especially effective are combinations of neutral endopeptidase inhibitors [57, 103, 169] and soybean trypsin inhibitors (*i.e.*, serine protease inhibitors) [103, 166]. In normal airways, VIP is degraded largely by NEP, but in the presence of airway inflammation, tryptic enzymes released from mast cells acquire greater significance [170].

10. VIP as a Regulator of Airways Function

10.1. Airways Smooth Muscle Tone and Reactivity

10.1.1. VIP and the NANC inhibitory system: VIP is a likely co-transmitter of the NANC component of neurogenic relaxation of the airways, the dominant component of autonomic relaxation, especially in guinea pigs, cats and humans [171–173], though not in dogs [174]. Supporting a transmitter role for VIP in NANC relaxation of ASM, is the following evidence:

a) VIP fulfills the criteria of a neurotransmitter [175].
b) It is present in the vagus nerve [176], where the signals for NANC airway relaxation are transmitted, and is released *in vivo* by electrical field stimulation of the vagi in the presence of cholinergic and adrenergic blockade [177].
c) It mimicks the electrophysiological changes in airways smooth muscle elicited by NANC nerve stimulation [45, 46].
d) Prolonged incubation of ASM with VIP, which produces desensitization of its relaxant effect, also reduces NANC relaxation [45, 178].
e) Transmural electrical field stimulation of guinea pig tracheal segments, in the presence of adrenergic and muscarinic cholinergic receptor blockade, elicits relaxation of the airway segments, and release of VIP into the medium, in proportion to the magnitude of the relaxation [178].
f) In the same experiments, both the VIP release and the tracheal relaxation are markedly inhibited by tetrodotoxin, which blocks neurotransmission [178].
g) Airway relaxation is also greatly ($> 70\%$) reduced if the tracheal segments are pre-incubated with a specific anti-VIP antiserum [178].

Other observations have been cited against a major transmitter role for VIP in airway relaxation, especially the failure of α-chymotrypsin (α-CT), which cleaves VIP [180], to reduce the NANC response of human bronchial strips although it reduces that of guinea pig trachea [181, 182]. However, this apparent discrepancy may be due to a) the inability of α-CT, which has a molecular mass of 21.6 kDa, to penetrate fully into human broncial tissues, which are thicker than guinea pig trachea, and/or b) the inactivation of α-CT by endogenous human antiproteases, which are strongly expressed in excised human airway tissues, and their activity may be far greater than that of α-CT at the neural sites of peptide release [183].

10.1.2. Nitric oxide as a co-transmitter of NANC relaxation: Within the past few years, it has become increasingly apparent that, like VIP, NO is a major transmitter of airways and other smooth muscle relaxation. The two transmitters, acting in concert via different second messengers, collaborate to bring about the relaxation [40, 101, 157]. The relative contributions of VIP and NO, and the interactions between them in different tissues, are under active investigation [41, 101, 150–152, 122, 155, 184–186].

10.2. Secretion

The actions of VIP on water-electrolyte and mucus secretion in the upper and lower respiratory passages suggest a possible physiological role in the regulation of airway secretion, but such a role is difficult to establish without the use of specific and potent antagonists of VIP receptors in the airways.

10.3. Pulmonary Microvascular Permeability

As noted above, VIP, and the related peptide helodermin, can attenuate or prevent alveolar epithelial-microvascular leakage in several different experimental models. Evidence suggests that VIP not only protects the lungs when given exogenously in pharmacological concentrations, but may also be a physiological modulator of lung and airway injury: 1) Injured lungs release relatively large amounts of immunoreactive peptide (100 ng per ml of pulmonary perfusate); 2) VIP is released in other conditions of tissue injury, such as septic shock [187, 188], intestinal ischemia [189], and following PAF infusion is anesthetized dogs [Pakbaz et al., 1988]; 3) VIP is co-released with SP in response to acute capsaicin-induced airway inflammation in guinea pigs [190]; and 4) VIP mRNA in the lung and airways is rapidly and markedly increased upon application of paraquat [191].

10.4. Surfactant Synthesis and Secretion

VIP binds to adenylyl cyclase-linked receptors on cultured alveolar type II cells derived from human bronchoalveolar cell carcinoma [134]. Through these receptors, VIP may regulate type II cell functions, such as the production of surfactant or some of its constituents. Other compounds that promote cAMP levels in these cells stimulate the synthesis and secretion of surfactant [192–194].

10.5. Airways Smooth Muscle Cell Proliferation

As described above, VIP inhibits basal and histamine-stimulated proliferation of human ASM cells in culture. This anti-mitogenic effect is potentiated by forskolin and other cAMP-elevating agents, and its magnitude is correlated with cAMP levels. Through this action, VIP may modulate the proliferation of smooth muscle cells in human airways in vivo. Lack of this modulatory effect may lead to excessive hyperplasia of ASM cells, because of the unbalanced action of endogenous mitogens.

11. VIP in Airway Pathophysiology and Disease

11.1. Airway Hyperreactivity and Inflammation

As summarized above, VIP protects the lungs against acute injury in different experimental models, and may be a physiological modulator of injury. Preliminary evidence suggests that VIP may also modulate airway inflammation. These observations, together with its potent airway relaxant, anti-bronchoconstrictor, and anti-inflammatory activities, suggest that VIP serve as an endogenous modulator of airway hyperreactivity and airway inflammation. The latter features are now widely recognized as important components of bronchial asthma.

11.2. Possible Role in Airway Diseases

The physiological and pharmacological properties of VIP have led to the hypothesis that its decreased biological activity, due to deficient expression of the peptide or its receptors, or its excessive degradation or neutralization, may contribute to the pathogenesis of two of the major and more common disorders of the airways and lungs: cystic fibrosis and bronchial asthma [179, 195].

11.2.1. Cystic fibrosis (CF): Some of the manifestations of CF, notably the decreased Cl-permeability in sweat gland ducts [196], seemed consistent with a deficiency of VIP in exocrine glands. Several observations are consistent with the possibility that CF may be causally related to a deficiency of VIP: a) The peptide richly innervates all exocrine organs [36], the principal organs affected in this disorder; b) it influences all major exocrine function, stimulating water, intestinal and bronchial Cl-, pancreatic HCO3-, and macromolecular secretion, and increasing blood flow [197]; c) it binds to specific receptors on exocrine glands [198]; and d) VIP innervation of sweat glands in CF patients is significantly decreased around the acini and virtually absent around the ducts [199]. VIP was recently found to activate a sustained, cAMP-mediated Cl^- conductance with similar kinetic properties to other cAMP-activated Cl^- channels, including the CF Transmembrane Regulator (CFTR) [70]. No direct link has been established, however, between CF and a deficiency of VIP biosynthesis, binding or metabolism.

11.2.2. Bronchial asthma: A related hypothesis, linking the pathogenesis of bronchial asthma to a postulated lack of VIP innervation of airways smooth muscle, is based on evidence that a) VIP is a co-transmitter of NANC relaxation of airways; b) the NANC system is the dominant, if not exclusive, relaxant system in human airways [172]; and c) a lack of

NANC relaxation might explain the airway hyperreactivity of bronchial asthma [179, 200]. Support for a pathogenetic role for VIP deficiency in asthma comes from the observations that: a) VIP-containing nerves are selectively lacking in airways of human asthmatics [92]; and b), VIP inhibits airway inflammation and airways smooth muscle multiplication, two major features of asthma.

12. Therapeutic Potential

12.1. Asthma and Airway Inflammation

VIP presents a number of features that offer promise for its potential usefulness in the management of asthma and airway inflammation. These features include: a) its presence naturally in nerves supplying the airways; b) potent relaxation of large and small airways, including lung parenchymal strips, from human subjects [53]; c) promotion of water and Cl^- transport in airway epithelium, rendering bronchial secretions easier to mobilize and eliminate; anti-inflammatory activity in the lung and airways [200]; and d) inhibition of ASM cell proliferation.

12.1.1. Clinical trials in asthma: Despite the excellent potential of VIP as an anti-asthma agent, the few clinical trials reported in this disease have not been as successful as might have been predicted. Given by i.v. infusion (1, 3 or 6 pmol/kg · min^{-1} for 15 min), VIP had no effect on specific airway conductance in six normal subjects [54], but in another study on seven atopic asthmatics, it caused significant bronchodilation (improvement in forced expiratory air flow) and attenuated histamine-induced bronchoconstriction [200]. In both studies, a moderate tachycardia and decrease in arterial blood pressure were observed at the higher dose rates. Given as an aerosol to six atopic subjects with mild asthma, VIP (total dose 100 μg, or approximately 1.4 μg/kg) reduced bronchial reactivity to histamine but did not significantly increase specific airway conductance on its own [203]. VIP pretreatment also did not prevent exercise-induced bronchoconstriction in six adult asthmatics [204]. In a third study, on 10 of 15 moderately severe asthmatics, 5 μg/kg of sterile VIP by inhalation, resulted in significant improvement in spirometric data or specific airway conductance, but the observed bronchodilatation was less than that elicited by inhalation of the β-agonist salbutamol [205].

12.1.2. VIP analogues that resist enzymatic degradation: The above results suggested that inhaled VIP was rapidly degraded by airway peptidases at or near airway epithelium. Two possible approaches seemed possible: a) to combine VIP with one or more selective peptidase inhibitors that would protect it against enzymatic degradation, or b) identify a related peptide that has similar bronchial-relaxant activity but is more resistant to inactivation by airway proteases. The former approach was validated by the demonstration that the addition of phosphoramidon and other protease inhibitors (see above), potentiates and prolongs VIP-induced tracheal relaxation *in vitro*. Alternatively, one naturally occurring peptide and several synthetic analogs of VIP have been characterized that are more resistant to enzymatic digestion. These peptides include helodermin, originally isolated from the lizard Gila monster *Heloderma* but also present in mammalian tissues. It exhibits close sequence similarity to VIP, but has a carboxy-terminal extension that may account for its greater resistance to enzymatic degradation [59] and thus for its longer-lasting tracheal relaxation [60]. The synthetic analogues are more potent than VIP as ASM relaxants, are longer-acting, or both [206, 207]. The special advantages of these VIP analogs in the management of asthma have yet to be tested in clinical trials.

12.2. Acute Lung Injury

The promising results in experimental models, demonstrating protection of the lungs and airways by VIP and helodermin, suggest the possible usefulness of these and related peptides in the clinical counterparts of these models, respectively, the acute lung injury of the adult respiratory distress syndrome (ARDS), and the airway inflammation of bronchial asthma.

12.3. Pulmonary Hypertension

Because of their vasodilator activity, VIP and related peptides may prove helpful in the management of pulmonary hypertension. VIP was recently administered directly into the pulmonary artery of 5 patients with primary pulmonary hypertension (PPH), a rare and fatal disease for which medical management is largely unsatisfactory. Relief of pulmonary hypertension in this disorder has depended on the use of pulmonary vasodilators, so far with mixed results. VIP was infused at $100-400$ picomoles $kg^{-1} hr^{-1}$ for up to 150 min, during right heart catheterization. A beneficial haemodynamic response (increased cardiac output, decreased pulmonary vascular resistance, or both) was noted in 3 of the 5 patients; the other two showed no significant improvement. The pulmonary vasodilation occurred without a decline in arterial blood O_2 saturation or systemic blood pressure, or other side effects [208]. This preliminary trial suggests that VIP may exert a greater vasodilator effect on the human pulmonary circulation than on the systemic circulation, and may therefore be useful in the relief of pulmonary hypertension in PPH.

12.4. Lung Cancer

VIP inhibits the growth and multiplication of small cell lung cancer (SCLC) cell lines NCI-H345 and NCI-H69, and the closely related peptide helodermin inhibits the proliferation of NCI-H345 cells with even higher efficacy. The VIP-induced suppression of cell proliferation is enhanced in the presence of other cAMP-promoting agents (isobutyl-methyl xanthine, forskolin and 8-bromo-cAMP), or of an anti-bombesin monoclonal antibody. VIP also inhibits the growth of explants of SCLC tumours in athymic nude mice [209]. This evidence raises the hope that VIP, or a related peptide, may have therapeutic potential in the management of this highly malignant lung cancer. As a naturally occurring peptide, VIP is likely to be less toxic than conventional chemotherapeutic agents.

13. Summary and Conclusions; Future Perspectives

The study of neuropeptides in the airways is a young and vigorous field that was given a major boost by the discovery and characterization of VIP. Since then, much has been learned about the localization and distribution of VIP, its biological activities, biosynthesis, receptors and possible physiological and pathophysiological significance. The existence of receptor subtypes for VIP and related peptides has been suspected for some time, but only recently have these receptors been characterized and cloned. To date, specific antagonists for VIP receptors in airways and lung are still lacking, making it difficult to ascertain the postulated physiological roles of the peptide.

Considerable data from *in vitro* studies and animal experiments suggest that VIP is an important regulator of ASM tone, airway reactivity, secretion, smooth muscle proliferation and pulmonary and bronchial blood flow. It is now necessary to determine to what extent our present knowledge applies to human airway function and dysfunction, and how this knowledge might be used to improve our ability to prevent and manage airways diseases.

Acknowledgments

The author's research cited here was supported in part by NIH Grant HL 30450 and by funds from the Department of Veterans Affairs (VA). SIS was a Medical Investigator of the VA.

References

1. Said SI. Vasoactive substances in the lung. In Proceedings of the Tenth Aspen Emphysema Conference, Aspen, Colorado, June 7–10, 1967. US Public Health Service Publication 1787, 1967: 223–8.
2. Said SI, Mutt V. Polypeptide with broad biological activity: Isolation from small intestine. Science 1970; 169: 1217–8.
3. Said SI, Rosenberg RN. Vasoactive intestinal polypeptide: abundant immunoreactivity in neural cell lines and normal nervous tissues. Science 1976; 192: 907–8.
4. Said SI. Polypeptide-containing neurons and their function in airway smooth muscle. In Coburn RF, editor. Airway Smooth Muscle in Health and Disease. New York: Plenum, 1989; 55–76.
5. Boomsma JD, Said SI. The role of neuropeptides in asthma. Chest 1992; 101: 389S–92S.
6. Dey RD, Shannon WA, Jr, Said SI. Localization of VIP-immunoreactive nerves in airways and pulmonary vessels of dogs, cats, and human subjects. Cell Tissue Res 1981; 220: 231–8.
7. Uddman R, Sundler F. VIP nerves in human upper respiratory tract. Otorhinolaryngology 1979; 41: 221–6.
8. Lundberg JM, Lundblad L, Martling C-R, Saria A, Stjärne P, Änggård A. Coexistence of multiple peptides and classic transmitters in airway neurons: functional and pathophysiologic aspects. Am Rev Resp Dis 1987; 136: S16–S22.
9. Lundberg JM, Saria A. Polypeptide-containing neurons in airway smooth muscle. Ann Rev Physiol 1987; 49: 557–72.
10. Dey RD, Altemus JB. Distribution of cholinergic and VIP-containing neurons in ferret tracheal plexus. Am Rev Resp Dis 1991; 143: A362.
11. Dey RD, Altemus JB, Michalkiewicz M. Distribution of VIP- and SP-containing nerves originating from neurons of airway ganglia in cat bronchi. J Comp Neurol 1991; 304: 330–40.
12. Springall DR, Polak JM, Howard L, Power RF, Krausz T, Manisckam S, et al. Persistence of intrinsic neurones and possible phenotypic changes after extrinsic denervation of human respiratory tract by heart lung transplantation. Am Rev Resp Dis 1990; 141: 1538–46.
13. Pakbaz H, Berisha H, Absood A, Foda HD, Said SI. VIP in sensory nerves of the lung: Capsaicin-induced release of immunoreactive vasoactive intestinal peptide [VIP] from guinea pig lungs. Am Rev Resp Dis 1993; 147: A477.
14. Dey RD, Hoffpair J, Said SI. Co-localization of VIP- and SP-containing nerves in cat bronchi. Neuroscience 1988; 24: 275–81.
15. Dey RD, Altemus JB, Zervos I, Hoffpair J. Origin and colocalization of CGRP- and SP-reactive nerves in cat airway epithelium. J App Physiol 1990; 68: 770–8.
16. Cutz E, Chan W, Track NS, Goth A, Said SI. Release of vasoactive intestinal polypeptide in mast-cells by histamine liberators. Nature 1978; 275: 661–2.
17. Goetzl EJ, Sreedharan SP, Turck CW. Structurally distinctive vasoactive intestinal peptides from rat basophilic leukemia cells. J Biol Chem 1988; 263: 9083–6.
18. Wershil Bk, Turck CW, Sreedharan SP, Yang J, An S, Galli SJ, Goetzl EJ. Variants of vasoactive intestinal peptide in mouse mast cells and rat basophilic leukemia cells. Cell Immunol 1993; 151: 369–78.
19. Aliakbari J, Sreedharan SP, Turck CW, Goetzl EK. Selective localization of vasoactive intestinal peptide and substance P in human eosinophils. Biochem Biophys Res Comm 1987; 148: 1440–5.
20. Luts A, Uddman R, Absood A, Håkanson R, Sundler F. Chemical coding of endocrine cells of the airways: presence of helodermin-like peptides. Cell Tissue Res 1991; 265: 425–33.
21. Cardell LO, Uddman R, Luts A, Sundler F. Pituitary adenylate cyclase activating peptide (PACAP) in guinea-pig lung: distribution and dilatory effects. Regulatory Peptides 1991; 36: 379–90.
22. Moller K, Zhang Y-Z, Håkanson R, Luts A, Sjölund B, Uddman R, et al. Pituitary adenylate cyclase activating peptide is a sensory neuropeptide: immunocytochemical and immunochemical evidence. Neuroscience 1993; 57: 725–32.

23. Itoh N, Obata K, Yanaihara N, Okamoto H. Human preprovasoactive intestinal polypeptide contains a novel PHI-27-like peptide, PHM-27. Nature 1983; 304: 547–9.
24. Gozes I, Nakai H, Byers M, Avidor R, Weinstein Y, Shani Y, et al. Sequential expression in the nervous system of the VIP and c-myc genes located on the human chromosomal region 6q24. Somatic Cell and Mol Genetics 1987; 13: 305–13.
25. Linder ST, Barkhem A, Norberg H, Persson H, Schalling M, Hökfelt T, Magnusson G. Structure and expression of the gene encoding the vasoactive intestinal peptide precursor. Proc Nat Acad Sci 1987; 84: 604–9.
26. Ohsawa K, Hayakawa Y, Nishizawa M, Yamagami T, Yamamoto H, Yanaihara N, Okamoto H. Synergistic stimulation of VIP/PHM-27 gene expression by cyclic AMP and phorbol esters in human neuroblastoma cells. Biochem Biophys Res Comm 1985; 132: 885–91.
27. Uddman R, Luts A, Sundler F. Nerves fibres containing peptide histidine isoleucine [PHI] in the respiratory tract. Arch Otorhinolaryngol 1985; 242: 189–93.
28. Fruness JB, Bornstein JC, Murphy R, Pompolo S. Roles of peptides in transmission in the enteric nervous system. Trends Neurosci 1992; 15: 66–71.
29. Shimosegawa T, Foda HD, Said SI. Opioid peptides in guinea pig and rat lungs: Their sources and colocalization with VIP and PHI. Am Rev Resp Dis 1989; 139: A470.
30. Dey RD, Zhu W. Origin of galanin in nerves of cat airways and colocalization with vasoactive intestinal peptide. Cell Tissue Res 1994; 273: 193–200.
31. Bowden JJ, Gibbins IL. Vasoactive intestinal peptide and neuropeptide Y co-exist in non-adrenergic sympathetic neurons to guinea pig trachea. J Auton Nervous Sys 1992; 38: 1–20.
32. Lundberg JM, Hökfelt T, Schultzberg M, Uvnas-Wallensten K, Kohler C, Said SI. Occurrence of vasoactive intestinal polypeptide (VIP)-like immunoreactivity in certain cholinergic neurons of the cat: evidence from combined immunohistochemistry and acetylcholine esterase staining. Neuroscience 1979; 4: 1539–59.
33. Laitinen A, Partanen M, Hervonen A, Pelto-Huikko M, Laitinen LA. VIP-like immunoreactive nerves in human respiratory tract. Histochemistry 1985; 82: 313–9.
34. Kummer K, Fischer A, Mundel P, Mayer B, Hoba B, Philippin B, et al. Nitric oxide synthase in VIP-containing vasodilator nerve fibres in the guinea pig. Neuroreport 1992; 3: 653–5.
35. Dey RD, Mayer B, Said SI. Colocalization of vasoactive intestinal peptide and nitric oxide synthase in neurons of ferret trachea. Neuroscience 1993; 54: 839–43.
36. Lundberg JM, Änggård A, Emson P, Fahrenkrug J, Hökfelt T, Mutt M. Vasoactive intestinal polypeptide in cholinergic neurons of exocrine glands: functional significance of coexisting transmitters of vasodilation and secretion. Proc Nat Acad Sci 1980; 77: 1651–5.
37. Ellis JL, Farmer SG. Modulation of cholinergic neurotransmission by vasoactive intestinal peptide and peptide histamine isoleucine in guinea-pig tracheal smooth muscle. Pulm Pharmacol 1989; 2: 107.
38. Martin JG, Wang A, Zacour M, Biggs DF. The effects of vasoactive intestinal polypeptide on cholinergic neurotransmission in an isolated innervated guinea pig tracheal preparation. Resp Physiol 1990; 79: 111–21.
39. Said SI. Neuropeptides as modulators of injury and inflammation. Life Sciences 1990; 47: PL-19–PL-21.
40. Said SI. Nitric oxide and vasoactive intestinal peptide as co-transmitters of smooth muscle relaxation. News Physiol Sci 1992; 7: 181–3.
41. Grider JR, Murthy KS, Jin J-G, Makhlouf GM. Stimulation of nitric oxide from muscle cells by VIP: prejunctional enhancement of VIP release. Am J Physiol 1992; 262: G774–8.
42. Piper PJ, Said SI, Vane JR. Effects on smooth muscle preparations of unidentified vasoactive peptides from the intestine and lung. Nature 1970; 225: 1144–6.
43. Said SI, Kitamura S, Yoshida T, Preskitt J, Holden LD. Humoral control of airways. Ann NY Acad Sci 1974; 221: 103–14.
44. Wasserman MA, Griffin RL, Malo PE. Comparative in vitro tracheal-relaxant effects of porcine and hen VIP. In Said SI, editor. Vasoactive Intestinal Peptide. New York: Raven Press, 1982; 177–84.
45. Ito Y, Takeda K. Non-adrenergic inhibitory nerves and putative transmitters in the smooth muscle of cat trachea. J Physiol 1982; 330: 497–511.

46. Cameron AC, C.T. Kirkpatrick CT, Kirkpatrick MCA. The quest for the inhibitory neurotransmitter in bovine tracheal smooth muscle. Quart J Exp Physiol 1983; 68: 413–26.
47. Hamasaki Y, Saga TM, Mojarad M, Said SI. VIP counteracts leukotriene D4-induced contractions of guinea pig trachea, lung and pulmonary artery. Trans Ass Am Physicians 1983; 96: 406–11.
48. Said SI. Influence of neuropeptides on airway smooth muscle. Am Rev Resp Dis 1987; 136: S52–S58.
49. Said SI. Vasoactive intestinal peptide in the lung. Ann NY Acad Sci 1988; 527: 450–64.
50. Boomsma JD, Foda HD, Said SI. Vasoactive intestinal peptide (VIP) reverses endothelin-induced contractions of guinea pig trachea and pulmonary artery. Am Rev Resp Dis 1990; 141: A485.
51. Altiere RJ, Diamond L. Comparison of vasoactive intestinal peptide and isoproterenol relaxant effects in isolated cat airways. J Appl Physiol 1984; 56: 986–92.
52. Hand JM, Laravuso RB, Will JA. Relaxation of isolated guinea pig trachea, bronchi and pulmonary arteries produced by vasoactive intestinal peptide [VIP]. Eur J Pharmacol 1984; 98: 279–84.
53. Saga T, Said SI. Vasoactive intestinal peptide relaxes isolated strips of human bronchus, pulmonary artery, and lung parenchyma. Trans Ass Am Physicians 1984; 97: 304–10.
54. Palmer JB, Cuss FMC, Barnes PJ. VIP and PHM and their role in non-adrenergic inhibitory responses in isolated human airways. Am J Physiol 1986; 61: 1322–8.
55. Said SI, Geumei A, Hara N. Bronchodilator effect of VIP *in vivo*: Protection against bronchoconstriction induced by histamine or prostaglandin F2α. In Said SI, editor. Vasoactive Intestinal Peptide. New York: Raven Press, 1982: 185–91.
56. Diamond L, Szarek JL, Gillespie MN, Altiere RJ. *In vivo* bronchodilator activity of vasoactive intestinal peptide in the cat. Am Rev Resp Dis 1983; 128: 827–32.
57. Liu L-W, Sata T, Kubota E, Paul S, Said SI. Airway relaxant effect of vasoactive intestinal peptide (VIP): Selective potentiation by phosphoramidon, an enkephalinase inhibitor. Am Rev Resp Dis 1987; 135: A86.
58. Sharaf H, Said SI. Tracheal relaxant response to vasoactive intestinal peptide [VIP]: Influence of airway epithelium and peptidases. FASEB J 1993; 7: A686.
59. Liu L-W, Trotz M, Erdös EG, Said SI. Vasoactive intestinal peptide [VIP] and helodermin degradation by airway enzymes. Am Rev Resp Dis 1991; 143: A618.
60. Foda HD, Said SI. Helodermin, a C-terminally extended VIP-like peptide, evokes long-lasting tracheal relaxation. Biomed Res 1989; 10: 107–10.
61. Yoshihara S, Ichimura T, Yanaihara N. Lasting inhibitory effect of helodermin inhalation on guinea pig airway contraction. Biomed Res 1992; 13 [Suppl.2]: 361–71.
62. Bolin DR, Cottrell J, Michalewsky J, Garippa R, O'Neill N, Simko B, *et al*. Degradation of vasoactive intestinal peptide in bronchial alveolar lavage fluid. Biomed Res 1992; 13: 25–30.
63. Ito O, Tachibana S. Vasoactive intestinal polypeptide precursors have highly potent bronchodilatory activity. Peptides 1991; 12: 131–7.
64. Nathanson I, Widdicombe JH, Barnes PJ. Effect of vasoactive intestinal peptide on ion transport across dog tracheal epithelium. J Appl Physiol 1983; 55: 1844–48.
65. Stoff JS, Rosa R, Hallac R, Silva P, Epstein FH. Hormonal regulation of active chloride ion transport in the dogfish rectal gland. Am J Physiol 1979; 237: F138–44.
66. Lowry RJ, Schreiber JH, Ernst SA. Vasoactive intestinal peptide stimulates ion transport in avian salt gland. Am J Physiol 1987; 252: C670–6.
67. Gerstberger R, Gray A. Fine structure, innervation and functional control of avian salt glands. Int Rev Cytol 1993; 144: 129–215.
68. Cliff WH, Frizzell RA. Separate Cl⁻ conductances activated by cAMP and Ca²⁺ in Cl-secreting epithelial cells Proc Nat Acad Sci 1980; 87: 4956–60.
69. Huang SJ, Fu W, Chung YW, Zhou TS, Wong PYD. Properties of cAMP-dependent and Ca²⁺-dependent whole cell Cl⁻ conductances in rat epididymal cells. Am J Physiol 1993; 264: C794–C802.
70. Martin SC, Shuttleworth TJ. Vasoactive intestinal peptide stimulates a cAMP-mediated Cl⁻ current in avian salt gland cells. Regulatory Peptides, 1994; 52: 205–14.
71. Peatfield AC, Barnes PJ, Bratcher C, Nadel JA, Davis B. Vasoactive intestinal peptide stimulates tracheal submucosal gland secretion in ferret. Am Rev Resp Dis 1983; 128: 89–93.

72. Elgavish A, Pillion DJ, Meezan E. Evidence for vasoactive intestinal peptide receptors in apical membranes from tracheal epithelium. Life Sci 1989; 44: 1037–42.
73. Coles SJ, Said SI, Reid LM. Inhibition by vasoactive intestinal peptide of glycoconjugate and lysozyme secretion by human airways *in vitro*. Am Rev Resp Dis 1981; 124: 531–6.
74. Sakai N, Tamaoki J, Kobayashi K, Kanemura T, Isono K, Takeyama K, *et al.* Vasoactive intestinal peptide stimulates ciliary motility in rabbit tracheal epithelium: modulation by neutral endopeptidase. Regulatory Peptides 1991; 34: 33–41.
75. Malm L, Undler SF, Uddman R. Effects of vasoactive intestinal polypeptide on resistance and capacitance vessels in the nasal mucosa. Acta Otolaryngology 1980; 90: 304–8.
76. Widdicombe JG. Pulmonary and respiratory tract receptors. J Exp Biol 1982; 100: 41–57.
77. Nandiwada PA, Kadowitz PJ, Said SI, Mojarad M, Hyman AL. Pulmonary vasodilator responses to vasoactive intestinal peptide in the cat. J Appl Physiol 1985; 58: 1723–8.
78. Sata T, Misra HP, Kuboto E, Said SI. Vasoactive intestinal polypeptide relaxes pulmonary artery by an endothelium-independent mechanism. Peptides 1986; 7(Suppl.): 225–7.
79. Laitinen LA, Laitinen A, Salonen RO, Widdicombe JG. Vascular actions of airway neuropeptides. Am Rev Resp Dis 1987; 136: S59–64.
80. Laitinen LA, Laitinen A, Widdicombe JG. Effects of inflammatory and other mediators on airway vascular beds. Am Rev Resp Dis 1987; 135: S67–70.
81. Said SI. VIP and nitric oxide: physiological co-transmitters with antagonistic roles in inflammation. Biomed Res 1993. In press.
82. Krco CJ, Gores A, Go VLW. Gastrointestinal regulatory peptides modulate *in vitro* immune reactions of mouse lymphoid cells. Clin Immunol Immunopathol 1986; 39: 308–18.
83. Ottaway CA. Selective effects of vasoactive intestinal peptide on the mitogenic response of murine T cells. Immunology 1987; 62: 291–7.
84. Sun L, Ganea D. Vasoactive intestinal peptide inhibits interleukin [IL]-2 and IL-4 production through different molecular mechanisms in T cells activated via the T cell receptor/CD3 complex. J Neuroimmunol 1993; 48: 59–70.
85. Rola-Pleszczynski M, Bolduc D, St.-Pierre S. The effects of vasoactive intestinal peptide on human natural killer cell function. J Immunol 1985; 135: 2569–73.
86. Wiik P. Vasoactive intestinal peptide inhibits the respiratory burst in human monocytes by a cyclic AMP-mediated mechanism. Regulatory Peptides 1989; 25: 187–97.
87. Litwin DK, Wilson AK, Said SI. Vasoactive intestinal polypeptide inhibits rat alveolar macrophage phagocytosis and chemotaxis. Regulatory Peptides 1992; 40: 63–74.
88. Undem BJ, Dick EC, Buckner CK. Inhibition by vasoactive intestinal peptide of antigen-induced histamine release from guinea-pig minced lung. Eur J Pharmacol 1983; 88: 247–9.
89. Cox CP, Linden J, Said SI. VIP elevates platelet cyclic AMP [cyclic AMP] levels and inhibits *in vitro* platelet activation induced by platelet-activating factor (PAF). Peptides 1984; 5: 325–8.
90. O'Dorisio MS, Hermina N, Balcerzak SP, O'Dorisio TM. Vasoactive intestinal polypeptide stimulation of adenylate cyclase in purified human leukocyte. J Immunol 1981; 127: 2551–4.
91. Maruno K, Said SI. Inhibition of human airway smooth muscle cell proliferation by vasoactive intestinal peptide (VIP). In Rosselin G, editor. VIP, PACAP, and Related Regulatory Peptides. River Edge, NJ: World Scientific, 1994: 587–92.
92. Ollerenshaw S, Jarvis D, Woolcock A, Sullican C, Scheibner T. Absence of immunoreactive vasoactive intestinal polypeptide in tissue from the lungs of patients with asthma. New Eng J Med 1989; 320: 1244–8.
93. Hossain S. Quantitative measurement of bronchial muscle in men with asthma. Am Rev Resp Dis 1973; 107: 99–109.
94. James AL, Pare PD, Hogg JC. Mechanisms of airway narrowing in asthma. Am Rev Resp Dis 1989; 139: 242–6.
95. Wiggs BR, Bosken CH, Paré PD, James A, Hogg JC. A model of airways narrowing in asthma and in chronic obstructive pulmonary disease. Am Rev Resp Dis 1992; 145: 1251–8.

96. Wiggs BR, Moreno R, Hogg JC, Hilliam C, Paré PD. A model of the mechanics of airway narrowing. J Appl Physiol 1990; 69: 849–60.
97. Lundberg JM, Martling C-R, Hökfelt T. Airways, oral cavity and salivary glands: classical transmitters and peptides in sensory and autonomic motor neurons. In Bjorklund A, Hökfelt T, Owman C, editors. Handbook of Chemical Neuroanatomy vol. 6. Peipheral Nervous System. Amsterdam: Elsevier, 1988: 391–444.
98. Ward JK, Belvisi MG, Fox AJ, Miura M, Tadjkarimi S. Modulation of cholinergic neural bronchoconstriction by endogenous nitric oxide and vasoactive intestinal peptide in human airways in vitro. J Clin Invest 1993; 92: 736–43.
99. Barnes PJ. Neuropeptides and asthma. Am Rev Resp Dis 1991; 143: S28–S32.
100. Young HM, Furness JB, Shuttleworth CWR, Bredt DS, Snyder SH. Co-localization of nitric oxide synthase immunoreactivity and NADPH diaphorase staining in neurons of the guinea-pig intestine. Histochemistry 1992; 97: 375–8.
101. Li CG, Rand MJ. Nitric oxide and vasoactive intestinal polypeptide mediate nonadrenergic, noncholinergic inhibitory neuro-transmission to smooth muscle of the rat gastric fundus. Eur J Pharmacol 1990; 191: 303–9.
102. Sharaf HH, Said SI. VIP relaxation of airway, pulmonary artery and aortic vascular smooth muscle is unaffected by blockade of nitric oxide synthase. Am Rev Resp Dis 1992; 145: A382.
103. Lilly CM, Martins MA, Drazen JM. Peptidase modulation of vasoactive intestinal peptide pulmonary relaxation in tracheal superfused guinea pig lungs. J Clin Invest 1993; 91: 235–43.
104. Tare M, Parkington HC, Coleman HA, Neild TO, Dusting GJ. Hyperpolarization and relaxation of arterial smooth muscle caused by nitric oxide derived from the epithelium. Nature 1990; 346: 69–71.
105. Standen NB, Quayle JM, Davies NW, Brayden JE, Huang Y, Nelson MT. Science 1989; 245: 177–80.
106. Said SI. Vasoactive intestinal peptide [VIP] and related peptides as anti-asthma and anti-inflammatory agents. Biomed Res 1992; 13 [Suppl. 2]: 257–62.
107. Berisha B, Pakbaz H, Absood A, Said SI. Nitric oxide as a mediator of oxidant lung injury due to paraquat. Proc Nat Acad Sci, 1994; 91: 7445–9.
108. Pakbaz H, Berisha H, Absood A, Foda HD, Said SI. Nitric oxide mediates oxidant tissue injury caused by paraquat and xanthine oxidase. Ann NY Acad Sci 1994; 723: 422–5.
109. Paul S, Said SI. Characterization of receptors for vasoactive intestinal peptide solubilized from the lung. J Bio Chem 1987; 262: 158–62.
110. Patthi S, Simerson S, Veliçelebi G. Solubilization of rat lung vasoactive intestinal peptide receptors in the active state. Characterization of the binding properties and comparison with membrane-bound receptors. J Biol Chem 1988; 263: 19363–9.
111. Luis J, Said SI. Characterization of VIP- and helodermin-preferring receptors on human small cell carcinoma cell lines. Peptides 1990; 11: 1239–44.
112. Leroux P, Vaudry H, Fournier A, St.-Pierre S, Pelletier G. Characterization and localization of vasoactive intestinal peptide receptors in the rat lung. Endocrinology 1984; 114: 1506–12.
113. Lazarus SC, Basbaum CB, Barnes PJ, Gold WM. Mapping of VIP receptors by use of an immunocytochemical probe for the intracellular mediator cyclic AMP. Am J Physiol 1986; 251: C115–9.
114. Muller J-M, El Battari A, Ah-Kye E, Luis J, Ducret F, Pichon J, Marvaldi J. Internalization of the vasoactive intestinal peptide [VIP] in human adenocarcinoma cell line [HT29]. Eur J Biochem 1985; 152: 107–14.
115. Omary MB, Kagnoff MF. Identification of nuclear receptors for VIP on a human colonic adenocarcinoma cell line. Science 1987; 238: 1578–81.
116. Paul S, Ebaid M. Vasoactive intestinal peptide: Its interactions wth calmodulin and catalytic antibodies. Neurochem Int 1993; 23: 197–214.
117. Couvineau A, Voisin T, Guijarro L, Laburthe M. Purification of vasoactive intestinal peptide receptor from porcine liver by a newly designed one-step affinity chromatography. J Biol Chem 1990; 265: 13386–90.
118. Ishihara T, Shigemoto R, Mori K, Takahashi K, Nagata S. Functional expression and tissue distribution of a novel receptor for vasoactive intestinal polypeptide. Neuron 1992; 8: 811–19.

119. Sreedharan SP, Patel DR, Huang J-X, Goetzl EJ. Cloning and functional expression of a human neuroendocirne vasoactive intestinal peptide receptor. Biochem Biophys Res Comm 1993; 193: 546–53.

120. Lutz EM, Sheward WJ, West KM, Morrow JA, Fink G, Harmar AJ. The VIP2 receptor: Molecular characterisation of a cDNA encoding a novel receptor for vasoactive intestinal peptide. FEBS Letters 1993; 334: 3–8.

121. Bodanszky M, Klausner YS, Said SI. Biological activities of synthetic peptides corresponding to fragments of and to the entire sequence of the vasoactive intestinal peptide. Proc Nat Acad Sci 1973; 70: 382–4.

122. Chakder S, Rattan S. The entire vasoactive intestinal polypeptide molecule is required for the activation of the vasoactive intestinal polypeptide receptor: Functional and binding studies on opossum internal anal sphincter smooth muscle. J Pharmacol Exp Ther 1993; 266: 392–9.

123. Turner AJ, Neuropeptides and their peptidases. Ellis Horwood Series in Biomedicine, Chichester, England, 1987.

124. Goosens J-F, Pommery N, Lohez M, Pommery J, Helbecque N, Cottelle P, Lhermitte M, Henichart J-P. Antagonistic effect of a vasoactive intestinal peptide fragment, vasoactive intestinal peptide [1-11], on guinea pig trachea smooth muscle relaxation. Mol Pharmacol 1992; 41: 104–9.

125. Pandol SJ, Dharmsathaphorn K, Schoeffield MS, Vale Y, Rivier J. Vasoactive intestinal peptide receptor antagonist [4Cl-D-Phe^6, Leu^{17}] VIP. Am J Physiol 1986; 250: G553–7.

126. Grider JR, Rivier JR. Vasoactive intestinal peptide [VIP] as transmitter of inhibitory motor neurons of the gut: evidence from the use of selective VIP antagonists and VIP antiserum. J Pharmacol Exp Ther 1990; 253: 738–42.

127. Waelbroeck M, Robberecht P, Coy DH, Camus J-C, De Neep P, Christophe J. Interaction of growth-hormone-releasing factor (GRF) and 14 GRF analogs with vasoactive intestinal peptide (VIP) receptors of rat pancreas. Discovery of (N-Ac-Tyr, D-Phe2)-GRF(1-29)-NH2 as a VIP antagonist. Endocrinology 1985; 116: 2643–59.

128. Gozwa I, Meltzer E, Rubinrout S, Brenneman DE, Fridkin M. Vasoactive intestinal peptide potentiates sexual behavior: inhibition by novel antagonist. Endocrinology 1989; 125: 2945–9.

129. Christophe J. Type I receptors for PACAP (a neuropeptide even more important than VIP?). Biochim Biophys Acta 1993; 1154: 183–99.

130. Ganz P, Sandrock AW, Landis SC, Leopold J, Gimbrone Jr, MA, Alexander RW. Vasoactive intestinal peptide: vasodilation and cyclic AMP generation. Am J Physiol 1986; 250: H755–60.

131. Shreeve SM, DeLuca AW, Diehl NL, Kermode JC. Molecular properties of the vasoactive intestinal peptide receptor in aorta and other tissues. Peptides 1992; 13: 919–26.

132. Robberecht P, Conlon P, Gardner JD. Interaction of porcine vasoactive intestinal peptide with dispersed pancreatic acinar cells from the guinea pig: structural requirements for effects of VIP an secretin on cellular cyclic AMP. J Biol Chem 1976; 251: 4635–9.

133. Dupont C, Laburthe M, Broyart JP, Bataille D, Rosselin G. Cyclic AMP production in isolated colonic epithelial crypts: a highly sensitive model for the evaluation of vasoactive intestinal peptide action in human intestine. Eur J Clin Invest 1980; 10: 67–76.

134. Sakakibara H, Kouichiro S, Said SI. Characterization of vasoactive intestinal peptide [VIP] receptors on rat alveolar macrophages. Am J Physiol, 1994; 267: 256–62.

135. Maruno K, Said SI. Inhibition of human airway smooth muscle cell proliferation by vasoactive intestinal peptide (VIP). Am Rev Resp Dis 1993; 147: A253.

136. Gozes I, Brenneman DE. VIP: molecular biology and neurobiological function. Mol Neurobiol 1989; 3: 201–26.

137. Francis SH, Noblett BD, Todd BW, Wells JN, Corbin JD. Relaxation of vascular and tracheal smooth muscle by cyclic nucleotide analogs that preferentially activate purified cGMP-dependent protein kinase. Mol Pharmacol 1988; 34: 506–17.

138. Meisheri KD, Rüegg JC. Dependence of cyclic-AMP induced relaxation on Ca^{2+} and calmodulin in skinned smooth muscle of guinea pig Taenia coli. Pflügers Archives 1983; 399: 315.

139. Lincoln TM, Cornwell TL, Taylor AE. cAMP-dependent protein kinase mediates the reduction of Ca^{2+} by cAMP in vascular smooth muscle cells. Am J Physiol 1990; 258: C399–C407.

140. De Lanerolle P, Paul RJ. Myosin phosphorylation/dephosphorylation and regulation of airway smooth muscle contractility. Am J Physiol 1991; 261: L1–L14.

141. Gerthoffer WT. Regulation of the contractile element of airway smooth muscle. Am J Physiol 1991; 261: L15–L28.

142. Audigier S, Barberis C, Jard S. Vasoactive intestinal polypeptide increases inositol phosphate breakdown in the rat superior cervical ganglion. Ann NY Acad Sci 1988; 527: 579–81.

143. Malhotra RK, Wakade TD, Wakade AR. Vasoactive intestinal polypeptide and muscarine mobilize intracellular Ca^{2+} through breakdown of phosphoinositides to induce catecholamine secretion. J Biol Chem 1988; 263: 2123–6.

144. Tatsuno I, Yada T, Vigh S, Hidaka H, Arimura A. Pituitary adenylate cyclase activating polypeptide and vasoactive intestinal peptide cytosolic free calcium concentration in cultured rat hippocampal neurons. Endocrinology 1992; 131: 73–81.

145. Gressens P, Hill JM, Gozes I, Fridkin M, Brenneman DE. Growth factor function of vasoactive intestinal peptide in whole cultured mouse embryos. Nature 1993; 362: 155–8.

146. Sand O, Chen B, Li Q, Karlsen HE, Bjoro T, Haug E. Vasoactive intestinal peptide [VIP] may reduce the removal rate of cytosolic Ca^{2+} after transient elevations in clonal rat lactotrophs. Acta Physiol Scand 1989; 137: 113–23.

147. Andersson M, Carlquist M, Maletti M, Marie JC. Simultaneous solubilization of high-affinity receptors for VIP and glucagon and of a low-affinity binding protein for VIP, shown to be identical to calmodulin. FEBS Letters 1993; 318: 35–40.

148. Stallwood D, Brugger CH, Baggenstoss BA, Stemmer PM, Shiraga H, Landers DF, *et al.* Identity of a membrane-bound vasoactive intestinal peptide-binding protein with calmodulin. J Biol Chem 1992; 267: 19617–21.

149. Said SI. Vasoactive intestinal peptide: its interactions with calmodulin and catalytic antibodies. Neurochem Int 1993; 23: 197–214.

150. Grider JR. Interplay of VIP and nitric oxide in regulation of the descending relaxation phase of peristalsis. Am J Physiol 1993; 264: G334–40.

151. He XD, Goyal RJ. Nitric oxide involvement in the peptide VIP-associated inhibitory junction potential in the guinea-pig ileum. J Physiol 1993; 461: 485–99.

152. Huizinga JD, Tomlinson J, Pintin-Quezada J. Involvement of nitric oxide in nerve-mediated inhibition and action of vasoactive intestinal peptide in colonic smooth muscle. J Pharmacol Exp Ther 1992; 260: 803–8.

153. Jin J-G, Murthy KS, Grider JR, Makhlouf GM. Activation of distinct cAMP- and cGMP-dependent pathways by relaxant agents in isolated gastric muscle cells. Am J Physiol 1993; 264: G470–7.

154. Murthy KS, Zhang K-M, Jin J-G, Grider JR, Makhlouf GM. VIP-mediated G protein-coupled Ca^{2+} influx activates a constitutive NOS in dispersed gastric muscle cells. Am J Physiol 1993; 265: G660–71.

155. Grider JR, Jin J-G. Vasoactive intestinal peptide release and L-citrulline production from isolated ganglia of the myenteric plexus: evidence for regulation of vasoactive intestinal peptide release by nitric oxide. Neurosci 1993; 54: 521–6.

156. Spessert R. Vasoactive intestinal peptide stimulation of cyclic guanosine monophosphate formation: further evidence for a role of nitric oxide synthase and cytosolic guanylate cyclase in rat pinealocytes. Endocrinology 1993; 132: 2513–7.

157. Lilly CM, Stamler JS, Gaston B, Meckel C, Loscalzo J, Drazen JM. Modulation of vasoactive intestinal peptide pulmonary relaxation by NO in tracheally superfused guinea pig lungs. Am J Physiol 1993; 265: L410–5.

158. Murad F, Mittal CK, Arnold WP, Katsuki S, Kimura H. Guanylate cyclase: activation by azide, nitro compounds, nitric oxide, and hydroxyl radical and inhibition by hemoglobin and myoglobin. Adv Cyclic Nucl Res 1978; 9: 145–58.

159. Orange RP, Austen WG, Austen KF. Immunological release of histamine and slow-reacting substance of anaphylaxis from human lung. I. Modulation by agents influencing cellular levels of cyclic 3′,5′-adenosine monophosphate. J Exp Med 1971; 136–48.

160. Katsuki S, Murad F. Regulation of adenosine cyclic 3′,5′-monophosphate and guanosine cyclic 3′,5′-monophosphate levels and contractility in bovine tracheal smooth muscle. Mol Pharmacol 1977; 13: 330–41.

161. Pfitzer G, Hofmann F, DiSalvo J, Ruegg JC. cGMP and cAMP inhibit tension development in skinned coronary arteries. Pflügers Arch 1984; 401: 277–80.
162. Anderson MP, Gregory RJ, Thompson S, Souza DW, Paul S, Mulligan RC, *et al.* Demonstration that CFTR is a chloride channel by alteration of its anion channel selectivity. Science 1991; 253: 202–5.
163. Kume H, Hall IP, Washabau RJ, Takagi K, Kotlikoff M. β-adrenergic agonist regulate KCa channels in airways smooth muscle by cyclic AMP-dependent and -independent mechanisms. J Clin Invest 1994; 93: 371–9.
164. Kase H, Wakui M, Petersen OH. Stimulatory and inhibitory actions of VIP and cyclic AMP on cytoplasmic Ca^{2+} signal generation in pancreatic acinar cells. Pflügers Archives 1991; 419: 668–70.
165. Caughey GH, Leidig F, Viro NF, Nadel JA. Substance P and vasoactive intestinal peptide degradation by mast-cell tryptase and chymase. J Pharmacol Exp Ther 1988; 244: 133–7.
166. Tam EK, Caughey GH. Degradation of airway neuropeptides by human lung tryptase. Am J Resp Cell Mol Biol 1990; 3: 27–32.
167. Goetzl EJ, Sreedharan SP, Turck CW, Bridenbaugh R, Malfroy B. Preferential cleavage of amino- and carboxyl-terminal oligopeptides from vasoactive intestinal polypeptide by human recombinant enkephalinase [neutral endopeptidase, EC 3.4.24.11]. Biochem Biophys Res Comm 1989; 158: 850–4.
168. Franconi G, Graf PD, Lazarus SC, Nadel JA, Caughey GH. Mast-cell tryptase and chymase reverse airway smooth muscle relaxation induced by vasoactive intestinal peptide in the ferret. J Pharmacol Exp Ther 1989; 248: 947–51.
169. Hachisu M, Hiranuma T, Tani S, Iizuka T. Enzymatic degradation of helodermin and vasoactive intestinal polypeptide. J Pharmacobiol Dyn 1990; 14: 126–31.
170. Lilly CM, Kobzik L, Hall AE, Drazen JM. Effects of chronic airway inflammation on the activity and enzymatic inactivation of neuropeptides in guinea pig lungs. J Clin Invest 1994; 93: 2667–74.
171. Coburn RF, Tomita T. Evidence for nonadrenergic inhibitory nerves in guinea pig trachealis muscle. Am J Physiol 1973; 224: 1072–80.
172. Richardson J, Beland J. Nonadrenergic inhibitory nervous system in human airways. J Appl Physiol 1976; 41: 764–71.
173. Kubota E, Hamasaki Y, Sata T, Said SI. Autonomic innervation of pulmonary artery: evidence for a NANC inhibitory system. Exp Lung Res 1988; 14: 349–58.
174. Russell JA. Nonadrenergic inhibitory innervation in canine airways. J Appl Physiol 1980; 48: 16–22.
175. Said SI. Peptides common to the nervous system and the gastrointestinal tract. In Martini L, Ganong WF, editors. Frontiers in Neuroendocrinology Vol. 6, New York: Raven Press, 1980: 293–331.
176. Lundberg JM, Hökfelt T, Nilsson G, Terenius L, Rehfeld J, Edle R, Said SI. Peptide neurons in the vagus, splanchnic, and sciatic nerves. Acta Physiol Scand 1978; 104: 499–501.
177. Irvin CG, Boileau R, Tremblay J, Martin RR, Macklem PT. Bronchodilation: Non-cholinergic, nonadrenergic mediation demonstrated *in vivo* in the cat. Science 1980; 207: 791–2.
178. Venugopalan GS, Said SI, Drazen JM. Effect of vasoactive intestinal peptide on vagally mediated tracheal pouch relaxation. Resp Physiol 1984; 56: 205–16.
179. Matsuzaki Y, Hamasaki Y, Said SI. Vasoactive intestinal peptide: A possible transmitter of non-adrenergic relaxation of guinea pig airways. Science 1980; 210: 1252–3.
180. Mutt V, Said SI. Structure of the porcine vasoactive intestinal octacosapeptide: the amino acid sequence: use of kallikrein in its determination. Eur J Biochem 1974; 42: 581–9.
181. Diamond L, Altiere RJ. Airway nonadrenergic noncholinergic inhibitory nervous system. In: Kaliner MA, Barnes P, editors. The Airways: Neural Control in Health and Disease. In Lenfant C, executive editor. Lung Biology in Health and Disease. New York: Marcel Dekker, 1988; 343–94.
182. Belvisi M, Stretton D, Verleden GM, Yacoub M, Barnes PJ. Nitric oxide is the endogenous transmitter of bronchodilator nerves in humans. Eur J Pharmacol 1992; 210: 221–2.

183. Bai TR, Bramley AM. Effect of an inhibitor of nitric oxide synthase on neural relaxation of human bronchi. Am J Physiol 1993; 264: L425–30.
184. Boeckxstaens GE, Pelckmans PA, De Man JG, Bult H, Herman AG, Van Maercke YM. Evidence for differential release of nitric oxide and vasoactive intestinal polypeptide by nonadrenergic noncholinergic nerves in the rat gastric fundus. Arch Int Pharmacodyn Ther 1992; 318: 107–15.
185. Lei YH, Barnes PJ, Rogers DF. Inhibition of neurogenic plasma exudation in guinea pig airways by CP-96,345 a new non-peptide NK-1 receptor antagonist. Br J Pharmacol 1992; 105: 261–2.
186. Fisher JT, Anderson JW, Waldron MS. Nonadrenergic noncholinergic neurotransmitter of feline trachealis: VIP or nitric oxide? J Appl Physiol 1993; 74: 31–9.
187. Brandtzaeg P, Oktedalen O, Kierulf P, Opstad PK. Elevated VIP and endotoxin plasma levels in human gram-negative septic shock. Regulatory Peptides 1989; 24: 37–44.
188. Revhaug A, Lygren I, Jenssen TG, Giercksky K-E, Burhol PG. Vasoactive intestinal peptide in sepsis and shock. Ann NY Acad Sci 1988; 527: 536–45.
189. Modlin IM, Bloom SR, Mitchell S. Plasma vasoactive intestinal polypeptide (VIP) levels and intestinal ischaemia. Experienta 1978; 34: L535–6.
190. Pakbaz H, Liu L-W, Foda HD, Berisha H, Said SI. Vasoactive intestinal peptide (VIP) as a modulator of PAF-induced lung injury. Clin Res 1988; 36: 626A.
191. Said SI. VIP and messenger plasticity. Trends Neurosci 1994; 17: 339.
192. Chander A, Fisher AB. Regulation of lung surfactant secretion. Am J Physiol 1990; 258: L241–53.
193. Mendelson CR, Boggaram V. Hormonal control of the surfactant system in fetal lung. Ann Rev Physiol 1991; 53: 415–40.
194. Wright JR. Regulation of pulmonary surfactant secretion and clearance. Ann Rev Physiol 1991; 53: 395–414.
195. Said SI. Vasoactive intestinal peptide and the lung. In Bloom SR, Polak JM, Lindenlaub E, editors. Systemic Role of Regulatory Peptides. New York: F.K. Schattauer Verlag, 1982: 293–300.
196. Quinton PM, Bijman J. Higher bioelectrical potentials due to decreased chloride absorption in the sweat glands of patients with cystic fibrosis. New Engl J Med 1983; 308: 1185–9.
197. Said SI, Heinz-Erian P. VIP and exocrine function: possible role in cystic fibrosis. In Mastella G, Quinton PM, editors. Cellular and Molecular Basis of Cystic Fibrosis. San Francisco, CA: San Francisco Press, 1988: 355–61.
198. Amiranoff B, Rosselin G. VIP receptors and control of cyclic AMP production. In Said SI, editor. Vasoactive Intestinal Peptide. New York: Raven Press, 1982: 307–22.
199. Heinz-Erian P, Flux M, Dey RD, Said SI. Deficient vasoactive intestinal peptide innervation in sweat glands of cystic fibrosis patients. Science 1985; 229: 1407–8.
200. Richardson JB, Nerve supply to the lungs. Am Rev Resp Dis 1979; 119: 785–802.
201. Said SI. VIP as a modulator of lung inflammation and airway constriction. Am Rev Resp Dis 1991; 143: S22–S24.
202. Morice A, Unwin RJ, Sever PS. Vasoactive intestinal peptide causes bronchodilatation and protects against histamine-induced bronchoconstriction in asthmatic subjects. Lancet 1983; 11/26: 1225–6.
203. Barnes PJ, Dixon CMS. The effect of inhaled vasoactive intestinal peptide on bronchial reactivity to histamine in humans. Am Rev Resp Dis 1984; 130: 162–6.
204. Bundgaard A, Enehjelm SD, Aggestrup S. Pretreatment of exercise-induced asthma with inhaled vasoactive intestinal peptide (VIP). Eur J Resp Dis 1983; 64: 427–9.
205. Mojarad M, Grode TL, Cox CP, Kimmel G, Said SI. Differential responses of human asthmatics to inhaled vasoactive intestinal peptide (VIP). Am Rev Resp Dis 1985; 131: A281.
206. Bolin DR, Cottrel J, Garippa R, O'Neill N, Simko B, O'Donnell M. Structure-activity studies of vasoactive intestinal peptide [VIP]: cyclic disulfide analogs. Int J Peptide Prot Res 1993; 41: 124–32.
207. Jaeger E, Remmer HA, Abdel-Razek TT, Said SI. Structure activity studies on VIP-11, synthesis of analogues modified at positions Arg[12], Arg[14]-Lys[15], Met[17] and Lys[20]-Lys[21], including a potent VIP/PHM-hybrid. In Rosselin G, editor. VIP, PACAP, and Related Regulatory Peptides. River Edge, NJ: World Scientific, 1994: 89–92.

208. Pavlou TA, Bergofsky EH, Dervan JP, Absood A, Said SI. Infusion of vasoactive intestinal peptide improves hemodynamics in primary pulmonary hypertension. Am Rev Resp Dis 1993; 147: A536.
209. Maruno K, Absood A, Said SI. *In vivo* inhibition of human small cell lung carcinoma (SCLC) tumors by vasoactive intestinal peptide (VIP). Am Rev Resp Dis 1994; 149: A174.

Airways Smooth Muscle: Peptide Receptors, Ion Channels and Signal Transduction
ed. by D. Raeburn and M. A. Giembycz
© 1995 Birkhäuser Verlag Basel/Switzerland

CHAPTER 5
Atrial Natriuretic Peptides

Neil C. Thomson

Department of Respiratory Medicine, Western Infirmary, Glasgow, Scotland, UK.

1   Introduction
2   Natriuretic Peptides
2.1   Type A Natriuretic Peptides
2.1.1 α-Atrial Natriuretic Peptide
2.1.2 Urodilatin
2.2   Type B Natriuretic Peptides
2.2.1 Brain Natriuretic Peptide
2.3   Type C Natriuretic Peptides
2.3.1 C-Type Natriuretic Peptide
2.4   Analogues
3   Natriuretic Peptide Receptors
3.1   ANP-$_A$ Receptor
3.2   ANP-$_B$ Receptor
3.3   ANP-$_C$ Receptor
3.4   Natriuretic Peptide Receptors in Airways Smooth Muscle
3.5   Natriuretic Peptide Antagonists
4   Signal Transduction
4.1   Guanylyl Cyclases
4.1.1 Particulate Guanylyl Cyclases
4.1.2 Soluble Guanylyl Cyclases
4.2   Cyclic Guanosine 5′-Monophosphate (cGMP)
5   Functional Effects
5.1   Animal Airways Smooth Muscle
5.1.1 Guinea-Pig
5.1.2 Rat
5.1.3 Bovine
5.2   Human Airways Smooth Muscle
5.3   Influence of Proteases on Human Airways Smooth Muscle Responses
6   Summary
      References

1. Introduction

In 1981 de Bold and colleagues demonstrated that the injection of atrial,
but not ventricular, extract into rats caused a natriuresis and diuresis
[1]. It is now known that this extract contained a peptide, atrial
natriuretic peptide (ANP) or atrial natriuretic factor, which is one of a
family of hormones known to have an important role in salt and water
homeostasis [2–6]. Natriuretic peptides are produced primarily in the
heart but are also released in other tissues including the kidneys and

central nervous system. Circulating ANP has effects on the kidney causing natriuresis and diuresis and on vascular tissue causing vasodilatation. ANP actions also include inhibition of the release or effect of several hormones including aldosterone, angiotensin II and endothelin [5]. Recent *in vivo* studies in man have shown that ANP has important actions on airways function including bronchodilatation and on the modification of bronchial reactivity to direct and indirect challenges [7–15]. This chapter reviews what is known of the natriuretic peptides, their receptors and signal transduction pathways and summarise their functional effects on airways smooth muscle.

Natriuretic Peptides

2.1. Type A Natriuretic Peptides

2.1.1. α-atrial natriuretic peptide: The structural features of the ANP gene are similar between species. In man the gene is located on chromosome 1, band p36 [4, 6, 16]. The transcription of the gene causes mRNA to encode an amino-acid precursor peptide, preproANP. The size of preproANP differs between species and can contain between 149 and 153 amino acids *e.g.*, dog = 149, man = 151, rabbit = 153. In man the preproANP is rapidly converted to a 126-amino-acid prohormone (proANP$_{1-126}$) and this is the principal form of ANP that is stored in atrial granules [4–6]. The biologically active hormone, the C-terminal peptide, α-ANP$_{99-126}$, is cleaved from proANP and leaves an inactive N-terminal fragment, ANP$_{1-98}$. The active hormone α-ANP$_{99-126}$ or α-ANP$_{1-28}$ (Figure 1) as it is often denoted, has a similar amino-acid sequence among species except at residue 110 which is methionine in humans and cows, but isoleucine in rats. A common 17-member central ring structure that has a disulphide bridge between cysteines 105 and 121 is found in all active forms of the natriuretic peptides [4] (Figure 1).

ANP is synthesised predominantly in the cardiac atria and the normal ventricle contains only small quantities of ANP mRNA. The main stimuli to affect atrial ANP gene expression and release is atrial stretch associated with pressure and volume overload [5]. A range of other factors may be involved in causing ANP secretion including exercise, hypoxia, and hormones such as glucocorticosteroids, catecholamines, angiotensin II and endothelin [5]. The finding that proANP or ANP immunoreactivity is present in other extracardiac tissues including brain, peripheral nerves, aorta, intestine and lung [3, 17] suggests that ANP may be synthesised and stored in these tissues.

α-ANP has an important role in salt and water homeostasis [2–6] and may be involved in the control of pulmonary vascular tone.

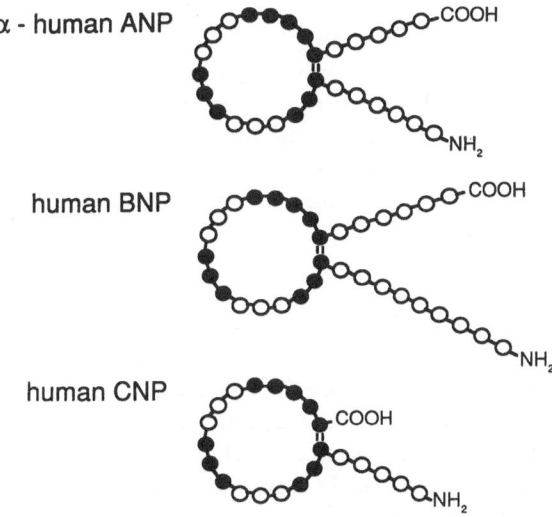

Figure 1. Human natriuretic peptides:- type A: α-ANP, type B: BNP, type C: CNP. Amino acids within the common 17 member central ring structure that are identical are represented by shaded areas. The parallel lines indicate a disulphide bridge between cysteine residues.

2.1.2. Urodilatin: Another form of ANP, urodilatin, was first identified in human urine and is synthesised in the kidney. It is the main natriuretic peptide in urine but it is not detectable in the blood [18]. Urodilatin is thought to be produced following atypical processing of proANP by renal tissue. The amino-acid sequence is the same as ANP except that it is extended by 4 additional residues, ANP_{95-126}. Urodilatin acts on the same receptors as ANP and activates guanylyl cyclase [19] to cause diuresis and natriuresis. Unlike ANP, however, it has been suggested that urodilatin is not metabolised by neutral endopeptidase [20–21].

2.2. Type B Natriuretic Peptides

2.2.1. Brain natriuretic peptide: Brain natriuretic peptide (BNP) was first isolated from porcine brain extracts [5, 6, 22]. Like ANP, human BNP contains a common 17-member central ring structure that has a disulphide bridge and differs from ANP in only 6 of the amino-acids within the ring (Figure 1). The N- and C-terminal extensions differ considerably from ANP in their length and amino-acid composition. There is marked variation in the amino-acid sequence of BNP between different species. In humans, the circulating hormone contains 32 amino-acids whereas a 46 amino-acid BNP is found in rodents (also called iso-ANP).

BNP is synthesised mainly in heart atrial tissue rather than in the central nervous system. The concentrations of BNP within the atrium are approximately 45 times lower than ANP. The normal ventricle produces levels of BNP 100 times lower than the atrium; during heart disease the ventricle becomes the main site of BNP synthesis. The reported actions of BNP include natriuretic, diuretic and hypotensive effects. The function of BNP may be related to pathological states such as heart failure.

2.3. Type C Natriuretic Peptides

2.3.1. C-type natriuretic peptide: C-type natriuretic peptide (CNP) was first isolated from porcine brain extracts [23] and like ANP and BNP it contains a common 17-member central ring structure that has a disulphide bridge (Figure 1). The N-terminal extension differs considerably from ANP and BNP in length and amino-acid composition. CNP has no C-terminal extension. Between species the amino-acid sequence of CNP is similar.

CNP is found principally within the central nervous system but is also detected in vascular endothelium, kidney and heart. CNP is a potent venodilator when compared with either ANP or BNP, but causes only a weak natriuretic effect. It acts on ANP-$_B$ receptors and may represent the natural ligand for this receptor.

2.4. Analogues

A number of synthetic ANP analogues have been developed in the hope of prolonging the therapeutic effect of ANP due to its short half-life. A synthetic ANP analogue NNC 70-0270 (Novo Nordisk A/S) demonstrated a prolonged action on animal and human arterial preparations *in vitro* [24]. Several truncated ANP analogues have also been assessed and shown to act as full agonists [25]. These analogues incorporated two portions of ANP_{1-28}, the 8 amino acids C-terminal to Cys^7, and two amino acids from the C-terminal (phenylalanine and arginine), into disulphide-bonded cyclic peptides. The most potent analogues have cyclohexylalinin at position 8. One such compound designated A-68828 was found to have a binding affinity approximately 1/400 of native ANP. A-68828 was less potent than ANP in stimulating the biosynthesis of cGMP in bovine cultured aortic endothelial cells but more potent in inhibiting ACTH-induced aldosterone release. Other analogues of $ANP_{105-126}$ were found to be less potent in various *in vitro* assays and in a binding assay, but to have higher activity *in vivo* suggesting reduced clearance and/or breakdown [26]. An oxidized analogue MetSO- α-hANP binds more selectively to the ANP-$_C$ receptor than the ANP-$_A$ receptor [27]. Novel natriuretic peptides with specific renal and/or cardiovascular activity are now being developed [28].

3. Natriuretic Peptide Receptors

Specific ANP receptors have been found in many tissues including blood vessels, where ANP causes vasodilatation, the kidney, where it causes natriuresis and diuresis, and the adrenals, where it inhibits aldosterone synthesis and release. Receptors are also found in the central nervous system and the choroid plexus [3]. Molecular cloning has identified three receptors:-

3.1. ANP-$_A$ Receptor (previously named GC-A, ANP-R$_1$ or ANP$_B$)

The human receptor consists of 1029 amino-acids (rat and mouse 1057 amino-acids) and has a molecular weight of 120 000–140 000. It has a transmembrane domain which divides the receptor in half (Figure 2). The intracellular region contains protein kinase and guanylyl cyclase

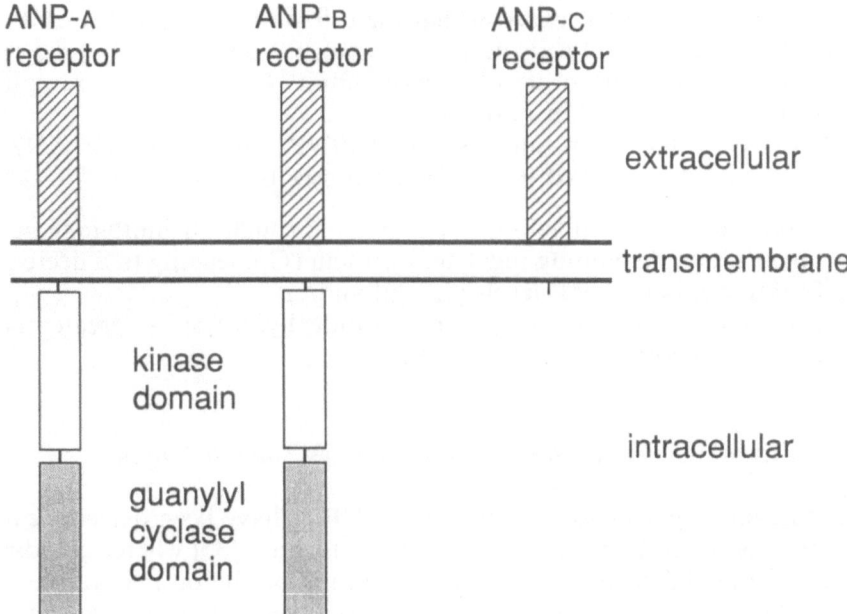

Figure 2. Natriuretic peptide receptors: ANP-$_A$ receptor, ANP-$_B$ receptor, ANP-$_C$ receptor.

domains. Ligand occupancy of the extracellular domain leads to activation of particulate guanylyl cyclase [29]. ANP is ~ 10 fold more potent than BNP at the ANP-$_A$ receptor [30].

3.2. ANP-$_B$ Receptor (previously named GC-B)

This receptor consists of 1025 amino acids. It has one transmembrane domain and like the ANP-$_A$ receptor contains intrinsic guanylyl cyclase activity (Figure 2). CNP is the most potent endogenous agonist at the ANP-$_B$ receptor with other members having at least 50–500 times lower affinity [30]. CNP may represent the natural ligand for the ANP-$_B$ receptor. It seems unlikely that BNP or ANP are the natural ligands for this receptor.

3.3. ANP-$_C$ Receptor (previously named ANP-R_2, CGC)

The ANP-$_C$ receptor, which has a molecular weight of 60 000–70 000 does not contain intrinsic guanylyl cyclase activity. It consists of 540 amino acids (human), exists as a disulphide-linked homodimer with one transmembrane spanning domain for each subunit (Figure 2). The receptor may be involved with the clearance of α-ANP. It has been proposed that circulating ANP binds to the ANP-$_C$ receptor to form a complex which is internalised within the cell, and subsequently ANP is degraded by lysosomal enzymes; the receptor is then recycled to the cell surface [31]. The rank order of binding affinity for the ANP-$_C$ receptor is ANP > CNP > BNP in both humans and rat tissues [30].

Recently it has been suggested that ANP binding to the ANP-$_C$ receptor may be involved in cellular responses by causing:

1) a reduction in adenylyl cyclase activity through an inhibitory guanine-nucleotide binding regulatory protein (G_i), leading to a decrease in intracellular cAMP levels [32, 33] and/or
2) a stimulatory effect on phosphoinositide hydrolysis – greater potency of atriopeptin I than ANP [34].

3.4. Natriuretic Peptide Receptors in Airways Smooth Muscle

Specific binding sites for radiolabelled ANP$_{1-28}$ have been detected over guinea-pig tracheal smooth muscle and epithelium [35] whereas binding sites for radiolabelled ANP$_{5-28}$ were not found on the bronchi of the rat [36]. However, studies on rat tracheal smooth muscle cultures have found binding sites for ANP and BNP [37]. The binding of each radiolabelled peptide was abolished by inclusion of excess amounts of the other unlabelled peptide, suggesting that ANP and BNP share binding sites in the trachea. Studies using a ring-deleted analogue C-ANP$_{4-23}$, which strongly competed for the natriuretic peptide binding sites suggests that the receptors were of the ANP-$_C$ or clearance receptor subtype [37]. The receptor subtype(s) in human airways smooth muscle is unknown.

3.5. Natriuretic Peptide Antagonists

HS-142-1 is a non-peptide antagonist which blocks cGMP production elicited by natriuretic peptides; it prevents ANP, BNP and CNP binding to the guanylyl cyclase-linked ANP receptor [38]. L-α-aminosuberic acid[^{7,23}]-β-ANP$_{7-28}$ acts as an antagonist (pA$_2$ = 7.5) at the ANP receptor.

4. Signal Transduction

4.1. Guanylyl Cyclases

Guanylyl cyclase proteins are present in the particulate and soluble fractions of cell homogenates and act as receptors for various ligands [reviewed in 39, 40].

4.1.1. Particulate guanylyl cyclases: Particulate guanylyl cyclases act as plasma membrane receptors for different ligands including the natriuretic peptides ANP, BNP and CNP. The extracellular region of particulate guanylyl cyclase contains one single, putative transmembrane domain flanked by peptide-binding domains; the intracellular region contains protein kinase-like and cyclase catalytic domains. Particulate guanylyl cyclases exist in at least two forms: guanylyl cyclases-A (ANP-$_A$ receptor) and guanylyl cyclases-B (ANP-$_B$ receptor) although other forms may exist. The ANP-$_A$ receptor and ANP-$_B$ receptor are identical across the extracellular domain and show 91% and 71% homology within the distal cyclase and protein kinase-like domains respectively (Figure 2). The function of the protein kinase-like region is unclear. It has been suggested that ANP may promote ATP to bind to the kinase-like domain which then causes activation of the cyclase region. An additional regulatory protein may be involved in this process.

A heat stable enterotoxin from *Escherichia coli* stimulates a guanylyl cyclase receptor (GC-C) in the intestines of several species. This receptor has a different extracellular region from the ANP-$_A$ and ANP-$_B$

receptor and is not activated by natriuretic peptides. An endogenous ligand has not yet been identified.

4.1.2. Soluble guanylyl cyclases: Soluble guanylyl cyclases (GC-S) are activated by nitric oxide, nitrosothiols, nitroglycerin and sodium nitro-prusside and may mediate the vasodilators effects of acetylcholine, bradykinin and substance P. The lung contains soluble guanylyl cyclase which has a molecular weight of 150 000 and exists as a heterodimer of an α-subunit (mol. weight 82 000) and a β-subunit (mol. weight 70 000). Soluble guanylyl contains haem; nitric acid is thought to bind to this site, which then results in enzyme activation. The peptide regions show similarities in structure between the α- and β-two subunits. The cyclase domains are identical between the subunits and with particulate guany-lyl cyclase. Whether only one or both soluble subunits require stimula-tion to cause activation of cyclase catalytic domains is not established. There are many heterodimeric guanylyl cyclases receptors and these can be tissue-specific, which suggests that there may be specific controlling influences for each subtype of soluble guanylyl cyclase.

4.2. Cyclic Guanosine 5′-Monophosphate (cGMP)

Cyclic GMP is widely believed to be the second messenger for ANP in the cell [41]. Little is known, however, about the signal transduction pathways in airways smooth muscle. In several animal models a direct dose relaxant effect of ANP has been demonstrated and work on bovine isolated and guinea pig trachea suggests that this effect is likely to be mediated by cGMP [42, 43]. Studies on rat pituitary tumour cells have demonstrated that ANP stimulates voltage-activated potassium chan-nels [44]. This effect is preceded by an increase in cGMP production and requires cGMP-dependent protein kinase activity. It has been suggested that activation of protein phosphatase by cGMP-dependent protein kinase could explain the inhibitory action of natriuretic peptides on electrical activity. However, the neuromodulatory action of ANP may be independent of increases in cGMP, and rather it may act via the ANP-$_C$ receptor to suppress adenylyl cyclase activity and neurotrans-mission [33].

5. Functional Effects

5.1. Animal Airways Smooth Muscle

5.1.1. Guinea-pig: In guinea pig isolated airway preparations, atriopep-tins (I (ANP$_{1-21}$), II (ANP$_{5-27}$), III (ANP$_{5-28}$)) and ANP have a direct relaxant effect [35, 45–48]. Atriopeptin III, the most potent of the atriopeptins, is 16 times less potent (EC_{50} $4.7 \pm 3.0 \times 10^{-8}$ M) in relaxing resting tracheal smooth muscle tone than isoprenaline [35]. The relaxant response to the atriopeptins is not influenced by H_2 receptor blockade, β-adrenoceptor blockade or cyclooxygenase inhibition. Removal of the epithelium [48] or the addition of the neutral endopeptidase inhibitors phosphoramidon [48] or thiorphan [47] increases the potency of atriopeptins and ANP. In contrast to these findings, however, O'Donnell *et al.* reported that epithelial cell removal did not alter the response to atriopeptins in relaxing resting tracheal smooth muscle tone [35].

In tracheal tissue ANP causes relaxation of intrinsic tone as well as tone induced by carbachol, methacholine, arachidonic acid, histamine, 5-HT or leukotriene D_4 [14, 42, 46]. ANP may be more effective in relaxing methacholine- than leukotriene D_4-induced tone [45]. ANP is 2–3 fold more potent than the atriopeptins in relaxing guinea-pig tracheal smooth muscle [48]. The potency of ANP has been reported as similar to that of isoprenaline when relaxing methacholine-induced tone [45], although others found that ANP was 20-fold less potent in relaxing carbachol contracted tracheal tissue [46]. ANP is 2–3 fold less potent when compared with isoprenaline in relaxing arachidonic acid-, histamine- and 5-HT-induced tone [46]. ANP has no effect on cholinergic contraction induced by electrical field stimulation [49].

The mechanism of the relaxant response to ANP has not been established but is not due to the generation of cyclooxygenase products [45]. In contrast to the potent effects of ANP on the trachea, it is relatively inactive at relaxing parenchymal strips contracted by methacholine or leukotriene D_4 [45].

5.1.2. Rat: In rat tracheal tissue atriopeptins cause a weak relaxation of intrinsic tone as well as tone induced by carbachol [48]. Removal of the epithelium or the addition of the neutral endopeptidase (NEP) inhibitor phosphoramidon decreases the potency of the atriopeptins [48]. Thus in contrast to the finding in the guinea pig, the epithelium in the rat trachea may convert ANF_{5-27} to a more active relaxant peptide [48].

5.1.3. Bovine: In bovine isolated tracheal tissue ANP, atriopeptin II and atriopeptin III have a direct relaxant effect on tone induced by carbachol (5×10^{-8} M), histamine or 5-HT [43]. The relaxant effects observed were much less when higher concentrations of carbachol were used. The potency of ANP was intermediate between isoprenaline and sodium nitroprusside. Another study in bovine bronchial tissue found that ANP caused a slight relaxation of methacholine-contracted tissue and evoked a significant rightward shift of the cumulative concentration-response curve to methacholine [50]. The ability of phosphoramidon to enhance an inhibitory effect of ANP would point to a role for NEP in bovine airways.

5.2. Human Airways Smooth Muscle

In a preliminary experiment human ANP caused a direct, if low potency, relaxant effect on tone induced with methacholine (3×10^{-6} M) in human isolated bronchi [51]. A more detailed study confirmed these earlier findings [50] and showed that ANP conferred protection against methacholine-induced contraction of isolated bronchi (Figures 3 and 4).

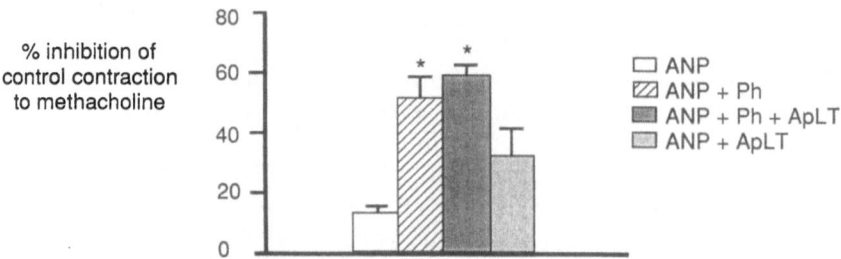

Figure 3. Effect of the addition of ANP(10^{-6} M) alone, and with a combination of phosphoramidon (Ph) alone or in combination with aprotinin (Ap), leupeptin (L) and soybean trypsin (T) (each at 20 μg/ml) inhibitor on human tissue precontracted with methacholine (n = 6). Values are mean ± SEM. * p < 0.05: ANP alone versus ANP + Ph or ANP + Ph + ApLT.

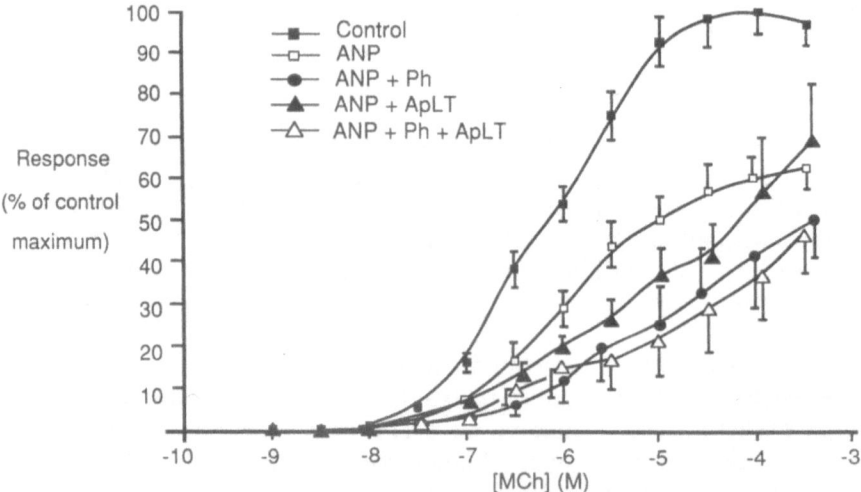

Figure 4. Effect of preincubation with ANP(10^{-6} M) alone, and with a combination of phosphoramidon (Ph) alone or in combination with aprotinin (Ap), leupeptin (L) and soybean trypsin inhibitor (T) (each at 20 μg/ml) on the cumulative concentration response curves to methacholine compared with controls in human bronchial rings (n = 6). Values are mean ± SEM.

The mechanism of ANP-induced relaxation of human airways smooth muscle tone has not been established.

In contrast to our results, several studies performed on isolated human bronchial tissue have failed to show that ANP or atriopeptins can reduce resting bronchial tone or relax carbachol or histamine induced contractions [47, 48, 52]. Several factors may explain these differing conclusions. Firstly, atriopeptins, which are less potent than human ANP, were used in 2 of these studies [48, 52]. Secondly, in the one study which used human ANP [47] the concentration administered was lower than we used. Thirdly, as there is an inverse relationship between the level of airway tone and the potency of relaxant agonists [53], the higher concentrations of carbachol used by Fernandes and coworkers [48] and of histamine used by Labat and colleagues [52] may have masked a relaxant effect of ANP.

5.3. Influence of Proteases on Human Airways Smooth Muscle Responses

Two principle mechanisms have been proposed for the inactivation of ANP: degradation by the enzyme NEP [54] and binding to a non-guanylyl cyclase clearance receptor (ANP-$_C$ receptor) [55]. There is evidence of clearance of ANP across the lung [56] although the main site for removal of circulating ANP appears to be the kidney [57]. NEP has been localised to the lung and it appears to be present in high concentrations in airways epithelium [58] although it is also found in submucosal glands, airways smooth muscle and nerves [59]. The widespread distribution of NEP within the airways suggests that it may play a role in modulating the effect of ANP on airways smooth muscle.

Phosphoramidon is an NEP inhibitor [60] which was found to potentiate ANP-induced relaxation of methacholine-induced tone and the ability of ANP to evoke a significant rightward shift of the cumulative concentration-response curve to methacholine [50] (Figures 3 and 4). This would suggest that NEP is important in the breakdown of ANP in the airways. NEP has been demonstrated in the lung and in higher concentrations in airway epithelium [58] in comparison with pulmonary artery and lung parenchyma. It could be speculated that degradation of ANP by this enzyme in vivo would limit any potential bronchodilator response to ANP when it is administered by inhalation. This may explain the absence of a significant bronchodilator effect when ANP is given in low dose (1 mg) by inhalation to asthmatic patients [12, 14] while there is a very significant bronchodilator response when it is given intravenously [7, 8, 10]. In a high dose (5 mg), inhalation of ANP does have a bronchodilator effect [15], and it could be postulated that at this dose the clearance mechanisms are saturated allowing the bronchodilator effects to be seen. It is possible that analogues of ANP which are more resistant to NEP may offer the prospect of being a novel group of bronchodilator drugs.

Recent studies examining the interaction of ANP with salbutamol in human bronchial rings have found that the combination of ANP (in the presence of phosphoramidon) and salbutamol evoked a greater effect than either agent alone, both in reversing and protecting against metha-choline-evoked contraction [61]. Such combinations may be of benefit in the treatment of patients, allowing lower doses of drug to be used.

Airway inflammation is associated with the release of various proteases including mast cell tryptase and chymase [62]. To investigate the possible role of these and other proteases in modulating the effect of ANP on bronchial smooth muscle we examined the effect of the protease inhibitors leupeptin, aprotinin and soybean trypsin inhibitor on the airway response to ANP [50]. A similar cocktail of protease inhibitors in combination with the NEP inhibitor phosphoramidon has been found to potentiate VIP induced bronchodilatation [63]. However, this combination of inhibitors has no effect on the action of ANP on human isolated bronchial tissues (Figures 3 and 4). These results suggest that these proteases are unlikely to be important in regulating the effects of ANP on smooth muscle tone.

6. Summary

ANP, one of a family of natriuretic peptides, is a 28 amino acid hormone secreted by the cardiac atria and isolated lung tissue and has an important role in salt and water homeostasis. Molecular cloning has identified three receptors in different tissues: ANP-$_A$ receptor, ANP-$_B$ receptor, ANP-$_C$ receptor, but the receptor subtypes in human or animal airways smooth muscle are unknown. It is thought that ANP acts by stimulating particulate guanylyl cyclase to cause the generation of cGMP. ANP causes relaxation of human and animal airways smooth muscle although there is variability in the potency between species. The ability of phosphoramidon to enhance the relaxant effect of ANP would point to a role for neutral endopeptidase in modulating its action on human bronchial tissue. It is possible that analogues of ANP which are more resistant to neutral endopeptidase may offer the prospect of being a novel group of bronchodilator drugs.

References

1. de Bold AJ, Borenstein HB, Veress AT, Sonnenberg H. A rapid and potent natriuretic response to intravenous injection of atrial myocardial extract in rats. Life Sci 1981; 28: 89–94.

2. Sagnella GA, MacGregor GA. Cardiac peptides and the control of sodium excretion. Nature 1984; 309: 666–9.
3. Gutkowska J, Nemer M. Structure, expression and function of atrial natriuretic factor in extraatrial tissues. Endocrine Rev 1989; 10: 519–36.
4. Rosenzweig A, Seidman CE. Atrial natriuretic factor and related peptide hormones. Ann Rev Biochem 1991; 60: 229–55.
5. Ruskoaho H. Atrial natriuretic peptide: synthesis, release, and metabolism. Pharmacol Rev 1992; 44: 479–602.
6. Koller KJ, Goeddel DV. Molecular biology of the natriuretic peptides and their receptors. Circulation 1992; 86, 4: 1081–8.
7. Hulks G, Jardine A, Connell JMC, Thomson NC. Bronchodilator effect of atrial natriuretic peptide in asthma. Brit Med J 1989; 299: 1081–2.
8. Chanez P, Mann C, Bousquet J, Chabrier PE, Godard P, Braquet P, Michel F-B. Atrial natriuretic factor (ANF) is a potent bronchodilator in asthma. J Allergy Clin Immunol 1990; 86: 321–4.
9. Hulks G, Jardine A, Connell JMC, Thomson NC. Effect of atrial natriuretic factor on bronchomotor tone in the normal human airway. Clin Sci 1990; 79: 51–5.
10. Hulks G, Jardine A, Connell JMC, Thomson NC. Influence of elevated plasma levels of atrial natriuretic factor on bronchial reactivity in asthma. Am Rev Resp Dis 1991; 143: 778–82.
11. McAlpine LG, Hulks G, Thomson NC. Effect of atrial natriuretic peptide given by intravenous infusion on bronchoconstriction induced by ultrasonically nebulized distilled water (FOG). Am Rev Resp Dis 1992; 146: 912–5.
12. Hulks G, Thomson NC. Inhaled atrial natriuretic peptide and asthmatic airways. Brit Med J 1992; 304: 1156.
13. Angus RM, McCallum MJA, Thomson NC. The bronchodilator, cardiovascular and cyclic guanylyl monophosphate (cGMP) response to high dose infused atrial natriuretic peptide in asthma. Am Rev Resp Dis 1993; 147: 1122–5.
14. Angus RM, MacCallum MJA, Thomson NC. Inhaled atrial natriuretic peptide inhibits methacholine-induced bronchoconstriction in asthma. Clin Exp Allergy 1994; 24: 784–8.
15. Hulks GH, Thomson NC. High dose inhaled atrial natriuretic factor is a bronchodilator in asthmatic subjects. Eur Resp J 1994; 7: 1593–97.
16. Yang-Feng TL, Floyd-Smith G, Nemer M, Drouin J, Francke U. The pronatriodilatin gene is located on the distal short arm of human chromosome 1 and on mouse chromosome 4. Am J Hum Genet 1985; 37: 1117–28.
17. Vesely DL, Palmer PA, Giordano AT. Atrial natriuretic factor prohormone peptides are present in a variety of tissues. Peptides 1992; 13: 165–70.
18. Kentsch M, Ludwig D, Drummer C, Gerzer R, Müller-Esch G. Haemodynamic and renal effects of urodilatin in healthy volunteers. Eur J Clin Invest 1992; 22: 319–25.
19. Heim JM, Kiefersauer S, Fülle H-J, Gerzer R. Urodilatin and beta-ANF. Binding properties and activation of particulate guanylate cyclase. Biochem Biophys Res Commun 1989; 163: 37–41.
20. Gagelmann M, Hock D, Forssmann W-G. Urodilatin (CDD/ANP[95-126]) is not biologically inactivated by a peptidase from dog kidney cortex membrane in contrast to atrial natriuretic peptide/cardiodilatin (α-hANP/CDD[99-126]). FEBS Lett 1988; 233: 249–54.
21. Abassi ZA, Tate J, Hunsberger S, Klein H, Trachewsky D, Keiser HR. Pharmacokinetics of ANP and urodilatin during cANF receptor blockade and neutral endopeptidase inhibition. Am J Physiol 1992; 263: E870–6.
22. Sudoh T, Kangawa K, Minamino N, Matsuo H. A new natriuretic peptide in porcine brain. Nature 1988; 332: 78–81.
23. Sudoh T, Minamino N, Kangawa K, Matsuo H. C-type natriuretic peptide (CNP): A new member of natriuretic peptide family identified in porcine brain. Biochem Biophys Res Commun 1990; 168: 863–70.
24. Weis J. Effects on isolated arteries from rat, rabbit and man of NNC 70-0270, a synthetic atrial natriuretic factor analogue with a prolonged action. Eur J Pharmac 1991; 205: 17–20.
25. Holleman WH, Budzik GP, Devine EM, Pollock DM, Opgenorth A, Thomas AM et al. Truncated atrial natriuretic factor analogs retain agonist activity. Can J Physiol Pharmacol 1991; 69: 1622–7.

26. Schiller PW, Mazlak LA, Nyugen TM-D, Godin J, Garcia R, Delean A, Cantin M. Superactive analogs of the atrial natriuretic peptide (ANP). Biochem Biophys Res Commun 1987; 143: 499–505.
27. Koyama S, Terai T, Inoue T, Inomata K, Tamura K, Kobayashi Y et al. An oxidized analog of α-human atrial natriuretic polypeptide is a selective agonist for the atrial-natriuretic-polypeptide clearance receptor which lacks a guanylate cyclase. Eur J Biochem 1992; 203: 425–32.
28. Gardner DG. Designer natriuretic peptides. J Clin Invest 1993; 92: 1606–7.
29. Chinker M, Garbers DL, Chang MS, Lowe DG, Chin H, Goeddel DV, Schulz S. A membrane form of guanylate cyclase is an atrial natriuretic peptide receptor. Nature 1989; 338: 78–83.
30. Suga S-I, Nakao K, Hosoda K, Mukoyama M, Ogawa Y, Shirakami G, Arai H, Saito Y, Kambayashi Y, Inouye K, Imura H. Receptor selectivity of natriuretic peptide family, atrial natriuretic peptide, brain natriuretic peptide, and C-type natriuretic peptide. Endocrinology 1992; 130: 229–39.
31. Nussenzveig DR, Lewicki JA, Maack T. Cellular mechanisms of the clearance function of type C receptors of atrial natriuretic factor. J Biol Chem 1990; 265: 20952–8.
32. Anand-Srivastava MB, Sairam MR, Cantin M. Ring-deleted analogs of atrial natriuretic factor inhibit adenylate cyclase/cAMP system. J Biol Chem 1990; 265: 8566–72.
33. Drewett JG, Ziegler RJ, Trachte GJ. Neuromodulatory effects of atrial natriuremic peptides correlate with an inhibition of adenylate cyclase but not an activation of guanylate cyclase. J Pharmacol Exp Ther 1992; 260: 689–96.
34. Hirata M, Chang CH, Murad F. Stimulatory effects of atrial natriuretic factor on phosphoinositide hydrolysis in cultured bovine aortic smooth muscle cells. Biohem Biophys Acta 1989; 1010: 346–51.
35. O'Donnell M, Garippa R, Welton AF. Relaxant activity of atriopeptins in isolated guinea pig airway and vascular smooth muscle. Peptides 1985; 6: 597–601.
36. Bianchi CJ, Gutkowska G, Thibault R, Garcia J, Genest J, Cantin M. Radioautographic localization of ^{125}I-atrial natriuretic factor (ANF) in rat tissues. Histochemistry 1985; 82: 441–52.
37. James S, Burnstock G. Atrial and brain natriuretic peptides sharing binding sites on cultured cells from the rat trachea. Cell Tissue Res 1991; 265: 555–65.
38. Morishita Y, Sano Y, Kase H, Yamada K, Inagami T, Matsuda Y. Hs-142-1, a novel atrial natriuretic peptide (ANP) antagonist, blocks ANP-induced renal responses through a specific interaction with guanylyl cyclase-linked receptors. Eur J Pharm 1992; 225: 203–7.
39. Schulz S, Yuen PST, Garbers DL. The expanding family of guanylyl cyclase. Trends Pharmacol Sci 1991; 12: 116–20.
40. Garbers DL. The guanylyl cyclase – receptor family. Can J Physiol Pharmacol 1991; 69: 1618–21.
41. Murad F, Leitman DC, Bennett BM, Molina C, Waldeman SA. Regulation of guanylate cyclase by atrial natriuretic factor and the role of cyclic GMP in vasodilation. Am J Med Sci 1987; 294: 139–43.
42. Watanabe H, Suzuki K, Takagi K, Satake T. Mechanism of atrial natriuretic polypeptide and sodium nitroprusside-induced relaxation in guinea-pig tracheal smooth muscle. Arzneim-Forsch 1990; 40: 771–6.
43. Ishii K, Murad F. ANP relaxes bovine tracheal smooth muscle and increases cGMP. Am J Physiol 1989; 256: C495–C500.
44. White RE, Lee AB, Shcherbatko AD, Lincoln TM, Schonbrunn A, Armstrong DL. Potassium channel stimulation by natriuretic peptides through cGMP-dependent dephosphorylation. Nature 1993; 361: 263–6.
45. Hamel R, Ford-Hutchinson AW. Relaxant profile of synthetic atrial natriuretic factor on guinea pig pulmonary tissues. Eur J Parmacol 1986; 121: 151–5.
46. Potvin W, Varma DR. Bronchodilator activity of atrial natriuretic peptide in guinea pigs. Can J Physiol Pharmacol 1989; 67: 1213–8.
47. Candenas M-L, Naline E, Puybasset L, Devillier P, Advenier C. Effect of atrial natriuretic peptide and of atriopeptins on the human isolated bronchus. Comparison with the reactivity of the guinea-pig isolated trachea. Pulmonary Pharmacol 1991; 4: 120–5.

48. Fernandes LB, Preuss JMH, Goldie RG. Epithelial modulation of the relaxant activity of atriopeptides in rat and guinea-pig tracheal smooth muscle. Eur J Pharmacol 1992; 219: 187–94.
49. Verleden GM, Pype JL, Demedts M. The effect of atrial natriuretic peptide on neurotransmission in guinea-pig airways *in vitro*. Am Rev Resp Dis 1992; 145: A43.
50. Angus RM, Nally JE, McCall R, Young LC, Thomson NC. Modulation of the effect of atrial natriuretic peptide in human and bovine bronchi by phosphoramidon. Clin Sci 1994; 86: 291–5.
51. Hulks G, Crabb KG, McGrath JC, Thomson NC. *In vitro* effects of atrial natriuretic factor and sodium nitroprusside on bronchomotor tone in human bronchial smooth muscle. Am Rev Resp Dis 1991; 143: A344.
52. Labat C, Norel X, Benveniste J, Brink C. Vasorelaxant effects of atrial peptide II on isolated human pulmonary muscle preparations. Eur J Pharmacol 1988; 150: 397–400.
53. Van den Brink FG. The model of functional interaction II. Experimental verification of a new model: the antagonism of β-adrenoceptor stimulants and other antagonists. Eur J Pharmacol 1973; 22: 279–86.
54. Erdos EG, Schulz WW. Neutral endopeptidase 24.11 (enkephalinase) and related regulators of peptide hormones. FASEB J 1989; 3: 145–51.
55. Maack T, Suzuki M, Almeida FA, Nussen-Zveig D, Scarborough RM, McEnroe GA, Lewicki JA. Physiological role of silent receptors of atrial natriuretic factor. Science 1987; 238: 675–8.
56. Akaike MD, Ishikura F, Nagata S, Kimura K, Miyatake K. Direct secretion from left atrium and pulmonary extraction of human atrial natriuretic peptide. Am Heart J 1992; 123: 984–9.
57. Sato F, Kamoi K, Wakiya Y, Ozawa T, Arai O, Ishibashi M, Yamaji T. Relationship between plasma atrial natriuretic peptide levels and atrial pressure in man. J Clin Endocrinol Metab 1986; 63: 823–7.
58. Johnson AR, Ashton J, Schulz WW, Erdos EG. Neutral metalloendopeptidase in human lung tissue and cultured cells. Am Rev Resp Dis 1985; 132: 564–8.
59. Nadel JA. Neutral endopeptidase modulates neurogenic inflammation. Eur Resp J 1991; 4: 745–54.
60. Ura N, Carretero O, Erdos EG. Role of endopeptidase 24.11 in kinin metabolism *in vitro* and *in vivo*. Kidney Int 1987; 32: 507–13.
61. Nally JE, Clayton RA, Thomson NC, McGrath JC. The interaction of α-human atrial natriuretic peptide (ANP) with salbutamol, sodium nitroprusside and isosorbide dinitrate in human bronchial smooth muscle. Br J Pharmacol 1994; 113: 1328–32.
62. Tam EK, Caughey GH. Degradation of airway neuropeptides by human lung tryptase. Am J Resp Cell Mol Biol 1990; 3: 27–32.
63. Tam EK, Franconi GM, Nadel JA, Caughey GH. Protease inhibitors potentiate smooth muscle relaxation induced by vasoactive intestinal peptide in isolated human bronchi. Am J Resp Cell Mol Biol 1990; 2: 449–52.

Airways Smooth Muscle: Peptide Receptors, Ion Channels and Signal Transduction
ed. by D. Raeburn and M. A. Giembycz

CHAPTER 6
Platelet-Derived Growth Factor, Transforming Growth Factor-β and Connective Tissue Growth Factor

Jason Kelley

University of Vermont College of Medicine, Given, Burlington, Vermont, USA

1   Introduction
2   Basic Mechanisms of Cytokine Action
2.1 Cytokine Peptide Structure
2.2 Platelet-Derived Growth Factor: A Prototypic Mitogenic Cytokine
2.3 PDGF Receptors and Signal Transduction
3   Cytokines in Lung Biology
3.1 Macrophage Production of PDGF
3.2 Release of PDGF by Airway Structural Cells
3.3 Connective Tissue Growth Factor (CTGF)
4   Cytokines and Smooth Muscle Cells: Lessons from Vascular Biology
5   Cytokines as Physiological Regulators
6   Cytokines and Scarring of Airways
7   Future Directions
     References

1. Introduction

Cell-to-cell communication determines the compostition and hence function of the airways of the lung in health. Conversely, altered balances of cytokines, either primary or secondary are thought to play a determining role in acute and chronic pathological disorders. The protein cytokines are a unique group of extracellular signalling molecules with critical roles in these processes. Cytokines (also termed growth factors) are pleiotrophic molecules produced ubiquitously and which act on a wide array of target cells. They modulate a wide array of cellular functions including growth (stimulation or inhibition), differentiation, survival and a wide range of phenotypic expression. Although generally small multimeric proteins, the cytokines are considerably larger than other signalling molecules such as the kinins, prostaglandins, leukotrienes and other molecules involved in modulating cell phenotype.

Like other signalling molecules, cytokines act by binding to one or more specific high affinity surface receptors on target cells. These membrane receptors in turn are linked to a complex array of intracellular signalling pathways. Cellular activities triggered by cytokines – migration, proliferation, change in phenotype – often require macromolecule synthesis and are therefore characterized by fairly long response times. Unlike more immediate agonists, the results of cytokine actions become apparent within hours rather than seconds or minutes.

The potential roles of cytokines in airways disease are only beginning to be examined. Specific clinical disorders in which pathological remodelling is mediated through imbalances in cytokines include such disparate phenomena as the hyperplasia of smooth muscle in a variey of airways disorders and the development of peribronchial fibrosis in asthma.

This chapter discusses the relevant physiology of cytokines, focussing on the available studies related to cytokine involvement in diseases of the airways. Because a general review of all cytokines is beyond the scope of this chapter, emphasis has been placed on the role of the predominantly mitogenic cytokine platelet-derived growth factor (PDGF) and the fibrogenic cytokine transforming growth factor-β (TGF-β). The very important subject of cytokine biology of bronchial epithelial cells has been reviewed recently and is not covered here [1]. For the same reason, in-depth discussion of the roles of cytokines released by migratory inflammatory cells is only alluded to. The emphasis of this chapter is on smooth muscle cells and fibroblasts of the airways as key target cells of cytokine action.

However, exciting recent insights into smooth muscle cell regulation by cytokines have come from studies of these cells in vessel walls. Because of their obvious significance these enlightening studies are included in this review.

2. Basic Mechanisms of Cytokine Action

2.1. Cytokine Peptide Structure

As proteins, cytokines are relatively small molecules with molecular weights in the range 5–35 kDa (for review, see [2]). Many cytokines are multimers of several identical or related gene products. They are often glycosylated and covalently linked by disulphide bonds. In general, their individual peptides do not have biological activity when chemically separated, indicating that tertiary structure is essential for receptor binding. Many but not all secreted cytokines exhibit hydrophobic leader sequences which direct transmembrane secretion.

Prior to the 1970's, investigators had sought for decades for solid evidence of secreted molecules with biological activity as mitogens. The discovery of the mitogenic cytokines only became possible when cells could be cultured in defined medium containing little or no serum. Observable phenotypic changes of cultured cell lines could then be elicited and the responsible proteins isolated using standard purification techniques. After purification of adequate amounts of cytokine, the amino acid and nucleic acid sequences could be deduced.

Elucidation of nucleic acid sequence has allowed development of recombinant cytokines for use as biological probes. cDNA probes have been hybridized to RNA as well as tissue sections for quantitation and localization of gene expression. Appropriate cDNA probes have also served to search cDNA and genomic libraries for related peptides. In many instances, this strategy has elucidated whole cytokine gene families. Indeed, many of the more than 100 cytokines discovered to date can be shown to be members of one or another of these super-families which often also contain proteins with functions related to growth and development. There are generally close amino acid homologies and function of cytokines across species. Hence, receptors of one species often respond to cytokine purified from multiple other species. However, such a generalization does not hold for all cytokines and there are some intriguing exceptions. Moreover, cross-species studies indicate that cytokines do not always elicit the same repertoire of phenotypical changes.

2.2. Platelet-Derived Growth Factor: A Prototypic Mitogenic Cytokine

Platelet-derived Growth Factor (PDGF) serves as a prototype of the growth modulating cytokines [3]. PDGF is a potent mitogenic and chemotactic protein which is a cationic glycoprotein with molecular weights ranging from 28–32 kDa. PDGF was originally discovered as a soluble serum factor the origin of which was traced to the α granules of platelets which lysed during collection of blood [4]. As with most growth factors, it is now recognized to have multiple forms. Hence, the PDGFs are members of a family of related but distinct gene products which act on cells through related cell surface receptors. The cellular targets of PDGF action in the lung and other organs – cells with cognate PDGF surface receptors – include fibroblasts and smooth muscle cells as well as certain epithelial and endothelial cells. At the whole tissue level, PDGF promotes granulation tissue formation; when added exogenously to wounds purified PDGF can overcome impaired wound healing in such settings as diabetes [5].

The now classical studies with serum-derived PDGF indicated that it acts during the early hours of the cell cycle. Pledger and colleagues [6] found that it acts to render serum-starved BALB/c 3T3 murine cells competent to begin the G_0/G_1 transition into the mitotic cycle, a process which initiates cell replication. Later acting factors such as Insulin-like Growth Factor (IGF-1), promote progression into the later stages of the cell cycle. It bears pointing out that although studies of cell cycle kinetics based upon the BALB/c 3T3 cells have been extremely fruitful in modelling the cell cycle, subsequent studies have cast doubt that such precise timing of PDGF action applies to all mammalian cells [7].

In addition to its role as a mesenchymal cell mitogen, PDGF serves as a chemoattractant molecule and may have weak effects as an inducer of extracellular matrix formation; however, it plays little or no role as a pro-inflammatory mediator. Like other cytokines, its actions are often synergistically modulated by the presence of certain other cytokines. As mentioned above, PDGF acts synergistically with IGF-I or epidermal growth factor (EGF) to induce mesenchymal cells to divde.

Sequencing of proteins purified from platelet granules revealed the presence of two distinct but related peptide chains termed PDGF-A and PDGF-B [8, 9]. Both were shown to be linked into dimers by mulitple disulfide bonds. These PDGFs peptides can exist in three different dimeric molecular forms composed of pairs of A and B chains: PDGF-AA, -AB, and -BB. All three dimeric isoforms exhibit biological activity on target cells which express appropriate receptor subtypes [10–13]. Hence the PDGF molecules represent a family fo allied mitogenic peptides.

2.3. PDGF Receptors and Signal Transduction

Receptors for various cytokines of the lung have been reviewed recently [14]. Cytokine receptors exhibit classical extracellular, transmembrane, and intracellular domains. The extracellular domain of the receptor binds the cytokine ligand, whereupon the intracellular portion initiates its signalling cascade. The cytoplasmic domains of receptors are linked biochemically to a complex array of intracellular signalling pathways. All cells express a heterogeneous population of high affinity receptors for various cytokines; this array may change during the life of the cell as well as in response to external or internal stimuli.

Extracellular PDGF acts on a particular subset of neighbouring cells by interacting with specific high affinity surface receptors. These receptors are glycoproteins integrally located within cell membranes. The exact phenotypic response of cells to PDGF depends on the physiological availability of the cytokine in the extracellular space. It also requires the presence of the appropriate cognate receptor which is linked to intracellular second messenger signalling pathways. When a single cell both secretes PDGF and responds to it by virtue of having the appropriate complement of PDGF receptors, the response pattern has been described as *autocrine regulation* of cell growth. Such cells, if constitutively producing PDGF, are theoretically capable of unregulated autonomous growth such as occurs in malignant transformation. When the PDGF-secreting cell is different from, but presumably closely adjacent to, the cells exhibiting PDGF receptors, the response pattern is said to be *paracrine regulation*.

A third model of cytokine signalling, so-called *juxtacrine secretion*, has been proposed. In this process active cytokine ligands are bound to external membranes of effector cells and act only when the effector cell comes into direct contact with congnate receptors on target cells [15]. Classically, cytokines are thought to act extracellularly, binding and activating receptors present in cell membranes. However, interactions between PDGF and its receptors have also been shown to take place within intracellular sites [16, 17]. This phenomenon has been termed *intracrine regulation*.

Studies of cytokine receptors have led to several unexpected and unique findings regarding receptor biology. For example, exposure of cells to one particular cytokine often alters their subsequent responses to other cytokines. This comes about through several mechanisms, including alteration of the specific population of receptors which the cells exhibits on its surface ("receptor transmodulation"). As just one example of this process, exposure of mesenchymal cells to PDGF alters the receptor binding of epithermal growth factor [18]. Moreover, several cytokines are known to up-regulate their own receptors, making target cells increasingly sensitive to their own action. Such modulation has the effect of providing a positive feedback loop which enhances cytokine potency.

Just as there are several forms of PDGF ligand, there is also more that one form of PDGF receptor. These receptors differ in their specificity for the multiple isoforms of PDGF [19]. High affinity binding of PDGF requires association of two receptor subunits: an α subunit that can bind either a PDGF-B or a PDGF-A chain, and a β subunit that can bind only the PDGF-B peptide. The α- and β subunits are similar in size but can be distinguished by binding specificity, as well as by anti-receptor monoclonal antibodies. Upon exposure to PDGF ligand, these subunits rapidly form reversible complexes on the cell surface. Both the absolute and relative numbers of these two PDGF receptor subunits vary on different cell types. This observation presumably accounts for the differences in mitogenic potency of the different PDGF isoforms when tested in different cell lines.

A study by Ferns and colleagues [20] confirms that regulation of receptor populations directly controls cell phenotype and response to cytokines in the local environment. Fibroblast lines without PDGF receptors were transfected with PDGF receptor cDNAs and examined for the ability of the three dimeric forms of PDGF to induce chemotaxis.

As predicted by this model, receptor-negative cells transfected with the β receptor subunit gene were only responsive to PDGF-BB, whereas cells expressing the α receptor subunit were equally responsive to all three dimeric forms (PDGF-AA, PDGF-AB, and PDGF-BB). In a parallel study of human arterial smooth muscle cells expressing both PDGF receptor subunits endogenously, the same investigators found that recombinant PDGF-AA could elicit a chemotactic response. However, the two smooth muscle cell isolates examined differed from each other in their chemotactic response to PDGF-AA. This difference correlated closely with their ability to respond mitogenically to this PDGF dimeric form. Of importance, the magnitude of both chemotactic and mitogenic responses was related to the proportion of the two receptor subunit species at the cell surface. It is generally acknowledged that PDGF-BB is a more potent mitogenic/chemotactic cytokine than PDGF-AA. These data would suggest that this difference in potency merely relates to the balance of receptors present on the cells under study.

PDGF acts by stimulating   dimerization of high affinity receptor subunits [21]. Association of receptor subunits on the cell surface gives rise to three possible receptor dimers: αα, αβ, and ββ. An early event following binding of PDGF to its receptor is activation of intrinsic receptor tyrosine kinase and auto-phosphorylation of tyrosine residues on the intra-cytoplasmic domain of the receptor [22]. Receptor chain phosphorylation can occur via inter-subunit or "trans-phosphorylation".

In addition to acting through different receptors, PDGF-AA and PDGF-BB exhibit other biological differences. In most cells studied, PDGF-BB is the more potent of the two homodimers; however, it is more poorly secreted than PDGF-AA [23]. The response to heterodimeric PDGF-AB is intermediate between the responses to homodimers.

Expression of PDGF receptors is tightly controlled, probably because they are so important to cell growth. Moreover, the α and β subunits of the receptor have been shown to be under independent regulation [24]. In a study of Gronwald et al. [24], exposure of 3T3 cells to TGF-β dampened their ability to bind different PDGF isoforms. PDGF-AA binding was most significantly down-regulated. The loss of PDGF-AA binding was accompanied by selective loss in the ability of 3T3 cells to undergo mitosis in response to PDGF-AA, but not to PDGF-BB. This sort of differential receptor phenotype regulation may allow cells to fine tune their response to the different isoforms of PDGF.

One aspect of this model which has only recently been resolved is whether PDGF-AB can bind to cells that express only PDGF β receptor and, if so, whether PDGF-AB can act as an agonist or an antagonist. Seifert and colleagues turned to cell lines derived from Patch mutant mouse embryos. The patch mutation involves deletion of the PDGF α receptor gene without alteration in expression of PDGF β receptors

[25]. Comparison between the binding and response properties of mutant and wild type 3T3 cell lines allowed these investigators to define the contribution that PDGF α receptors make to the ability of these cells to respond to PDGF-AB. They found that PDGF-AB binds to PDGF α receptor-negative 3T3 cells and can induce DNA synthesis, PDGF β receptor dimerization and phosphorylation on tyrosine residues in a temperature-dependent manner.

3. Cytokines in Lung Biology

3.1. Macrophage Production of PDGF

Despite its recognition as a platelet product, it is clear that PDGF is produced by multiple cell types and used by them as a secreted signalling molecule. Macrophages produce and secrete a large number of cytokines (for review, see [26]). Based on early studies of the unpurified mitogenic activities released by macrophages and monocytes, the aggregate activity was originally described broadly as "macrophage-derived growth factor." Once individual cytokines could be isolated and distinguished, however, it became clear that macrophages and monocytes produce a number of specific mitogenic, pro-inflammatory, anti-inflammatory and chemokinetic cytokines [27].

PGDF is one such key mitogenic product of macrophages, accounting for much of the growth promoting activity produced by lavage-derived human macrophages [7]. PDGF-B mRNA can be detected in human lung macrophages, macrophages derived from cultured monocytes [28, 29], and in macrophages derived from human monocytes induced to differentiate *in vitro* [30]. Regardless of the cellular source, macrophage PDGF appears to act primarily as a potent fibroblast mitogen [31]. Macrophage PDGF-B is also chemotactic for fibroblasts [32] and causes fibroblast mediated wound contraction [33]. There is mounting evidence for the importance of PDGF-B in pulmonary fibrosis in human patients. Alveolar macrophages from patients with pulmonary fibrosis have an increased rate of transcription of PDGF [34], an increased amount of PDGF-B mRNA [35] and exhibit spontaneous secretion of PDGF proteins [36].

Using *in situ* hybridization, Antoniades and colleagues have examined lung biopsies of patients with fibrotic lung disease to localize PDGF-B gene expression [37]. The cells appearing to express PDGF mRNA most vigorously appeared to be macrophages and alveolar pneumocytes. This study also documented local increases of PDGF-B-containing protein by immuno-staining. Vignaud and colleagues [38] have further demonstrated that PDGF-B is prominent in interstitial macrophages in lung biopsies from patients with lung fibrosis. This finding is true even in areas with little fibrosis, suggesting PDGF accumulation precedes focal remodelling. Such a finding argues for a causal role for PDGF in the pathogenesis of fibrosis.

In both clinical studies and in experimental models, PDGF-B-activity increases in the lungs within days of acute injury. In studies of patients with adult respiratory distress syndrome, Snyder and colleagues [39] have reported that lavage fluid contains soluble PDGF-related peptides. However, they described anaomalous molecular weights for this PDGF-like material (See below). In a similar vein, there is an increase in the steady-state level of PDGF-B mRNA and PDGF-like protein in the lungs of rats exposed to hyperoxia [40].

Fabisiak and colleagues found evidence of increased PDGF-B (c-*sis*) gene expression after oxidant injury induced by exposure of rats to 85% oxygen [41]. Low levels of PDGF-B mRNA were present at baseline in rats exposed only to room air. Steady-state levels of PDGF-B mRNA rose 2.5-fold within 3 days and remained elevated for at least a week. The rise in mRNA preceded evidence of generalized DNA synthesis and cell proliferation, in accord with a mitogenic role for PDGF in this model. In further investigations focusing on actin gene expression, these investigators have tracked the changes in actin during the response to hyperoxia. Cytoplasmic (β, γ) actin mRNA (2.1 kb) doubled in exposed lungs. Additionally, an mRNA which appeared to be a 1.7 kb muscle-specific α-actin was detected on day 7, consistent with hyperplasia of smooth muscle and/or myofibroblasts.

The triggers responsible for inducing macrophage PDGF-B gene expression in the setting of chronic inflammation and fibrosis are not known. Several possible mechanisms can be speculated upon. First, lymphocytes are prominent cells in these chronic inflammatory responses. This has led investigators to focus on the production of interferon-γ (IFN-γ) in these settings. IFN-γ is a product of T-cells of the TH_1 phenotype [42] and its presence has been reported in some patients with lung fibrosis [43]. IFN-γ clearly induces increased steady state levels of PDGF-B mRNA in lung macrophages [35]. Moreover, antigen-stimulated lymphocytes release a factor which causes macrophages to increase steady state levels of PDGF-B mRNA. This response is prevented by antibodies to IFN-γ [35].

Similar results have been observed in bleomycin-induced pulmonary fibrosis. Following bleomycin instillation in rats, cultures of lymphocytes obtained by lavage release one or more factors capable of inducing alveolar macrophages to release fibroblasts mitogens [44]. Thus, one hypothesis to explain the transition from chronic inflammation to fibrosis would be that antigen-stimulated TH_1-lymphocyte activation mediates the release of IFN-γ.

Physicochemical stimuli can also induce macrophages to release PDGF-like peptides as well. Cellular adherence to a substratum has been shown to cause PDGF gene activation in monocytes [29]. Similarly, activation can occur when cells are exposed to inorganic particles such as silica or asbestos fibres. At least *in vitro* such inorganic particles induce PDGF secretion by macrophages [45]. Additionally, exposure to immune complexes *in vitro* increases secretion of PDGF-BB by alveolar macrophages [46].

While most investigators have focussed on PDGF-B production by macrophages, PDGF-A subunits may also be produced by macrophages. An increase in PDGF-A transcription and mRNA abundance has been demonstrated in alveolar macrophages retrieved from subjects with pulmonary fibrosis. However, the abundance of PDGF-A mRNA is 10-fold less than the abundance of PDGF-B mRNA [34].

3.2. Release of PDGF by Airway Structural Cells

Like migratory inflammatory cells which traffic through the lung, resident structural cells of the mammalian lung also produce PDGF. For example, cultured lung fibroblasts themselves secrete a PDGF-like cytokine [47]. Fibroblasts cultured from rat lungs secrete *in vitro* a PDGF-like mitogenic cytokine which in turn causes proliferation of lung fibroblasts. The cytokine(s) responsible for this activity acts as a competence factor for murine BALB/c 3T3 cells in the classic bioassay for PDGF. By way of confirmation, anti-PDGF antibodies added to the fibroblast cultures blocks much, but not all, of this mitogenic activity. At the same time northern blots of whole cell RNA isolated from cultured fibroblasts detect mRNA for the PDGF-A chain in those cultures which secreted the PDGF-like activity but not in non-secreting lines. No PDGF-B chain gene expression was apparent in the lung fibroblast lines examined. Although developmental differences between foetal and adult cells might be expected based on what is known about the ontogeny of autocrine signalling, both the foetal and adult rat lung fibroblasts examined produce cytokine. Rat foetal lung fibroblasts produced significantly more PDGF-like cytokine than adult rat fibroblasts. Unlike the rat fibroblast lines examined, however, neither foetal nor adult human lines secreted a comparable activity. It is likely that these observations regarding PDGF secretion would extend to airways smooth muscle cells as well; however no such studies have been forthcoming at this time.

It is of interest that several cytokines act indirectly as mitogens for mesenchymal cells through the induction of PDGF. Transforming Growth Factor-β (TGF-β) acts as a bimodal growth modulator, stimulating proliferation at low concentrations (0.01–5 ng/ml) but repressing it at slightly higher concentrations (5–20 ng/ml) [48]. Initial studies suggested that TGF-β induces proliferation of connective tissue cells at low concentrations by stimulating PDGF-AA secretion, which then acts in an autocrine fashion on the secreting cell. In contrast, at higher concentrations of TGF-β, proliferation is decreased by down-regulation of PDGF α receptor and also be direct growth inhibitory effects. The same bimodal mitogenic response to varying concentrations of TGF-β have been seen in lung fibroblasts in the author's laboratory. Such behaviour reinforces the point that simple extrapolation from *in vitro* data to the whole organism is fraught with hazard.

3.3. Connective Tissue Growth Factor (CTGF)

Recently, investigators have called into question whether the PDGF peptides, described above, truly account for the mitogenic activity for smooth muscle cells and fibroblasts released by cell types other than platelets. While purified forms of native or recombinant PDGF dimers are clearly mitogenic and chemotactic when tested *in vitro*, many studies have ascribed mitogenic activity to these peptides based solely on circumstantial arguments: changing mitogenic activity released by cells in conjunction with altered mRNA levels for PDGFs have often been interpreted to indicate altered PDGF secretion. Missing from these analyses have been the rigorous isolation and sizing of the released cytokines.

A series of studies from the laboratory of G.R. Grotendorst point to the existence of several additional unrelated cytokine molecules which nevertheless act through the PDGF receptor [3, 49, 50]. These unique cytokines appear to differ from PDGF in being monomeric peptides. They have unique coding sequences quite distinct from the classical forms of PDGFs. They have been detected because they interact with one or both of the known PDGF receptors and also share antigenic determinants with the classic PDGFs. Hence they can be considered to be "mimitopes" of the PDGFs rather than true PDGF gene family members. The existence of such mimitopes would accord well with previous studies demonstrating that neutralising antibodies to PDGFs block the mitogenic activities of many cells.

Two such peptide cytokines have now been described. The first, a PDGF-like material secreted by peripheral blood monocytes, appears to be a unique product of such cells. It lacks inter-chain disulphide bridges and behaves as a 16 kDa monomer under non-reducing conditions. Moreover, it exhibits a different sensitivity pattern to formic acid as well as a different pattern of cyanogen bromide cleavage products compared to dimeric PDGF-A and -B chains [49]. Peptide moieties of this size have been noted in lavage fluid from subjects with ARDS [39], suggesting that it is released by alveolar and airway macrophages in acute lung injury.

The same investigators have also described a peptide of molecular weight higher than expected for authenic PDGF. This larger peptide has been named Connective Tissue Growth Factor (CTGF) [51]. A peptide of 34–38 kDa in size, CTGF was originally found in wound fluid along with the 16–17 kDa monocyte peptide but the cellular origin of this peptide in surgical wounds was not immediately established [50]. Additionally, PDGF related peptides in the 36–39 kDa range have been found to be secreted by foreskin fibroblasts upon treatment with TGF-β [52]. Indeed, Grotendorst has argued that CTGF rather than PDGF is the primary mitogenic peptide induced by TGF-β stimulation of these cells.

Undoubtedly, the current uncertainty on this issue is heightened by the observation that the release of these proteins is accompanied by increased PDGF-A gene expression in stimulated fibroblasts. A 38–40 kDa peptide which is immunoreactive with anti-PDGF antibodies but is a product of a gene distinct from the PDGF A and B genes has also been found to be produced by human umbilical vein endothelial cells [51]. Interestingly, CTGF supports normal wound healing in the absence of detectable authentic PDGF [50], suggesting that its efficacy rivals that described for PDGF.

To substantiate the role of CTGF-like peptides in lung remodelling, their presence in bleomycin-induced injury and remodelling in the rat has recently been sought [53]. Lung epithelial lining fluid harvested by lavage contains at least two platelet-derived growth factor-like peptides which appear sequentially over time. PDGF-like peptides from epithelial lining fluid were isolated and partially purified by cation exchange chromatography. Isolated peptides were then analyzed by immuno-blotting to determine their molecular weight and immunologic identity, probing the blots with polyclonal antibodies to PDGF-BB and PDGF-AA. PDGF-like peptides of two distinct size classes (38–40 kDa and 29 kDa) were present in alveolar fluid from all rats with lung injury induced by bleomycin but not control rats. The 38–40 kDa peptide was detected only with anti-PDGF-BB antibody. The 29 kDa peptide was detected only with anti-PDGF-AA antibody. The apparent amounts of these two peptides varied independently with time after exposure to bleomycin. The 38–40 kDa peptide peaked at 3–6 days. In contrast, the 29 kDa peptide was present at all times following injury, but with far less variation over time.

In parallel with these immuno-assays for PDGF-like molecules, it was found that there was abundant mitogenic activity for fibroblasts in concentrated epithelial lining fluid following the bleomycin-induced injury. The amount of growth promoting activity paralleled temporally the amount of the 38–40 kDa peptide detected on immuno-blots: at day 3 the mitogenic activity of chromatographically purified PDGF-like peptides was equivalent in potency to 12.1 ng/ml of authentic PDGF.

Anti-PDGF-BB antibodies blocked most but not all of the activity, whereas anti-PDGF-AA antibodies blocked only a small amount of the activity. This activity was consistent biologically and biochemically, and immunologically with Connective Tissue Growth Factor (CTGF).

Studies such as these suggest the CTGF may be a critical mitogenic activity and that earlier studies ascribing to PDGF the major role of mitogen for smooth muscle cells and fibroblasts in the lung must be revised. They clearly bring into question previous studies in which PDGF had been tentatively identified as the mitogen centrally responsible for focal cell proliferation.

A study of Bonner and colleagues showed that rat alveolar macrophages stimulted with chrysotile asbestos or carbonyl iron spheres *in vitro* produce PDGF peptides in the molecular weight range of 30–40 kDa and 16–18 kDa as determined by gel filtration chromatography [54]. Analysis by immunoassay suggested that the peptides in each molecular weight range consisted of a mixture of PDGF-AA, -AB, and -BB. While these authors ascribed the activity to authentic PDGF, a role for CTGF in these studies cannot be ruled out.

4. Cytokines and Smooth Muscle Cells: Lessons from Vascular Biology

While there is a paucity of information regarding the interaction of mitogenic cytokines for smooth muscle cells of airways, considerable information has come forward during the past decade regarding the functions of cytokines on vascular smooth muscle cells. It is more than likely that many of the insights from these studies will prove directly relevant to the biology of airways smooth muscle. These studies, therefore, bear review here.

Focal injury of vessel walls induces a prompt repair response involving coordinated modulation of cytokines and receptors. For example, rat smooth muscle cells of the carotid artery proliferate during the repair process that follows focal balloon catheter injury of vessel walls. Majesky and colleagues examined whether cells in damaged arteries might produce their own reparative growth factors. They hypothesized that such locally acting growth substances then might stimulate smooth muscle cell replication [55]. They found that the genes for both the PDGF-A and PDGF-B ligands as well as their receptor subunits are expressed in normal and injured carotid arteries. Moreover, they are independently regulated during the sequential stages of repair. Two phases of PDGF ligand and receptor gene expression were observed. Within hours of wounding, there was a large decrease in PDGF β receptor mRNA levels followed by a 10-fold increase in PDGF-A transcript levels. No change in PDGF α receptor or PDGF-B gene expression was found at these times. Two weeks after injury neointimal tissue had lower levels of PDGF α receptor mRNA and higher levels of PDGF β receptor mRNA (3- to 5-fold) than did restored media. *In situ* hybridization studies identified a sub-population of neointimal cells localized near the luminal surface with a different pattern of gene expression than the underlying carotid smooth muscle cells. Luminal vascular cells were strongly positive for PDGF-A and PDGF β receptor transcripts, but showed little or no hybridization for PDGF-B and PDGF α receptors. Immunohistochemical studies confirmed strongly positive staining for PDGF-A in cells along the luminal surface. Hence changes in PDGF ligand and receptor expression occur at specific times and locations in the injured carotid artery.

Because of their prominent role in the early phase of vascular wound repair, platelets had been thought to be a key local source of PDGF. Schwartz and colleagues showed that growth of certain smooth muscle cells depends on mechanisms independent of release of growth factors by platelets [56]. For their study, rat aortic smooth muscle cells were isolated under conditions which precluded their exposure to PDGF. Unlike cells from newborn animals or from injured vessels, smooth muscle cells so derived from adult rats did not produce PDGF, but did exhibit PDGF receptors. These PDGF-independent cells responded to PDGF when all other components of serum were absent. However, growth proceeded equally well in low concentrations of serum, independent of the presence of PDGF or similar growth factors detectable by growth-promoting assays using either 3T3 cells or smooth muscle cells whose growth is dependent on factors present in whole blood serum. This further emphasizes the importance of resident structural cells in producing locally acting mitogens.

As a consequence of its activities on smooth muscle cells, PDGF-like peptides appear to be central mediators of the wound healing process. Indeed, addition of recombinant PDGF-BB to wounds accelerates healing [57]. In an animal model of vascular neointimal smooth muscle proliferation induced by mechanical large vessel de-endothelialization in athymic rats, the application of anti-PDGF antibodies prevents vascular re-stenosis or occlusion [58]. During pathological skin fibrosis such as occurs in systemic sclerosis, there are increased steady-state levels of PDGF-B mRNA in skin macrophages [59] as well as a rise in cytoplasmic PDGF protein [60].

Jawien and colleagues questioned whether the *in vitro* observations that PDGF is a smooth muscle mitogen and chemoattractant reflect the situation *in vivo* [61]. They infused recombinant PDGF-BB into rats which had undergone focal injury to the carotid artery. Exogenously administered PDGF-BB produced a several-fold increase in medial cell proliferation and greatly increased the intimal thickening and the migration of smooth muscle cells from the media to the intima during the first week after injury.

Kraiss and colleagues have recently studied the expression and regional distribution of the PDGF protein isoforms and their receptors in a prosthetic vascular graft model [62]. Healing baboon synthetic grafts express PDGF mRNA in their neointimal layer. Both PDGF-A and PDGF-B were identified in macrophages within the interstices of the graft. However, the neointima contained predominantly PDGF-A antigen localized to both the endothelial surface and to the subjacent smooth muscle cell layers. Tissue extracts of neointima and graft material were active as mitogens for smooth muscle cells. Most of this proliferative activity was blocked by a neutralizing anti-PDGF antibody. Further, PDGF receptor subunit α and β proteins were detectable in the neointima and graft material. Moreover, the coexpression of ligand and receptor in the macrophage-rich matrix suggests that PDGF may participate in the foreign body response.

There are interesting interplays between the classical endocrine hormones and the mitogenic cytokines. The process of vascular intimal thickening in response to injury is inpaired in the absence of a functioning hypothalamic-pituitary axis. Fingerle and colleagues examined the cell kinetics of vascular intimal thickening in the balloon catheter model of injury in the rat [63]. In hypophysectomized rats the stimulation of thymidine labelling normally seen 2 days after injury of the carotid artery was almost completely eliminated. Total DNA content of the developing neointima two weeks after injury was only thirty percent of the values found in ballooned carotid arteries of normal rats. The proliferative response was not restored by treating hypophysectomized rats with recombinant human growth hormone. The investigators proposed two possible general explanations for the reduction of proliferation in hypophysectomized rats: smooth muscle cells may have lost the capacity to respond to the stimulus of injury; alternately, the smooth muscle response may require the presence of pituitary-dependent factors. Transplantation experiments were performed *in vivo* to distinguish between these possibilities. Carotid arteries in inbred Lewis rats were excised 1 hour after balloon injury to give platelets the opportunity to adhere. These vessels were than transplanted from rats into control animals and vice versa. Proliferation of smooth muscle cells in control-to-hypophysectomized transplants was markedly reduced compared to control-to-control transplants, whereas vessels from hypophysectomized rats increased their indices toward normal levels if transplanted into control animals.

5. Cytokines as Physiological Regulators

Protein cytokines are involved in normal tissue growth, homeostasis and senescence. All nucleated cells produce and secrete cytokines throughout their life span. Moreover, cytokines modulate multiple cellular functions in response to external challenge such as in inflammation and infection or wound repair. Although produced by a variety of cell types, the triggers that induce the production of a specific cytokine differ between cells. In the past several years, it has become apparent that many of the cytokines share regions of homologous nucleic acid sequences, making them members of larger cytokine gene families. Functionally, cytokines can be described as mitogenic, anabolic/ catabolic, anti- or pro-inflammatory. However, the pleiotropic nature of cytokines assures that such definitions invariably fail to cover all the functions of a single molecule. This phenomenon of nomenclature continues to confound the naming of newly discovered cytokines.

Most *in vitro* studies to date have not been able to take into account the obvious fact that cells and tissues are exposed to complex cytokine mixtures rather than to individual cytokines. Recent attention however has turned to understanding how cytokines interact. Interactions between cytokines *in vitro* and *in vivo* have invariably proven to yield results which could not be predicted based on knowledge of the separate actions of the individual cytokines involved. This complexity has led to the concept of cytokine networks [64]. Moreover, the constitutive expression of cytokines varies characteristically in detail in different organs in health [65].

Cytokine production by lung fibroblasts has recently been shown to be modulated directly by infectious agents. To demonstrate this, Fabisiak and colleagues isolated human lung fibroblasts from transbronchial biopsy specimens and assayed for the production of IL-6 and GM-CSF [66]. Fibroblasts were isolated from lung allografts, recipient lungs obtained at the time of transplant, and normal lung tissue removed during tumor resection. Several early-passage fibroblasts from transplant recipients contained mycoplasma-like organisms. Secreted IL-6 and GM-GSF were elevated over 50-fold in infected fibroblast lines but not the noninfected control lines. Removal of mycoplasma by antibiotic treatment of cultures reduced the production of these cytokines. Moreover, addition of TNF-β, a known inducer of IL-6 and GM-CSF secretion, had an enhanced effect in infected cultures. The observation that infected lung fibroblasts can induce inflammatory cytokines necessitates rigorous control of clinically obtained specimens. It seems likely that mitogenic cytokines may be similarly modulated by infectious agents; this possibility remains unexplored. This study should provide a cautionary note, particularly for investigators studying potentially infected airways cells from human subjects.

6. Cytokines and Scarring of Airways

Connective tissue is integral to the normal functioning of the airways and distal lung. Indeed, lung contains more connective tissue as a

proportion of total weight than any other soft tissue organ of the body
[67]. In pathological states, excess amounts of connective tissue can be
deposited at several levels in the airways: large and small airways as well
as in alveolar capillary membranes and airspaces. Excess peribronchial
deposition of matrix and basement membrane components may occur
in airways diseases and undoubtedly influences local physiology.

Roche and colleagues have reviewed the controversial issue of peribronchial basement membrane thickening in the small airways of asthmatics [68]. Their findings suggest that interstitial collagen deposition
accounts for the apparent basement membrane thickening. Using immunohistochemical and electron microscopic approaches, they found
that the subepithelial areas contained excess amounts of types I, III and
V collagens and fibronectin. Based on the spacial relationship between
areas of matrix deposition and cellular injury, the investigators surmised
that fibroblasts rather than damaged epithelial cells were responsible for
the local matrix changes. Unanswered in their studies are questions
regarding the physiological consequences of this process, which may be
important determinants of irreversible airway function in asthma.

The fibrogenic cytokines are assuredly important mediators of this
fibrosing process in small airways. In this regard, the members of the
TGF-β family are the most potent of the fibrogenic cytokines (For
review see [69–71]). The TGF-βs are pleiotrophic growth modulating
protein cytokines notable for their capacity to modulate a wide range of
cellular behaviours. At least five closely related TGF-β molecules, each
a distinct gene product, have now been described.

TGF-βs induce proliferation of certain target cells such as smooth
muscle cells and fibroblasts [72], albeit through an indirect mechanism
involving activation of PDGF-like molecules [52]. They inhibit proliferation of epithelial and endothelial cells, regulate multiple aspects of
connective tissue turnover and influence numerous control points in the
immune and haematopoietic systems. Upon binding of TGF-β to one of
its several cognate receptors, genes for a host of connective tissue matrix
proteins become up-regulated [73]. However, at least one of the matrix
proteins, decorin, appears to be a potent inactivator of TGF-β [74].
Hence an extracellular feedback loop is possible in response to TGF-β.
Clearly, other cytokines contribute indirectly to the fibrogenic process
by serving as chemoattractants for inflammatory cells, by promoting
maturation of proliferation in situ of cells involved in extracellular
matrix production, or through other mechanisms. Nevertheless, TGF-β
remains the most potent direct inducer of connective tissue production
among the cytokines.

Within wounds TGF-β stimulates granulation tissue formation in
vivo, promotes scar formation in wounds and enhances wound strength.
In this regard, however, it is of interest that TGF-β plays no role in

healing of foetal wounds. It is perhaps for this reason that foetal wounds heal without scar formation [75].

In addition to fibroblasts and smooth muscle cells, a variety of pulmonary cells have been shown to secrete TGF-β. These include among others macrophages [76], mast cells [77], human pulmonary endothelial cells exposed to bleomycin [78], and polymorphonuclear leukocytes [79]. All three major TGF-β isoforms (TGF-β1-3) are present in normal murine lung and appear to play an important role during development [80]. Immunohistochemical techniques detect TGF-β protein in the bronchial epithelium; TGF-β mRNA transcripts for each are present in fibroblasts and smooth muscle cells of the airway wall. Direct evidence for the specific involvement of TGF-β in peribronchial fibrosis does not exist. However, TGF-βs have clearly been implicated in diseases involving focal fibrotic derangement. These include skin, eye, liver and heart. The precise role of TGF-β as a fibrogenic agent in the distal lung has been examined in the context of spontaneous (idiopathic) pulmonary fibrosis.

Pulmonary macrophages store TGF-β in significant amounts [79]. This observation may explain why they so often stain positively in immuno-assays. Macrophage TGF-β production and secretion can be modulated by exposure of cells to other cytokines; by way of example, PDGF induces expression of TGF-β in a rat macrophage cell line [81]. Many aspects of TGF-β biology indicate it to be an anti-inflammatory cytokine. Recently, a TGF-β_1 deficient mouse strain has been developed in which unbridled inflammatory responses, particularly in lung and heart, lead to early mortality [82].

The amounts of TGF-β present are easily detected by immunohistochemistry in models of inflammation and remodelling process such as that induced by intratracheal instillation of bleomycin [83]. Khalil and colleagues measured tissue levels of TGF-β following instillation of bleomycin to rats. Easily detectable (microgram quantities) of TGF-β were present with a peak level seven days following a single bleomycin instillation. Interestingly, from the point of view of airway disease, TGF-β1 could be detected in bronchiolar epithelial cells and the adjacent matrix. Macrophages stained intensely positive at the same time as peak tissue levels of TGF-β (3–10 days). Whether bronchiolar epithelial cells and macrophages account quantitatively for the high tissue levels of extractable TGF-β following bleomycin could not be determined in this immunocytochemical study.

Several studies have probed the possible role of TGF-β in the genesis of pulmonary fibrosis in humans. In a study focusing on advanced remodelling associated with spontaneous pulmonary fibrosis, Khalil and colleagues [84] have confirmed in human subjects that bronchiolar epithelial cells appear to be a major source of TGF-β. Broekelmann and colleagues have delineated the sites of production of TGF-β1 mRNA in lung biopsy samples of patients with pulmonary fibrosis using the technique of *in situ* hybridization [85]. In biopsies from cases of chronic pulmonary fibrosis, TGF-β1 mRNA is localized to areas where there is intense macrophage aggregation. These areas are also rich in mRNA for fibronectin but not type I procollagen. However, attempts to detect cytokine protein using immunohistochemical techniques failed to demonstrate commensurate increases in stainable TGF-β1 protein in these macrophage aggregates. In contrast, areas such as alveolar buds populated predominantly by fibroblasts were relatively devoid of TGF-β mRNA but contained mRNAs for procollagen I and fibronectin. Such fibroblast-rich areas also stained positively for TGF-β protein. A cautionary note regarding these studies is warranted: only one isoform – TGF-β1 – of transforming growth factor was examined and it is certainly possible that other isoforms may prove important actors in the process as well.

In contrast to these histochemical studies, *in vitro* studies have suggested that TGF-β gene expression is not up-regulated in alveolar macrophages harvested from subjects with pulmonary fibrosis even though PDGF-B gene is [35]. Moreover, incubation of macrophages with the stimulant IFN-γ was capable of up-regulating PDGF-B gene expression but not that of TGF-β.

In certain diseases, specific isoforms of TGF-β are key mediators. Kulozik and colleagues [86] have shown by *in situ* hybridization that there is co-localization of TGF-β2 mRNA and proα1(I) collagen expression around dermal blood vessels in patients with the inflammatory stage of systemic sclerosis. There was no expression of either gene in the dermis of patients with the fibrotic stage of the same disorder or in control patients. TGF-β1 did not appear to play a role in this process.

There are likely to be important interactions between TGF-β and opposing cytokines such as TGF-α in fibrosing lung disorders. Such a pattern of opposing actions has been demonstrated in remodelling that occurs in other tissues. TGF-β and TGF-α have a number of opposing activities in modulating cell proliferation and phenotype. Castilla and colleagues examined liver remodelling in alcoholic cirrhosis [87]. They contrasted the tissue expression of TGF-β1 mRNA in this irreversible remodelling process of cirrhosis with that in normal or fatty livers. Simultaneous determinations of serum procollagen III peptide levels provided a surrogate marker of on-going fibrotic activity. Hepatic TGF-β1 mRNA levels correlated closely over time with tissue procollagen I mRNA and the serum procollagen III peptide levels. Tissue TGF-β1 mRNA levels were increased 2- to 14-fold above control. Moreover, treatment with IFN-α in 5 patients normalized the tissue levels of TGF-β1 mRNA. Of note was the observation that TGF-α mRNA was increased in all cases in which successful hepatic regeneration occurred. This latter observation is of particular interest, given that the latter cytokine often opposes the action of TGF-β in cultures of epithelial cells.

7. Future Directions

It is clear that the recognition of cytokines as signalling molecules has advanced our understanding of respiratory cell biology considerably during the past decade. An understanding of their biology has allowed recognition of their role in specific systemic and local pulmonary diseases such as hypereosinophilic syndrome [88] and asthma [89].

There is a need to understand the genetic bases for inter-individual variation in cytokine responses. For example, one study has indicated that certain individuals are more susceptible to up-regulation of cytokine production. Bitterman and colleagues found that unaffected relatives of subjects with pulmonary fibrosis exhibited enhanced insulin-like growth factor (IGF-I) release from lavage-derived macrophages [90]. The underlying genetic defect for this hyper-senstitive cytokine production state remains unclear.

Until further studies in such kindreds are carried out, it will not be clear whether other specific cytokines are also upregulated. However, a recent study of transplant patients lends support to the notion of important inter-individual differences in cytokine responsiveness [91]. Plasma TGF-β concentrations measured in women with advanced breast cancer could be used to predict the likelihood of development of liver of lung fibrosis after bone marrow transplantation.

Understanding of the structure and functions of cytokines has opened avenues for pharmacological intervention designed to augment their therapeutic benefits and prevent their untoward effects. Our evolving understanding of cytokine biology of the airways has provided insight into the action of known drugs. Strategies to block or augment cytokine actions have already entered clinical testing and practice.

Finally, much more needs to be learned about the potential and limits of cytokine action *in vivo*. Strategies which involve introduction of genetic material into tissues are in experimental use. Transgenic animals have been generated for studies of the *in vivo* effects of cytokines and cytokine receptors. In initial studies, transgenic animals have demonstrated the effects of excess production of specific cytokines.

References

1. Jetten AM. Growth and differentiation factors in tracheobroncial epithelium. Am J Physiol 1991; 260: L361–73.
2. Kelley J. State of the Art: Cytokines of the lung. Am Rev Resp Dis 1990; 141: 765–88.

3. Ross R, Raines EW, Bowen-Pope DF. The biology of platelet-derived growth factor. Cell 1986; 46: 155–69.
4. Raines EW, Bowen-Pope DF, Ross R. Platelet-derived growth factor. In Peptide Growth Factors and their Receptors. In: Sporn MB, Roberts AB, editors. Handbook of Experimental Pharmacology. Vol. 95. Heidelberg: Springer-Verlag, 1990: 173–262.
5. Grotendorst GR, Martin GR, Pencev D, Sodek J, Harvey AK. Stimulation of granulation tissue formation by platelet-derived growth factor in normal and diabetic rats. J Clin Invest 1985; 76: 2323–9.
6. Pledger WJ, Stiles CD, Antoniades HN, Scher D. An ordered sequence of events is required before BALB/c-3T3 cells become committed to DNA synthesis. Proc Natl Acad Sci USA 1977; 74: 4481–90.
7. Shimokado K, Raines EW, Madtes DK, Barrett TB, Benditt EP, Ross R. A significant part of macrophage-derived growth factor consists of at least two forms of PDGF. Cell 1985; 43: 277–86.
8. Antoniades HN, Hunkapiller MW. Human platelet-derived growth factor (PDGF): amino terminal amino acid sequence. Science 1983; 220: 963–5.
9. Johnsson A, Heldin C-H, Wasteson A, Westermark B, Deuel TF, Huang JS, et al. The c-sis gene encodes a precursor of the B-chain of platelet-derived growth factor. EMBO J. 1984; 3: 921–8.
10. Hart CE, Forstrom JW, Kelly JD, Seifert RA, Smith RA, Ross R, et al. Two classes of PDGF receptor recognize different isoforms of PDGF. Science 1988; 240: 1529–31.
11. Gronwald RGK, Grant FJ, Haldeman BA, Hart CE, O'Hara PJ, Hage FS, et al. Cloning and expression of a cDNA coding for the human platelet-derived growth factor receptor: evidence for more than one receptor class. Proc Natl Acad Sci USA 1988; 85: 3435–9.
12. Claesson-Welsh L, Hammacher A, Westermark B, Heldin C-H, Nistér M. Identification and structural analysis of the A type receptor for platelet-derived growth factor. J Biol Chem 1989; 264: 1742–7.
13. Claesson-Welsh L, Eriksson A, Westermark B, Heldin C-H. cDNA cloning and expression of the human A type PDGF receptor establishes similarity to the B type receptor. Proc Natl Acad Sci USA 1989; 86: 4917–21.
14. Shepherd VL. Cytokine receptors of the lung. Am J Resp Cell Mol Biol 1991; 5: 403–10.
15. Zimmerman GA, Lorant DE, McIntyre TM, Prescott SM. Juxtacrine intercellular signalling: Another way to do it. Am J Resp Cell Mol Biol 1993; 9: 573–7.
16. Keating MT, Williams LT. Autocrine stimulation of intracellular PDGF receptors in v-sis-transformed cells. Science 1988; 239: 914–6.
17. Bejcek BE, Li DI, Deuel T. Transformation by v-sis occurs by an internal autoactivation mechanism. Science 1989; 245: 1496–9.
18. Zachary I, Rozengurt E. Modulation of epidermal growth factor receptor by mitotic ligands: effects of bombesin and the role of protein kinase C. Cancer Surv 1985; 4: 729–65.
19. Seifert RA, Hart CE, Phillips PE, Forstrom JW, Ross R, Murray MJ, Bowen-Pope DF. Two different subunits associate to create isoform-specific platelet-derived growth factor receptors. J Biol Chem. 1989; 264: 8771–8.
20. Ferns GA, Sprugel KH, Seifert RA, Bowen-Pope DF, Kelly JD, Murray M, Raines EW, Ross R. Relative platelet-derived growth factor receptor subunit expression determines cell migration to different dimeric forms of PDGF. Growth Factors 1990; 3: 315–24.
21. Kelly JD, Haldeman BA, Grant FJ, Murray MJ, Seifert RA, Bowen-Pope DF, et al. Platelet-derived growth factor (PDGF) stimulates PDGF receptor subunit dimerization and intersubunit trans-phosphorylation. J Biol Chem 1991; 266: 8987-92.
22. Williams LT. Signal transduction by the platelet-derived growth factor receptor. Science 1989; 243: 1564-70.
23. LaRochelle WJ, Giese N, May-Siroff M, Robbins KC, Aaronson SA. Molecular localization of the transforming and secretory properties of PDGF A and PDGF B. Science 1990; 248: 1541–4.
24. Gronwald RGK, Seifert RA, Bowen-Pope DF. Differential regulation of expressing of two platelet-derived growth factor receptors by transforming growth factor-β. J Biol Chem. 1989; 264: 8120–5.
25. Seifert RA, van Koppen A, Bowen-Pope DF. PDGF-AB requires PDGF receptor alpha-subunits for high-affinity, but not for low-affinity, binding and signal transduction. J Biol Chem. 1993; 268: 4473–80.

26. Shaw RJ, Kelley J. Monocyte-Macrophages in Pulmonary Fibrosis. In: Phan SH, Thrall RS, editors. Pulmonary Fibrosis. New York: Marcel Dekker, 1994: 405–44.

27. Rappolee DA, Mark D, Banda MJ, Werb Z. Wound macrophages express TGF-α and other growth factors *in vivo*: analysis by mRNA typing. Science 1988; 241: 708–12.

28. Mornex J, Martinet Y, Yamauchi K, Bitterman PB, Grotendorst A, Chytil-Weir A, Martin GR, Crystal RG. Spontaneous expression of the cis-gene and release of platelet-derived growth factor-like molecule by human alveolar macrophages. J Clin Invest 1986; 78: 61–6.

29. Shaw RJ, Doherty DE, Ritter AG, Benedict SH, Clark AF. Adherence-dependent increase in human monocyte PDGF(B) mRNA is associated with increases in *c-fos, c-jun*, and EGR2 mRNA. J Cell Biol 1990; 111: 2139–48.

30. Pantazis P, Lanfrancone L, Pelicci PG, Dalla-Favera R, Antoniades HN. Human leukemia cells synthesize and secrete proteins related to platelet-derived growth factor. Proc Natl Acad Sci USA 1986; 83: 5526–30.

31. Beckmann MP, Betsholtz C, Heldin C-H, Westermark B, di Marco E, di Fiore PP, Robbins KC, Aaronson SA. Comparison of biological properties and transforming potential of human PDGF-A and PDGF-B chains. Science 1988; 241: 1346–9.

32. Nistér M, Hammacher A, Mellström K, Siegbahn A, Rönnstrand L, Estermark B, Heldin C-H. A glioma-derived PDGF A chain homodimer has different functional activities from a PDGF AB heterodimer purified from human platelets. Cell 1988; 52: 791–9.

33. Clark RAF, Folkvord JM, Hart CE, Murray MJ, McPherson JM. Platelet isoforms of platelet-derived growth factor stimulate fibroblasts to contract collagen matrices. J Clin Invest 1989; 84: 1036–40.

34. Nagaoka I, Trapnell BC, Crystal RG. Upregulation of platelet-derived growth factor-A and -B gene expression in alveolar macrophages of individuals with idiopathic pulmonary fibrosis. J Clin Invest 1990; 85: 2023–7.

35. Shaw RJ, Benedict SH, Clark RA, King TE, Jr. Pathogenesis of pulmonary fibrosis in interstitial lung disease. Alveolar macrophage PDGF(B) gene activation and up-regulation by interferon gamma. Am Rev Resp Dis 1991; 143: 167–73.

36. Martinet Y, Rom WN, Grotendorst GR, Martin GR, Crystal RG. Exaggerated spontaneous release of platelet-derived growth factor by alveolar macrophages from patients with idiopathic pulmonary fibrosis. N Engl J Med 1987; 317: 202–9.

37. Antoniades HN, Bravo MA, Avila RE, Galanopoulos T, Neville-Golden J, Maxwell M, Selman M. Platelet-derived growth factor in idiopathic pulmonary fibrosis. J Clin Invest 1990; 86: 1055–64.

38. Vignaud J-M, Allam M, Martinet N, Pech M, Plenat F, Martinet Y. Presence of platelet-derived growth factor in normal and fibrotic lung is specifically associated with interstitial macrophages, while both interstitial macrophages and alveolar epithelial cells express the *c-sis* oncogene. Am J Resp Cell Mol Biol 1991; 5: 531–8.

39. Snyder LS, Hertz MI, Peterson MS, Harmon KR, Marinelli WA, Henke CA, et al. Acute lung injury. Pathogenesis of intraalveolar fibrosis. J Clin Invest 1991; 88: 663–73.

40. Han RNN, Buch S, Freeman BA, Post M, Tanswell AK. Platelet-derived growth factor and growth-related genes in rat lung. II. Effect of exposure to 85% O_2. Am J Physiol 1992; 262: L140–6.

41. Fabisiak JP, Evans JN, Kelley J. Increased expression of PDGF-B (*c-sis*) mRNA in rat lung precedes DNA synthesis and tissue repair during chronic hyperoxia. Am J Resp Cell Mol Biol 1989; 1: 181–9.

42. Kay AB. Origin of type 2 helper T cells. N Engl J Med 1994; 330: 567–9.

43. Robinson BW, Rose AH. Pulmonary gamma interferon production in patients with fibrosing alveolitis. Thorax 1990; 45: 105–8.

44. Kovacs EJ, Kelley J. Lymphokine regulation of macrophage-derived growth factor secretion following pulmonary injury. Am J Pathol 1985; 121: 261–268.

45. Schapira RM, Osornio-Vargas AR. Brody AR. Inorganic particles induce secretion of a macrophage homologue of platelet-derived growth factor in a density- and time-dependent manner *in vitro*. Exp Lung Res 1991; 17: 1011–24.

46. Martinet Y, Rom WN, Grotendorst GR, Martin GR, Crystal RG. Exaggerated spontaneous release of platelet-derived growth factor by alveolar macrophages from patients with idiopathic pulmonary fibrosis. N Engl J Med 1987; 317: 202–9.

47. Fabisiak JP, Absher M, Evans JN, Kelley J. Spontaneous production of PDGF A-chain homodimer by rat lung fibroblasts *in vitro*. Am J Physiol 1992; 263: L185–93.

48. Battegay EJ, Raines EW, Seifert RA, Bowen-Pope DF, Ross R. TGF-β induces bimodal proliferation of connective tissue cells via complex control of an autocrine PDGF loop. Cell 1990; 63: 515–24.
49. Pencev D, Grotendorst GR. Human peripheral blood monocytes secrete a unique form of PDGF. Oncogene Res 1988; 3: 333–42.
50. Matsuoka J, Grotendorst GR. Two peptides related to platelet-derived growth factor are present in human wound fluid. Proc Natl Acad Sci USA 1989; 86: 4416–20.
51. Bradham DM, Igarashi A, Potter RL, Grotendorst GR. Connective tissue growth factor: a cysteine-rich mitogen secreted by human vascular endothelial cells is related to the SRC-induced immediate early gene product CEF-10. J Cell Biol 1991; 114: 1285–94.
52. Soma Y, Grotendorst GR. TGF-β stimulates primary human skin fibroblast DNA synthesis via an autocrine production of PDGF-related peptides. J Cell Physiol 1989; 140: 246–53.
53. Walsh J, Absher M, Kelley J. Variable expression of platelet-derived growth factor (PDGF) family proteins in acute lung injury. Am J Resp Cell Mol Biol 1993; 9: 637–44.
54. Bonner JC, Osornio-Vargas AR, Badgett A, Brody AR. Differential proliferation of rat lung fibroblasts induced by the platelet-derived growth factor (PDGF)-AA, -AA, -AB, and -BB isoforms secreted by rat alveolar macrophages. Am J Resp Cell Biol 1991; 5: 539–47.
55. Majesky MW, Reidy MA, Bowen-Pope DF, Hart CE, Wilcox JN, Schwartz SM. PDGF ligand and receptor gene expression during repair of arterial injury. J Cell Biol 1990; 111: 2149–58.
56. Schwartz SM, Foy L, Bowen-Pope DF, Ross R. Derivation and properties of platelet-derived growth factor-independent rat smooth muscle cells. Am J Pathol 1990; 136: 1417–28.
57. Pierce GF, Mustoe TA, Lingelbach J, Masakowski VR, Griffin GL, Senior RM, Deuel TF. Platelet-derived growth factor and transforming growth factor-β enhance tissue repair activities by unique mechanisms. J Biol Chem 1989; 109: 429–40.
58. Ferns GAA, Raines EW, Sprugel KH, Motani AS, Reidy MA, Ross R. Inhibition of neointimal smooth muscle accumulation after angioplasty by an antibody to PDGF. Science 1991; 253: 1129–32.
59. Olsen DR, Uitto J. Differentiation expression of type IV procollagen and laminin genes by fetal vs. adult skin fibroblasts in culture; determination of subunit mRNA steady-state levels. J Invest Dermatol 1989; 93: 127–31.
60. Gay S, Jones RE Jr, Huang GQ, Gay RE. Immunohistologic demonstration of platelet-derived growth factor (PDGF) and sis-oncogene expression in scleroderma. J Invest Dermatol 1989; 92: 301–3.
61. Jawien A, Bowen-Pope DF, Lindner V, Schwartz SM, Clowes AW. Platelet-derived growth factor promotes smooth muscle migration and intimal thickening in a rat model of balloon angioplasty. J Clin Invest 1992; 89: 507–11.
62. Kraiss LW, Raines EW, Wilcox JN, Seifert RA, Barrett TB, Kirkman TR, et al. Regional expression of the platelet-derived growth factor and its receptors in a primate graft model of vessel wall assembly. J Clin Invest 1993; 92: 338–48.
63. Fingerle J, Faulmuller A, Muller G, Bowen-Pope DF, Clowes MM, Reidy MA, Clowes AW. Pituitary factors in blood plasma are necessary for smooth muscle cell proliferation in response to injury in vivo. Arterioscler Thromb 1992; 12: 1488–95.
64. Kohase M, May LT, Tamm I, Vilcek J, Sehgal PB. A cytokine network in human diploid fibroblasts: interactions of β-interferons, tumor necrosis factor, platelet derived growth factor, and interleuken-1.Mol Cell Biol 1987; 7: 273–80.
65. Tovey MG, Content J, Gresser I, Gugneheim J, Blanchard B, Guymarho J, et al. Genes for interferon β_2 (interleukin-6, tumor necrosis factor, and interleukin-1) are exposed at high levels in the organs of normal individuals. J Immunol 1988; 141: 3106–10.
66. Fabisiak JP, Weiss RD, Powell GA, Dauber JH. Enhanced secretion of immune-modulating cytokines by human lung fibroblasts during in vitro infection with Mycoplasma fermentans. Am J Resp Cell Mol Biol 1993; 8: 358–64.
67. Kelley J. Collagen. In: Massaro D, editor. Lung Cells Biology. New York: Marcel Dekker, 1989: 821–66.
68. Roche WR, Beasley R, Williams JH, Holgate ST. Subepithelial fibrosis in the bronchi of asthmatics. Lancet 1989; 1: 520–4.

69. Kelley J. Transforming Growth Factor-β. In: Kelley J, editor. Cytokines of the Lung. New York: Marcel Dekker, 1992: 101–37.
70. McCartney-Francis NL, Wahl SM. Transforming growth factor-β: a matter of life and death. J Leukocyte Biol 1994; 55: 401–9.
71. Border WA, Ruoslahti E. Transforming growth factor-beta in disease: the dark side of tissue repair. J Clin Invest 1992; 90: 1–7.
72. Kelley J, Fabisiak JP, Hawes K, Absher M. Cytokine signalling in lung: transforming growth factor-β secretion by lung fibroblasts. Am J Physiol 1991; 260: L123–8.
73. Rossi R, Karsenty G, Roberts AB, Roche NS, Sporn MB, de Crombrugghe B. A nuclear 1 binding site mediates the transcriptional activation of a type 1 collagen promoter by transforming growth factor-beta. Cell 1988; 52: 405–414.
74. Border WA, Noble NA, Yamamoto T, Harper JR, Yamaguchi Y, Pierschbacher MD, Rouslahti E. Natural inhibitor of transforming growth factor-β protects against scarring in experimental kidney disease. Nature 1992; 360: 361–4.
75. Shah M, Foreman DM, Ferguson MWJ. Control of scarring in adult wounds by neutralising antibody to transforming growth factor β. Lancet 1992; 339: 213–4.
76. Assoian RK, Fleurdelys BE, Stevenson HC, Miller PJ, Madtes DK, Raines EW, Ross R, Sporn MB. Expression and secretion of type β transforming growth factor by activating human macrophages. Proc Natl Acad Sci USA 1987; 84: 6020–4.
77. Pennington DW, Lopez AR, Thomas PS, Peck C, Gold WM. Dog mastocytoma cells produce transforming growth factor-β_1 J Clin Invest 1992; 90: 35–41.
78. Phan SH, Gharaee-Kermani M, Wolber F, Ryan US. Stimulation of rat endothelial cell transforming growth factor-β production by bleomycin. J Clin Invest 1991; 87: 148–54.
79. Grotendorst GR, Smale G, Pencev D. Production of transforming growth factor beta by human peripheral blood monocytes and neutrophils. J Cell Physiol 1989; 140: 396–402.
80. Pelton RW, Johnson MD, Perkett EA, Gold LI, Moses HL. Expression of transforming growth factor-β1, -β2, and -β3 mRNA and protein in the murine lung. Am J Resp Cell Mol Biol. 1991; 5: 522–30.
81. Pierce GF, Mustoe TA, Lingelbach J, Masakowski VR, Griffin GL, Senior RM, Deuel TF. Platelet-derived growth factor and transforming growth factor-β enhance tissue repair activities by unique mechanisms. J Biol Chem 1989; 109: 429–40.
82. Kulkarni AB, Huh CG, Becker D, Geiser A, Lyght M, Flanders KC, et al. Transforming growth factor beta 1 null mutation in mice causes excessive inflammatory response and early death. Proc Natl Acad Sci USA 1993; 90: 770–4.
83. Khalil N, Bereznay O, Sporn M, Greenberg AH. Macrophage production of transforming growth factor β and fibroblast collagen synthesis in chronic pulmonary fibrosis. J Exp Med 1989; 170: 727–37.
84. Khalil N, O'Connor RN, Unruh HW, Warren PW, Flanders KC, Kemp A, Bereznay OH, Greenberg AH. Increased production and immunohistochemical localization of transforming growth factor-β in idiopathic pulmonary fibrosis. Am J Resp Cell Mol Biol 1991; 5: 155–62.
85. Broekelmann TJ, Limper AH, Colby TV, McDonald JA. Transforming growth factor-β_1 is present at sites of extracellular matrix gene expression in human pulmonary fibrosis. Proc Natl Acad Sci USA 1991; 88: 6642–6.
86. Kulozik M, Hogg A, Lankat-Buttgereit B, Krieg T. Co-localization of transforming growth factor β2 with α1(I) procollagen mRNA in tissue sections of patients with systemic sclerosis. J Clin Invest 1990; 86: 917–22.
87. Castilla A, Prieto J, Fausto N. Transforming growth factors β1 and α in chronic liver disease. N Engl J Med 1991; 324: 933–40.
88. Cogan E, Schandené L, Crusiaux A, Cochaux P, Velu T, Goldman M. Brief report: Clonal proliferation of type 2 helper T cells in a man with the hypereosinophilic syndrome. N Engl J Med 1994; 330: 535–8.
89. Robinson DS, Hamid Q, Ying S, Tsicipoulos A, Barkans J, Bentley AM, et al. Predominant T_{H2}-like bronchoalveolar T-lymphocyte population in atopic asthma. N Engl J Med 1992; 326: 298–304.
90. Bitterman PB, Rennard SI, Keogh BA, Wewers MD, Adelberg S, Crystal RG. Familial idiopathic pulmonary fibrosis. Evidence of lung inflammation in unaffected family members. N Engl J Med 1986; 314: 1343–7.
91. Ancher MS, Peters WP, Reisenbichler H, Petros WP, Jirtle RL. Transforming growth factor β as a predictor of liver and lung fibrosis after autologous bone marrow transplantation for advanced breast cancer. N Engl J Med 1993; 328: 1592–8.

Airways Smooth Muscle: Peptide Receptors, Ion Channels and Signal Transduction
ed. by D. Raeburn and M. A. Giembycz
© 1995 Birkhäuser Verlag Basel/Switzerland

CHAPTER 7
Voltage-Dependent and Receptor-Operated Calcium Channels

Ian W. Rodger

Merck Frosst Centre for Therapeutic Research, Pointe Claire-Dorval, Quebec, Canada

1   Introduction
2   Involvement of Calcium Ions (Ca^{2+}) in Airways Smooth Muscle Contraction
3   Plasma Membrane Calcium Ion Channels
3.1 Voltage-Dependent Calcium Channels
3.1.1 Structural Considerations
3.1.2 Pharmacological Considerations
3.2 Receptor-Operated Calcium Channels
4   Clinical Considerations
     References

1. Introduction

Contraction of airways smooth muscle is universally accepted as being the principal component of the acute phase of airflow limitation that characterizes an asthmatic attack. It is also acknowledged that the airway structure of asthmatic subjects is abnormal. The abnormality is, in all likelihood, a consequence of the profound inflammatory processes that thicken the airway wall and promote transudation of mucus and oedema fluid into the airway lumen. Coincident with, or consequent upon, these changes there occurs an increase in airways smooth muscle mass which is hyperresponsive to a wide range of provoking stimuli. Despite detailed knowledge of these fundamental changes in airways smooth muscle and its contractile state, some of which have been recognized for a considerable period of time, it is only relatively recently that the complex molecular mechanisms underlying contraction of airways smooth muscle have begun to be unravelled. Detailed discussion of several different aspects of this subject can be found elsewhere in this series. The specific objective of this chapter is to provide a brief overview of the pharmacology of the voltage-dependent and receptor-operated calcium channels present in airways smooth muscle cells. In this context, our attention will be focussed only on those calcium channels that exist within the plasmalemmal membrane of airways smooth muscle cells and not on those calcium release channels found in the sarcoplasmic reticulum within the cells. These latter channels are dealt with elsewhere.

2. Involvement of Calcium Ions (Ca^{2+}) in Airways Smooth Muscle Contraction

It is now almost universally accepted that two forms of excitation-contraction coupling exist: electromechanical and pharmacomechanical. Electromechanical coupling, as the term suggests, depends either on electrical depolarization of the plasma membrane which opens calcium ion channels leading to Ca^{2+} influx from the extracellular space, and a consequent increase in the intracellular concentration of free Ca^{2+}, or on a voltage-dependent release of Ca^{2+} from intracellular stores such as the sarcoplasmic reticulum (for a review see [1]). In marked contrast, the pharmacomechanical coupling mechanism is regarded as being voltage-independent. Thus, it may involve either extracellular Ca^{2+} influx via ligand-gated Ca^{2+} channels or release of activator Ca^{2+} from intracellular stores. The intracellular release mechanism is executed either via ligand-generated intracellular second messengers or via a direct action of a ligand on the intracellular sarcoplasmic reticular stores [1].

When a contractile agonist interacts with its specific cell surface receptors, the intracellular concentration of Ca^{2+} in airways smooth muscle cells rises abruptly from its resting level ($< 0.2\ \mu M$) to between 0.5 and 1 μM [2]. These activator Ca^{2+}, so called because they initiate the contractile sequence, can only be derived from two sources. They may be derived from the extracellular compartment where they reside in abundance or from intracellular stores such as those present in the sarcoplasmic reticulum. The relative contribution of activator Ca^{2+} from each of these sources is wholly dependent upon both the nature and concentration of the contractile agonist that is initiating contraction and the component of the contractile response being considered, *i.e.* the phasic or tonic element of contraction [2]. In addressing the remit of this chapter (calcium channels in the plasmalemma), discussion will focus only upon the movement of extracellular calcium ions into the cell.

In addition to the role that extracellular calcium ions play in either initiating or maintaining a contractile response, they are also known to play a fundamental role in replenishing the intracellular Ca^{2+} pool which is depleted by the action of ligands that induce contraction via either the inositol 1,4,5 trisphosphate (IP3)-dependent or cyclic ADP-ribose-dependent mechanisms.

3. Plasma Membrane Calcium Ion Channels

The resting membrane potential of airways smooth muscle cells lies somewhere between -45 mV and -60 mV. Under normal circumstances, most species (guinea pig and man being notable exceptions) do not exhibit spontaneous oscillations of the membrane potential, $i.e.$, the cells are electrically quiescent. A further characteristic feature is the cell membrane's remarkable propensity for outward electrical rectification consequent upon a depolarizing stimulus. This rectification behaviour is thought to be largely due to a voltage-dependent, Ca^{2+}-insensitive delayed rectifier K^+ current [3]. Additionally, G-protein regulated (both G_s and G_i), Ca^{2+}-activated K^+ channels of either low (about 90 pS) or high (250–290 pS) conductance are abundant in the airways smooth muscle plasma membrane [4, 5]. This rectification ability, in addition to limiting the magnitude of any depolarization, effectively prevents the membrane potential from attaining the threshold necessary to elicit opening of voltage-dependent Ca^{2+} channels. For a fuller description of the electrophysiological characteristics of airways smooth muscle calcium channels the reader is directed to the chapter by Small and Foster in an earlier volume [6].

Given that the cell membrane very effectively and efficiently partitions the intracellular and extracellular environments, it is self evident that activator Ca^{2+} originating in the extracellular domain can only gain admission to the cell once the membrane has been rendered permeable to them. This is achieved via the opening of Ca^{2+} channels in the plasma membrane through which the Ca^{2+} flow down their electrochemical and concentration gradients. Two types of Ca^{2+} channels have been proposed: voltage-dependent (VDC) and receptor-operated (ROC) [7].

3.1. Voltage-Dependent Calcium Channels

VDCs, as the term implies, possess a Ca^{2+} conductance that is directly proportional to the potential difference that exists across the plasma membrane. Thus, membrane depolarization increases both the probability of VDC opening and the duration of the open channel time. Numerous studies in many different laboratories have, using both electrophysiological and pharmacological techniques, differentiated several types of VDC. It is currently accepted that there are four main types of VDC [6, 8]. These are defined as a) the L-type (for long lasting), b) the T-type (for transient or tiny), c) the N-type (for neuronal which are neither L nor T) and d) the P-type (found in Purkinje cells). Of these four VDCs it is L-type Ca^{2+} channels that are the most well defined because of the availability of high affinity drugs with which to probe the channels and because of recent molecular genetic studies [8]. Furthermore, the L-type channels may be considered simply as those channels that are most sensitive to 1,4-dihydropyridine calcium channel antagonists. It is the L-type calcium channel that is most abundant in the plasmalemma of airways smooth muscle cells [6].

3.1.1. Structural considerations: The biochemical structure of the L-type calcium channel has been widely studied and 1,4-dihydropyridine agonists and antagonists have been used as probes to isolate the protein constituents of the channel. This calcium channel is a multimeric protein complex composed of five different polypeptide subunits each with different molecular masses (Figure 1) [8]. These subunits are referred to as α_1, α_2, β, γ and δ (Figure 1). The α_1 subunit (175 kD), which is encoded by three genes, apparently forms the ion selective pore and contains essential phosphorylation sites (an important characteristic of the L-type calcium channel is that in order to open when the membrane is depolarized it must first be phosphorylated) and binding sites for some calcium antagonists, for example dihydropyridines and phenylalkylamines (Figure 1) [8–11]. The protein has four repeating motifs each containing six putative transmembrane spanning regions

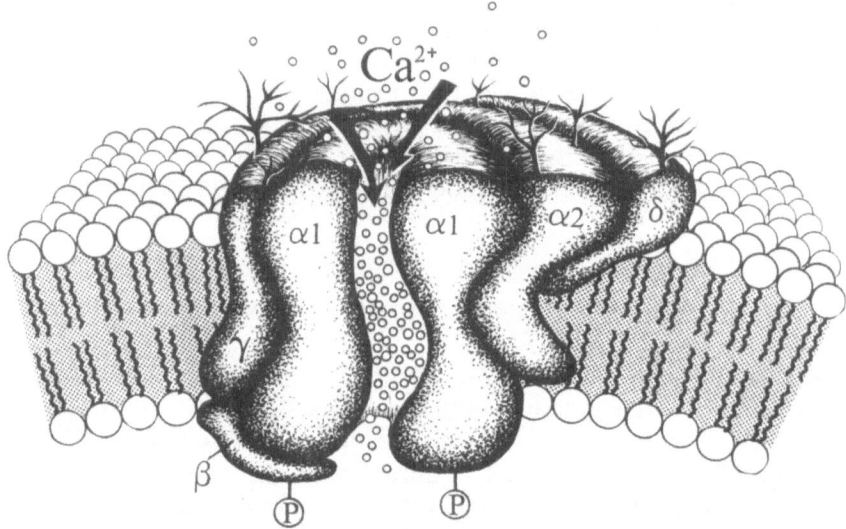

Figure 1. Diagrammatic representation of the structural organisation of the L-type calcium ion channel. The calcium channel is a pentameric protein complex. Each of the five different polypeptide subunits (α_1, α_2, β, γ, δ) has a different molecular mass. The α_1-subunit is depicted as the ion channel or pore. It contains the 1,4-dihydropyridine and phenylalkylamine calcium antagonist binding sites, essential regulatory (cyclic AMP-dependent) phosphorylation sites as well as the voltage sensor apparatus. Like the α_1-subunit, the β-subunit has similar phosphorylation sites for cyclic AMP-dependent protein kinase. The α_2, γ and δ subunits are involved in modulating channel conductance (see text for further details).

(termed S1–S6) of which the S4 spans constitute the putative voltage sensor [8]. The α_1 subunit closely resembles that found in sodium channels and is approximately 55% homologous with the sodium channel in the transmembrane spanning domain [8]. However, the cytosolic regions of the α_1 subunit are significantly different from those in the sodium channel. The α_1 subunit has been cloned from a variety of tissue types, including lung, and northern blot analysis has revealed specific messenger RNA (mRNA) transcript sizes that are tissue specific [8]. This indicates that, in all likelihood, distinct isoforms of the L-type calcium channel exist, most probably arising through the process of alternative splicing. A comparison of the deduced amino acids from the nucleotide sequences is also consistent with the existence of different isoforms.

Whilst the role of the other subunits has not been as well characterized as that of the α_1 subunit they do appear to exert modulatory control over the α_1 subunit. Thus, the α_2 subunit does not appear to possess binding sites for calcium antagonists nor does it appear to act as a channel in itself [10]. However, in conjunction with the δ subunit it is capable of increasing the calcium current following the injection of α_1 subunit mRNA in the *Xenopus* oocyte expression system [12]. The β subunit is encoded by at least three genes from which alternatively spliced variants are derived. Thus, several isoforms of the β-subunits exist. Functionally the β-subunit has been shown to increase both the rate of activation and inactivation of the channel and when co-expressed with the α_1 subunit it significantly increases current flow [12]. Additionally, the presence of the β subunit markedly enhances 1,4-dihydropyridine binding to the α_1-subunit (B_{max} increases some 12-fold) and markedly increases kinetics [13]. Although the function of the γ subunit of the L-type calcium channel is not yet clear it is known to be linked by sulphydryl bonds to the α_2 subunit.

3.1.2. Pharmacological considerations: There is a substantial literature of convincing evidence, from both electrophysiological and ion-flux/pharmacological studies, which supports the view that VDCs are both present in airways smooth muscle and that they are true L-type calcium ion channels [6, 14–19]. Furthermore, with regard to contractile events, there is overwhelming evidence demonstrating that VDCs are primarily responsible for contractions elicited by potassium chloride solution, tetraethylammonium and other depolarizing manoeuvers. Thus, for example, in bovine [20] canine [21–24] and guinea-pig [25, 26] airways smooth muscle, potassium chloride solution elicits a contraction that is accompanied by, and correlated with, both graded membrane depolarization and the influx of extracellular calcium ions as assessed by both the lanthanum technique [25–28] and through the use of fluorescent Ca^{2+} indicators [29]. Certain drugs, such as the 1,4-dihydropyridine

calcium channel agonists BAY K8644 and BAY R5417, that enhance
the open probability state of L-type calcium ion channels, have also
been shown to augment unitary cation currents (conductance ~ 25 pS),
extracellular $^{45}Ca^{2+}$ uptake and potassium chloride and calcium ion-
induced contractions of airways smooth muscle [18, 30–32]. Additionally, certain 1,4-dihydropyridine calcium antagonists (typified by
nifedipine, nitrendipine, and nisoldipine) have been shown to suppress
the slow wave activity elicited by potassium chloride and tetraethylammonium and in addition to abrogate the action potential discharge
evoked by tetraethylammonium. These effects occur coincident with an
inhibition of the associated $^{45}Ca^{2+}$ uptake and tension changes in
airways smooth muscle preparations [24, 27, 33–38].

Thus, the evidence attesting to the existence of true L-type VDCs in
airways smooth muscle is thoroughly convincing. It is entirely likely
that the opening of these channels is the mechanism underlying the
entry of extracellular calcium ions into airways smooth muscle cells in
response to a variety of depolarizing stimuli, for example tetraethylammonium and potassium chloride. In sharp contrast, however, there is no
real evidence to support the involvement of similar 1,4-dihydropyridine-
sensitive, L-type VDCs in the mechanisms underlying contraction *initiated* by a wide range of physiologically relevant agonists such as,
cholinomimetics, histamine, eicosanoids, tachykinins, endothelin and
bradykinin [19, 39–43]. The current consensus is that these agonists
elicit contraction of airways smooth muscle via induction of the release
of Ca^{2+} from intracellular stores. These conclusions are further rein-
forced by results from studies *in vitro* where the dependence of different
spasmogens upon extracellular Ca^{2+} has been investigated. In general,
as one would predict, contractions of airways smooth muscle elicited by
those spasmogens which utilize Ca^{2+} from the extracellular environment (for example KCl and tetraethylammonium) are substantially
more sensitive to reductions in the concentration of extracellular Ca^{2+}
than contractions induced by the more physiologically relevant agonists
mentioned above (reviewed in [19]).

Whilst VDCs may not be involved in admitting extracellular Ca^{2+} in
order to initiate the sequence of contractile events, recent evidence [44]
suggests that they may be intimately involved in the process of refilling
intracellular Ca^{2+} stores. It is activator Ca^{2+} from these stores, released
in response to IP_3 generated by agonist-receptor stimulations, that is
regarded as critical in initiating the contractile events in airways smooth
muscle [45]. In the study of Bourreau *et al.* [44], repetitive stimulation
of airways smooth muscle with acetylcholine, in the continuous presence
of nifedipine, resulted in a progressive loss in developed tension. This
was associated with a decrease in the content of the spasmogen-sensitive
intracellular Ca^{2+} stores. Similarly, agonist-sensitive internal Ca^{2+}
stores were readily depleted by successive or prolonged agonist stimula-

tion in a Ca^{2+}-free medium. Refilling of the empty Ca^{2+} stores after washout of the spasmogen required extracellular Ca^{2+}. This refilling process was decreased by nifedipine and increased by the calcium channel agonist BAY K8644 or by increasing the extracellular Ca^{2+} concentration. Refilling of the intracellular stores *during* an acetylcholine contraction in a Ca^{2+}-containing medium was similarly decreased by nifedipine and increased by BAY K8644. These results strongly suggest that replenishment of internal Ca^{2+} stores is, at least in part, dependent on the influx of extracellular Ca^{2+} via a 1,4-dihydropyridine-sensitive, L-type calcium channel.

Recently, it has been proposed that a calcium influx factor (CIF) is generated as a consequence of the depletion of activator Ca^{2+} from intracellular stores [46–48]. The CIF is thought to be responsible for gating the plasmalemmal "capacitative" Ca^{2+} refilling process. Detailed information on the CIF-activated process is not yet available but it will be interesting to know whether the process is sensitive to inhibition by 1,4-dihydropyridine calcium antagonists.

One final interesting feature of the L-type VDCs in airways smooth muscle is their modulation by β-adrenoceptor agonists. It is well recognized that the open state probability of the cardiac L-type calcium channel is enhanced by cyclic AMP-dependent phosphorylation mechanisms [49, 50] and also by the α subunit of G_s [51]. Given the extensive sequence identity between the cardiac and airways smooth muscle L-type channel proteins, it is perhaps not surprising to find that β-agonists modulate the airway L-type VDCs. The initial observation was reported by two independent laboratories who, using adult bovine tracheal smooth muscle cells, reported that isoprenaline increased the intracellular Ca^{2+} concentration apparently via a 1,4-dihydropyridine-sensitive pathway [52, 53]. Further examination of this phenomenon [54] showed that isoprenaline stimulated peak calcium currents in a dose-dependent manner via a β adrenoceptor mechanism. The isoprenaline effect was not mediated or caused by the stimulation of a K^+ or Na^+ current, by a decrease in intracellular concentrations of Ca^{2+} or H^+, or by stimulation of either the Na^+-H^+ or the Na^+-Ca^{2+} exchanger. Neither the basal nor isoprenaline-stimulated calcium current was affected by internal dialysis of the cell with cyclic AMP or its analogues or by the catalytic subunit of cyclic AMP-dependent protein kinase. In contrast, internal dialysis of the cells with guanosine 5'-O-(2-thiodiphosphate) (GDPβS) blocked the stimulation induced by isoprenaline whereas dialysis with guanosine 5'-O-(3-thiotriphosphate) (GTPγS; a stable analogue of GTP) induced an isoprenaline-like maximal increase in the calcium current. Thus, these results clearly illustrate that β-adrenoceptor stimulation augments the L-type VOC calcium current of isolated tracheal smooth muscle cells. Furthermore, since this effect occurs independent of either cyclic AMP or its dependent protein kinase it is in all likelihood, an effect mediated via a GTP/GDP regulated protein, probably G_s. Taken at face value this effect of β-adrenoceptor agonists is directly contrary to their well established ability to promote airways smooth muscle relaxation (presumably by lowering the intracellular Ca^{2+} concentration). Whether the effects on Ca^{2+} entry are a reflection of the non-physiological experimental conditions employed by Welling et al., [54] or an indication that different β-adrenoceptors subserve Ca^{2+} entry and airways smooth muscle relaxation remains to be determined.

3.2. Receptor-Operated Calcium Channels

ROCs are ion channels opened, or operated, by a receptor for a stimulant substance [7]. It is generally accepted that ROCs are not wholly selective for Ca^{2+}, have an ionic permeability determined by the controlling receptor, can be gated by either voltage-dependent or voltage-independent events and are not readily inhibited by organic Ca^{2+} antagonists such as the 1,4-dihydropyridines [7, 43]. In contrast to the extensive molecular biology effort that has been directed at the L-type calcium channels, little similar effort has been directed at the ROCs. The postulated existence of ROCs in airways smooth muscle, therefore, stems largely from pharmacological evidence which is of a more indirect nature.

In airways smooth muscle, most pharmacological agonists elicit contraction associated with graded membrane depolarization (the magnitudes differ substantially between different agonists) but without action potential discharge (see [55] for references). Furthermore, the same agonists can elicit contractions in fully depolarized airway preparations. It has also not been possible using the "lanthanum technique" to detect extracellular calcium ion influx in response to spasmogens during the period associated with tension development [27, 56]. Collectively, therefore, these data argue against both the existence of ROCs in airways smooth muscle and the involvement of VDCs, to any significant extent, in the actions underlying receptor stimulation by pharmacological agonists. This latter view is strengthened by the many observations that calcium antagonists, in concentrations that completely block the contractile and electrophysiological effects of KCl and of tetraethylammonium, fail to inhibit contractions elicited by a wide range of agonists in airway preparations from several species including man [24, 27, 34–38]. Additionally, BAY K8644 (the calcium channel agonist) does not potentiate contractions elicited by either acetylcholine or histamine in guinea-pig or human airway preparations [30, 31].

Given the body of evidence presented above, it is tempting and not unreasonable to dispel the notion that ROCs are present in the plasma membrane of airways smooth muscle cells. Notwithstanding, one has to be cognisant of the fact that the "lanthanum technique" is only useful for measuring fairly gross changes in extracellular calcium entry into cells and is ill equipped to measure the small changes which may be associated with calcium entry through ROCs. In this context, recent results [43, 62, 63], using Fura-2 fluorescence to measure intracellular Ca^{2+} concentrations in human and canine airways smooth muscle cells in culture, provide new insights into transmembrane Ca^{2+} movements. Through use of this very sensitive detection system Murray and Kotlikoff have demonstrated a receptor-activated calcium ion influx mechanism that is both 1,4-dihydropyridine insensitive and which has biophysical characteristics that are inconsistent with those of VDCs [43, 62]. Thus, in these cells contractile agonists activate sustained calcium influx that is decreased when the cell is depolarized and increased at more negative membrane potentials. These data are remarkably consistent with the earlier information described by Coburn [24] in which hyperpolarization manoeuvres increased contractions following exposure of canine tracheal smooth muscle to acetylcholine. The cation permeability of the influx pathway described by Murray and Kotlikoff [43] is not only distinct from that of VDCs in airways smooth muscle but also differs from that described in nonexcitable cells such as platelets and endothelial cells [64–66]. In human airways smooth muscle cells, Mn^{2+} permeability is not augmented by receptor activation; rather, all divalent cations examined including Ba^{2+} block calcium influx. Since the VDCs possess a high barium conductance and since depolarization actually decreases Ca^{2+} flux following agonist exposure, Ca^{2+} influx through VDCs is inconsistent with these observations. Furthermore, the fact that divalent cations block this pathway is suggestive that it is highly specific for Ca^{2+} and that a nonspecific cation channel is unlikely to be involved.

Precisely how ROCs are activated/regulated or, indeed, how important they are to the overall contractile process is not yet known. In terms of activation, however, recent studies in mast cells [67, 68] have supported the capacitative entry hypothesis [69, 70]. Thus, emptying of the agonist/IP_3-sensitive Ca^{2+} pool in the sarcoplasmic reticulum can create an enhanced permeability of the plasma membrane so permitting Ca^{2+} entry.

The implications from the above work are that intracellular activator calcium ions are responsible for the initiation of contraction but that the tonic or sustained phase of contraction is associated with a low level of extracellular calcium ion influx, via a mechanism that bears all the hallmarks of a ROC-mediated event. The fact that the compound SKF 96365, an inhibitor of ROC activity in nonexcitable cells, has been shown to block extracellular calcium ion entry in the experiments of Murray and Kotlikoff [43] simply confirms this conclusion. Whilst these recent findings are of a very exciting nature, they clearly require sub-stantiation. However, they may well signal for the first time not only the presence but also the active participation of ROCs in the excitation-con-traction coupling process in airways smooth muscle. Unambiguous definition of such a role, through the development of more selective inhibitors of Ca^{2+} influx via these ROCs, may very well lead to the development of exciting, novel, therapeutic agents for use in diseases such as asthma. The precise involvement of the CIF [46–48] in poten-tially regulating these ROCs, either as a means to support sustained contraction of airways smooth muscle cells or as a means of refilling the intracellular calcium store, is an exciting area that remains to be developed.

4. Clinical Considerations

In vascular smooth muscle inhibitors of the L-type calcium ion chan-nels, such as nifedipine, diltiazem and verapamil, have found substantial and impressive clinical utility in the treatment of a variety of cardiovas-cular diseases. Given that VDCs clearly exist in airways smooth muscle one might have anticipated that these L-type calcium channel antago-nists would have a similar clinical utility in asthma. This is not the case. Nifedipine, verapamil and diltiazem have all been investigated in asth-matic patients with very disappointing results. Whilst a weak inhibitory effect on bronchoconstriction induced by spasmogens, such as his-tamine, methacholine, exercise and cold air, has been reported, there is no convincing evidence that these drugs are either bronchodilator or that they have any significant effect on the clinical symptoms of asthma [71, 72].

The reasons for the stunning clinical ineffectiveness of these calcium antagonists in airways disease can now be rationalized given our under-standing of the mechanisms involved in contraction of airways smooth muscle [1, 45]. In airways smooth muscle the airway cellular events consequent upon receptor activation by a contractile agonist are now fairly well understood. In recent years, for example, we have gained substantial insight into receptor-G-protein interactions, G-protein acti-vation/regulation of certain critical enzymes such as phospholipase C, generation of second messengers such as IP_3 and diacylglycerol, intracel-lular Ca^{2+} release mechanisms and activation and regulation of the contractile proteins with respect to the **initial** generation of smooth muscle tension. In general, all these events are insensitive to L-type calcium channel antagonists. The events involved in the maintenance of developed tension, which equate with the sustained bronchospasm of an asthmatic attack, still remain to be elucidated. Although the mainte-nance of tone almost certainly depends upon Ca^{2+} entry from the

extracellular space, in all likelihood this entry process is via channels
(ROCs?) that are not sensitive to conventional L-type calcium antagonists. Additionally, the release of mediators from inflammatory cells is
not thought to be mediated via VDCs and there is no convincing
evidence that calcium antagonists have a significant effect on either
mediator release, activation of inflammatory cells or on the consequent
process of inflammation. Notwithstanding, the developing awareness of
ROCs in airways smooth muscle coupled with an appreciation of their
function, heralds a new era. Given that heterogeneity of ROCs is almost
certain to exist it holds out the possibility that new ROC-selective
calcium antagonists, selective for channels in respiratory tissue, may
well be developed. Such, new calcium antagonists could, potentially,
revolutionize asthma therapy.

References

1. Rodger IW. Airway smooth muscle: Signal transduction and contractile mechanisms. In:
   Holgate ST, Austen KF, Lichtenstein LM, Kay AB, editors. Asthma: physiology,
   immunopharmacology and treatment. London: Academic Press, 1993: 243–57.
2. Rodger IW, Small RC. The pharmacology of airway smooth muscle. In: Page CP, Barnes
   PJ, editors. Pharmacology of asthma. Handbook of experimental pharmacology; Vol. 98,
   Berlin: Springer Verlag, 1991: 107–41.
3. Kotlikoff MI. Potassium currents in canine airway smooth muscle cells. Am J Physiol
   1990; 259: L384–L395.
4. Kume H, Kotlikoff MI. Muscarinic inhibition of single K_{Ca} channels in smooth muscle
   cells by a pertussis-sensitive G-protein. Am J Physiol 1991; 261: C1204–C1209.
5. Saunders H-MH, Farley JM. Spontaneous transient outward currents and Ca^{++}-acti-
   vated K^+ channels in swine tracheal smooth muscle cells. J Pharmacol Exp Ther 1991;
   257: 1114–1120.
6. Small RC, Foster RW. The electrophysiology of calcium-channels in airways smooth
   muscle. In: Giembycz MA, Raeburn D, editors. Airways smooth muscle: Development,
   and regulation of contractility. Basel: Birkhauser Verlag, 1994: 137–61.
7. Bolton TB. Mechanisms of action of transmitters and other substances on smooth muscle.
   Physiol Rev 1979; 59: 606–718.
8. Spedding M, Paoletti R. Classification of calcium channels and the sites of action of drugs
   modifying channel function. Pharmacol Rev 1992; 44: 363–376.
9. Tanabe T, Takeshima H, Mikami A, Flockerzi V, Takahashi H, Kangawa K, et al.,
   Primary structure of the receptor for calcium channel blockers from skeletal muscle.
   Nature 1987; 328: 313–318.
10. Ellis SB, Williams ME, Ways NR, Brenner R, Sharp AH, Leung AT, et al., Sequence and
    expression of mRNAs encoding the α_1 and α_2 subunits of a DHP-sensitive calcium
    channel. Science 1988; 241: 1661–1664.
11. Armstrong D, Eckert R. Voltage-activated calcium channels that must be phosphorylated
    to respond to membrane depolarization. Proc Natl Acad Sci USA 1987; 84: 2518–2522.
12. Singer D, Biel M, Lotan I, Flockerzi V, Hofmann F, Dascal N. The roles of the subunits
    in the function of the calcium channel. Science 1991; 253: 1553–1556.
13. Varadi G, Lory P, Schultz D, Varadi M, Schwartz A. Acceleration of activation and
    inactivation by the β subunit of the skeletal muscle calcium channel. Nature 1991; 352:
    159–162.
14. Hisada T, Kurachi Y, Sugimoto T. Properties of membrane currents in isolated smooth
    muscle cells from guinea-pig trachea. Pflugers Arch 1990; 416: 151–161.
15. Kotlikoff MI. Ion channels in airway smooth muscle. In: Coburn RF, editor. Airway
    smooth muscle in health and disease. New York: Plenum Press, 1989: 169–182.

16. Marthan R, Martin C, Amédée T, Mironneau J. Calcium channel currents in isolated smooth muscle cells from human bronchus. J Appl Physiol 1989; 66: 1706–1714.
17. Kotlikoff MI. Calcium currents in isolated canine airway smooth muscle cells. Am J Physiol 1988; 254: C793–C801.
18. Worley JF III, Kotlikoff MI. Dihydropyridine-sensitive single calcium channels in airway smooth muscle cells. Am J Physiol 1990; 259: L468–L480.
19. Rodger IW. Calcium channels. Am Rev Resp Dis 1987; 136(4Pt2): S15–S17.
20. Kirkpatrick CT. Excitation and contraction in bovine tracheal smooth muscle. J Physiol (London) 1975; 244: 263–281.
21. Suzuki H, Morita K, Kuriyama H. Innervation and properties of the smooth muscle of the dog trachea. Jpn J Physiol 1976; 26: 303–320.
22. Coburn RF, Yamaguchi T. Membrane potential-dependent and -independent tension in the canine tracheal muscle. J Pharmacol Exp Ther 1977; 201: 276–284.
23. Farley JM, Miles PR. Role of depolarization in acetylcholine-induced contractions of dog trachealis muscle. J Pharmacol Exp Ther 1977; 201: 199–205.
24. Coburn RF. Electromechanical coupling in canine trachealis muscle: acetylcholine contractions. Am J Physiol 1979; 236: C177–C184.
25. Foster RW, Small RC, Weston AH. The spasmogenic action of potassium chloride in guinea-pig trachealis. Br J Pharmacol 1983; 80: 553–559.
26. Foster RW, Small RC, Weston AH. Evidence that the spasmogenic action of tetraethyl-ammonium in guinea-pig trachealis is both direct and dependent upon the cellular influx of calcium ion. Br J Pharmacol 1983; 79: 255–263.
27. Raeburn D, Rodger IW. Lack of effect of leukotriene D_4 on Ca-uptake in airway smooth muscle. Br J Pharmacol 1984; 83: 499–504.
28. Weiss GB, Pang IH, Goodman FR. Relationship between ^{45}Ca movements, different calcium components and responses to acetylcholine and potassium in tracheal smooth muscle. J Pharmacol Exp Ther 1985; 233: 389–394.
29. Takuwa Y, Takuwa N, Rasmussen H. Measurement of cytoplasmic free Ca^{2+} concentration in bovine tracheal smooth muscle using aequorin. Am J Physiol 1987; 253: C817–C827.
30. Allen SL, Foster RW, Small RC, Towart R. The effects of the dihydropyridine BAY-K 8644 in guinea pig isolated trachealis. Br J Pharmacol 1985; 86: 171–180.
31. Advenier C, Naline E, Renier A. Effects of BAY-K 8644 on contraction of the human isolated bronchus and guinea-pig isolated trachea. Br J Pharmacol 1986; 88: 33–39.
32. Marthan R, Armour CL, Johnson PRA, Black JL. The calcium channel agonist BAY-K 8644 enhances the responsiveness of human airway muscle to KCl and histamine but not to carbachol. Amer Rev Respir Dis 1987; 135: 185–189.
33. Richards IS, Kulkarni A, Brooks SM. Human fetal tracheal smooth muscle produces spontaneous electromechanical oscillations that are Ca^{2+}-dependent and cholinergically potentiated. Dev Pharmacol Ther 1991; 16: 22–28.
34. Kannan MS, Jager LP, Daniel EE, Garfield RE. Effects of 4-aminopyridine and tetra-ethylammonium chloride on the electrical activity and cable properties on canine tracheal smooth muscle. J Pharmac Exp Ther 1983; 227: 706–715.
35. Foster RW, Okpalugo BI, Small RC. Antagonism of Ca^{2+} and other actions of verapamil in guinea-pig isolated trachealis. Br J Pharmacol 1984; 81: 499–507.
36. Ahmad F, Foster RW, Small RC. Some effects of nifedipine in guinea-pig isolated trachealis. Br J Pharmacol 1985; 84: 861–869.
37. Baba K, Kawanishi M, Satake T, Tomita T. Effects of verapamil on the contractions of guinea-pig tracheal smooth muscle induced by Ca, Sr and Ba. Br J Pharmacol 1985; 84: 203–211.
38. Raeburn D, Roberts JA, Rodger IW, Thomson NC. Agonist-induced contractile responses of human bronchial muscle in vitro. Effects of Ca^{2+} removal, La^{3+} and PY108068. Eur J Pharmacol 1986; 121: 251–255.
39. Giembycz MA, Rodger IW. Electrophysiological and other aspects of excitation-contraction coupling and uncoupling in mammalian airway smooth muscle. Life Sci 1987; 41: 111–132.
40. Small RC, Foster RW. Airway smooth muscle: An overview of morphology, electrophysiology and aspects of the pharmacology of contraction and relaxation. In: Kay AB, editor. Asthma: clinical pharmacology and therapeutic progress. Oxford: Blackwell Scientific Publications, 1986: 101–113.

41. Kajita J, Yamaguchi H. Calcium mobilization by muscarinic cholinergic stimulation in bovine single airway smooth muscle. Am J Physiol 1993; 264: L496–L503.
42. Henry PJ. Endothelin-1 (ET-1)-induced contraction in rat isolated trachea: involvement of ET_A and ET_B receptors and multiple signal transduction systems. Br J Pharmacol 1993; 110: 435–441.
43. Murray RK, Kotlikoff MI. Receptor-activated calcium influx in human airway smooth muscle cells. J Physiol (London) 1991; 435: 123–144.
44. Bourreau J-P, Abela AP, Kwan CY, Daniel EE. Acetylcholine Ca^{2+} stores refilling directly involves a dihydropyridine-sensitive channel in dog trachea. Am J Physiol 1991; 261: C497–C505.
45. Rodger IW, Pyne NJ. Airway smooth muscle. In: Barnes PJ, Rodger IW, Thomson NC, editors. Asthma: basic mechanisms and clinical management. London: Academic Press, 1992: 59–84.
46. Clapham DE. A mysterious new influx factor? Nature 1993; 364: 763–764.
47. Randriamampita C, Tsien RY. Emptying of intracellular Ca^{2+} stores releases a novel small messenger that stimulates Ca^{2+} influx. Nature 1993; 364: 809–814.
48. Parekh AB, Terlau H, Stühmer W. Depletion of $InsP_3$ stores activates a Ca^{2+} and K^+ current by means of a phosphatase and a diffusable messenger. Nature 1993; 364: 814–818.
49. Hartzell HC, Méry P-F, Fischmeister R, Szabo G. Sympathetic regulation of cardiac calcium current is due exclusively to cAMP-dependent phosphorylation. Nature 1991; 351: 573–576.
50. Kameyama M, Hofmann F, Trautwein W. On the mechanism of β-adrenergic regulation of the Ca^{2+} in the guinea-pig heart. Pfugers Arch 1985; 405: 285–293.
51. Imoto Y, Yatani A, Reeves JP, Codina J, Birnbaumer L, Brown AM. α-Subunit of G_s directly activates cardiac calcium channels in lipid bilayers. Am J Physiol 1988; 255: H722–H728.
52. Felbel J, Trockur B, Ecker T, Landgraf W, Hofmann F. Regulation of cytosolic calcium by cAMP and cGMP in freshly isolated smooth muscle cells from bovine trachea. J Biol Chem 1988; 263: 16764–16771.
53. Takuwa Y, Takuwa N, Rasmussen H. The effects of isoproterenol on intracellular calcium concentration. J Biol Chem 1988; 263: 762–768.
54. Welling A, Felbel J, Peper K, Hofmann F. Hormonal regulation of calcium current in freshly isolated airway smooth muscle cells. Am J Physiol 1992; 262: L351–L359.
55. Rodger IW. Biochemistry of activation-contraction coupling. In: Busse W, Holgate ST, editors. Asthma and rhinitis. Cambridge USA: Blackwell Scientific, 1995.
56. Ahmed F, Foster RW, Small RC, Weston AH. Some features of the spasmogenic actions of acetylcholine and histamine in guinea-pig isolated trachealis. Br J Pharmacol 1984; 83: 227–233.
57. Jones TR, Davis C, Daniel EE. Pharmacological study of the contractile activity of leukotriene C_4 and D_4 on isolated human airway smooth muscle. Can J Physiol Pharmacol 1982; 60: 638–643.
58. Cerrina J, Advenier C, Renier A, Floch A, Duroux P. Effects of diltiazem and other Ca^{2+}-antagonists on guinea-pig tracheal muscle. Eur J Pharmacol 1983; 94: 241–249.
59. Advenier C, Cerrina J, Duroux P, Floch A, Renier A. Effects of five different organic calcium antagonists on guinea-pig isolated trachea. Br J Pharmacol 1984; 82: 727–733.
60. Baba K, Satake T, Takagi K, Tomita T. Effects of verapamil on the response of the guinea pig trachea muscle to carbachol. Br J Pharmacol 1986; 88: 441–449.
61. Roberts JA, Giembycz MA, Raeburn D, Rodger IW, Thomson NC. In vitro and in vivo effect of verapamil on human airway responsiveness to leukotriene D_4. Thorax 1986; 41: 12–16.
62. Murray RK, Fleischmann BK, Kotlikoff MI. Receptor-activated Ca influx in human airway smooth muscle: use of Ca imaging and perforated patch-clamp techniques. Am J Physiol 1993; 264: C485–C490.
63. Yang CM, Yo Y-L, Wang Y-Y. Intracellular calcium in canine cultured tracheal smooth-muscle cells is regulated by M_3 muscarinic receptors. Br J Pharmacol 1993; 110: 983–988.
64. Hallam TJ, Rink TJ. Agonists stimulate divalent cation channels in the plasma membrane of human platelets. FEBS Lett 1985; 186: 175–179.

65. Hallam TJ, Jacob R, Merritt JE. Evidence that agonists stimulate bivalent-cation influx into human endothelial cells. Biochem J 1988; 255: 179–184.
66. Merritt JE, Jacob R, Hallam TJ. Use of manganese to discriminate between calcium influx and mobilization from internal stores in stimulated human neutrophils. J Biol Chem 1989; 264: 1522–1527.
67. Penner R, Matthews G, Neher E. Regulation of calcium influx by second messengers in rat mast cells. Nature 1988; 334: 499–504.
68. Hoth M, Penner R. Depletion of intracellular calcium stores activates a calcium current in mast cells. Nature 1992; 355: 353–356.
69. Putney JW. A model for receptor-regulated calcium entry. Cell Calcium 1986; 7: 1–12.
70. Putney JW. Excitement about calcium signalling in inexcitable cells. Science 1993; 262: 676–678.
71. Barnes PJ. Clinical studies with calcium antagonists in asthma. Br J Clin Pharmacol 1985; 20 (Suppl. 2): 289S–298S.
72. Löfdahl C-G, Barnes PJ. Calcium channel blockade and asthma – the current position. Eur J Respir Dis 1985; 67: 233–237.

Airways Smooth Muscle: Peptide Receptors, Ion Channels and Signal Transduction
ed. by D. Raeburn and M. A. Giembycz
© 1995 Birkhäuser Verlag Basel/Switzerland

CHAPTER 8
High Conductance Calcium-Activated Potassium Channels

Gregory J. Kaczorowski[1] and Thomas R. Jones[2]

[1]*Department of Membrane Biochemistry and Biophysics, Merck Research Laboratories, Rahway, New Jersey, USA*
[2]*Department of Pharmacology, Merck Frosst Centre for Therapeutic Research, Pointe Claire-Dorval, Canada*

1 Introduction
2 Development of Probes for Maxi-K Channels
3 Molecular Characteristics of the Maxi-K Channel
4 Molecular Pharmacology of the Maxi-K Channel
5 Pharmacological Modulation of the Maxi-K Channel (Functional Studies)
6 Therapeutic Opportunities
  References

1. Introduction

Potassium channels represent a vast and diverse family of ion channel proteins which are found in many different types of tissues [1, 2]. These channels modulate the electrical excitability of certain cells, such as those derived from neuronal, endocrine and muscle sources. They also control the resting plasma membrane potential of a large variety of cells, regardless of whether the cells display electrically excitable or nonexcitable properties. K^+ channels are routinely categorized according to their biophysical and pharmacological properties. However, these proteins may also be sub-divided into two major classifications depending on whether they are activated by changes in membrane potential (voltage-sensitive channels), or by an interaction with small molecule modulators (ligand-gated channels). Unfortunately, when compared with other types of ion channels, such as with the members of the voltage-gated Na^+ and Ca^{2+} channel families [3], the biochemistry and molecular pharmacology of K^+ channels is still rather undeveloped.

Fortunately during the last few years through application of molecular biological techniques, the cDNA's for many different types of mammalian K^+ channels have been isolated and functionally expressed [4]. This work, which began with cloning of the voltage-gated K^+ channel that is responsible for the *Shaker* phenotpye in *Drosophila* [5], has led to testable ideas regarding the putative structure of K^+ channels. Interestingly, the predicted secondary structure of many voltage-gated K^+ channels that are homologous to *Shaker*, and the other related families of K^+ channels that were first cloned from *Drosophila* (*i.e.*, *Shaw*, *Shab* and *Shal*), display similarity with one of the four repeated regions present in Na^+ and Ca^{2+} channels [6]. This observation has led to the hypothesis that functional K^+ channels are tetramers formed by the association of either identical or dissimilar subunits [7–10]. This characteristic, together with alternative splicing of channel transcripts, is thought to give rise to the large diversity of K^+ channels that exists. Among the different families of K^+ channels which have recently been characterized by molecular biological techniques are the Ca^{2+}-activated K^+ channels [11–13]. These proteins, whose presumed transmembrane topology is related to those of the *Shaker*-like voltage-gated K^+ channel family (see below), are modulated by both membrane potential and the binding of Ca^{2+} to the intracellular surface of the channel. Several different sub-types of Ca^{2+}-activated K^+ channels have been identified and these have been conveniently classified according to their single channel conductance; large conductance (100–300 pS), intermediate conductance (40–100 pS) and small conductance (10–35 pS) channels. Despite the fact that all of these channels are gated by intracellular Ca^{2+}, the biophysical and pharmacological characteristics of the various sub-types are distinct. The most prevalent and hence best studied of the different types of Ca^{2+}-activated K^+ channels in terms of biochemical, biophysical, and pharmacological characterization is the high conductance or maxi-K (BK) channel.

Maxi-K channels are present in electrically excitable and nonexcitable cells [14, 15]. They are gated both by membrane depolarization and by micromolar concentrations of intracellular Ca^{2+}. The effects of these parameters are related because the channel's sensitivity to Ca^{2+} increases with membrane depolarization. Despite the unique feature of possessing a very high single channel conductance, these channels display a marked selectivity for K^+ over Na^+ or other ions. In excitable cells, maxi-K channels can contribute to the repolarization phase of the action potential. Hence, they are involved in certain types of neuroendocrine secretion and in excitation-contraction coupling processes in muscle. However, they are not present in cardiac tissue, but are present in cardiac vasculature. In nonexcitable cells, maxi-K channels can cause hyperpolarization of the plasma membrane when intracellular Ca^{2+} levels are elevated. In such cases, they have been implicated in the regulation of fluid secretion (*e.g.*, in exocrine gland and epithelial cells) or cell volume (*e.g.*, in red blood cells). These channels are prevalent in different types of smooth muscle and they are expected to affect the contractility of these tissues. For example, the myogenic activity of certain guinea pig smooth muscles (*e.g.*, bladder and taenia coli) appears to be affected by changes in maxi-K channel activity, while other types of spontaneously contracting smooth muscles isolated from this species (*e.g.*, portal vein, uterus) are not, even though the channel is present in each of these tissues [16]. Moreover, the role played by this channel in smooth muscle is not only tissue-dependent, but also species-dependent. Thus, the myogenic tone of rat portal vein is modulated by maxi-K channel activity [17], while this channel does not influence the contractile behaviour of the same tissue isolated from guinea pig. The contractility of quiescent smooth muscle such as guinea pig aorta and trachea is also controlled by the activity of the maxi-K channel. Indeed, activation of this conductance pathway appears to be responsible in part for smooth muscle relaxation brought about by various agents which elevate intracellular levels of cAMP or cGMP (*e.g.*, by the action of β-agonists, phosphodiesterase inhibitors, and activators of guanylyl cyclase; see below). In some cases, it has been possible to demonstrate regulation of maxi-K channels by phosphorylation, interaction with G-proteins, or by direct modulation via intracellular second messengers.

This review will focus on one particular type of potassium channel which has been identified in airways smooth muscle (*i.e.*, the large conductance calcium-activated (maxi-K or BK) channel). Other subtypes of potassium channels have been shown to be present in airways smooth muscle, namely small conductance Ca^{2+}-activated and delayed rectifier potassium channels, but will not be reviewed in this chapter. In addition, there is strong pharmacological evidence to support a role for metabolically controlled potassium channels such as the ATP-sensitive potassium (K_{ATP}) channels in the regulation and modulation of airway function [18, 19]. Again this channel will not be extensively reviewed in this chapter, except where important comparisons will be useful to the understanding of the pharmacology of maxi-K channels.

2. Development of Probes for Maxi-K Channels

Although only a few small molecules have been identified which are selective modulators of maxi-K channel activity, it is fortunate that a series of peptidyl inhibitors have been isolated from various scorpion venoms which have allowed an extensive characterization of this channel (for a review see 20). The first such agent was discovered by Miller *et al.* [21] who showed that crude venom from the scorpion *Leiurus quinquestriatus hebraeus* would block the activity of maxi-K channels reconstituted into planar lipid bilayers from skeletal muscle t-tubular membrane vesicles. This inhibitor which is a minor peptide component of the venom was named charybdotoxin (ChTX). In single channel recordings, the characteristic inhibitory pattern produced by the venom is the occurrence of prolonged silent periods which interrupt the pattern of normal channel gating activity. This behavior has been interpreted as resulting from binding of ChTX in the pore of the maxi-K channel to block the ion conduction pathway [21].

It was not until several years after the discovery of ChTX that the structure of this agent was elucidated. ChTX was first purified to homogeneity from *L. quinquestriatus hebraeus* venom by a combination of ion exchange and reversed phase HPLC chromatographic techniques, and its primary sequence was determined by Edman degradation [22]. Subsequently, several other groups also purified and characterized the protein structurally [23–25]. ChTX is a 37 amino acid peptide with a blocked N-terminus in the form of a pyroglutamic acid residue. It is highly positively charged with four lysine, three arginine and one histidine residue coexisting with only two acidic functionalities (a glutamic acid residue and the carboxy terminus). At physiological pH, ChTX displays a net positive charge of five, and this property is critical for its interaction with the maxi-K channel. Binding of ChTX occurs through an electrostatic interaction with negatively charged residues which form part of the channel's external pore [26]. Thus, the affinity of ChTX can be increased by lowering the ionic strength of the medium bathing the maxi-K channel, and lowered by methylation of carboxyl groups in the mouth of the channel with trimethyloxonium ion [27]. The high cysteine content of ChTX (6 cysteine residues in disulphide linkages) was a preliminary indication that this peptide is a highly compact structure.

Before ChTX could be developed as a pharmacological tool, it was important to produce large quantities of this peptide since it is a very minor component of *Leiurus quinquestriatus* venom. The chemical synthesis of ChTX was accomplished through solid phase synthetic techniques [28, 29]. After oxidation of the reduced peptide to its disulphide-containing form, ChTX was produced in good yield with full biological activity. The correct folding of synthetic ChTX was verified by comparing peptide maps of both synthetic and native toxins. This also led to the assignment of disulphide linkages in ChTX as Cys_{7-28}, Cys_{13-33} and Cys_{17-35} (Figure 1). Consistent with the prediction from these data that ChTX should have a very compact structure, are 2D-NMR spectroscopic results which have been used to elucidate the three dimensional structure of ChTX in solution [29, 30]. The peptide forms a surface of three antiparallel β sheets which are linked by disulphide bridges to a helix region composed of residues 10–18 lying behind the plane of the β sheet surface. These same structural features appear to be common motifs found in a number of different scorpion toxins [31].

ChTX has also been produced in high yield by biosynthetic means [32]. In this approach, the gene for a ChTX-fusion protein construct was incorporated into a plasmid which was then transfected into *E. coli*.

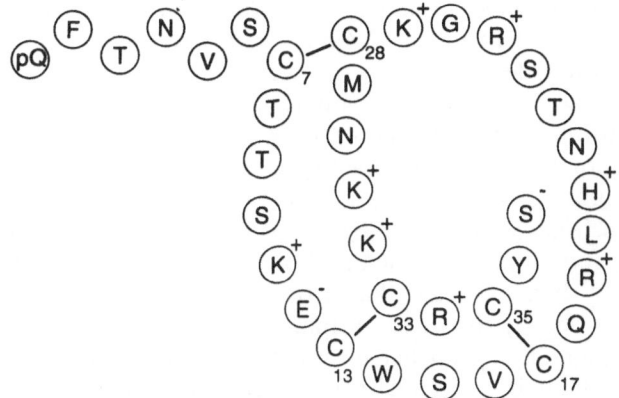

Figure 1. Secondary structure of charybdotoxin. The structure of charybdotoxin is depicted using the single letter code for the amino acids. The disulphide bonding pattern is also represented, as are the charges on the amino acid residues at physiological pH.

After expression and purification of the ChTX-fusion protein product, the toxin was cleaved from the fusion protein using factor Xa protease and purified by HPLC techniques following conversion of its N-terminal residue to pyroglutamic acid. The ChTX produced by this procedure folds into the proper conformation and is completely biologically active.

The production of ChTX by a biosynthetic route has allowed the construction of various mutants which should be useful for defining those residues on the toxin which make contact with the mouth of the maxi-K channel. Using the maxi-K channel from rat skeletal muscle reconstituted in planar lipid bilayers as an assay system, many ChTX mutants have been evaluated for their ability to affect channel activity. For example, the electrostatic interaction between ChTX and the channel was probed by modifying the charged residues on the toxin [33]. Point mutations at Lys_{11}, Glu_{12}, Arg_{19}, His_{21}, Lys_{31} and Lys_{32} had little effect on toxin affinity. In contrast, replacement of Arg_{25}, Lys_{27} or Arg_{34} by Gln drastically reduced the inhibitory activity of the mutant peptides. Most of the loss in potency was due to increases in toxin dissociation rates, with little effect on rates of association, indicating that close range interactions between the peptide and channel were disrupted. Furthermore, it was shown that Lys_{27} is the residue responsible for the interaction that occurs between toxin and K^+ entering the conduction pathway of the maxi-K channel from its inner surface [34]. Alterations in this residue suggest that Lys_{27} is close to a K^+ binding site at the external end of the ion conduction pathway and that occupation of this site by K^+ destabilizes ChTX binding [35].

Similar site-directed mutagenesis studies have been performed with other non-charged residues of ChTX and several hydrophobic amino acids (*e.g.*, Phe_2, Met_{29} and Tyr_{36}) were identified whose modification destabilizes the ChTX-maxi-K interaction [36]. Using this approach, point mutations in recombinant ChTX have identified 8 of 37 residues that are important for the interaction of toxin with the maxi-K channel. These residues cover *ca.* 25% of the peptide's surface and, given the three dimensional structure of ChTX as determined from NMR studies, a complete functional map of the molecular surface of ChTX has been deduced [37]. It is expected that a triangular area of the flat surface of the toxin formed by the three antiparallel β strands makes close contact with a complementary surface on the channel, perhaps spanning several subunits of what is expected to be a tetrameric protein complex. This interaction occludes the pore and blocks the ion conduction pathway through the maxi-K channel.

To develop the biochemistry and molecular pharmacology of the maxi-K channel, ChTX was radiolabeld to high specific activity with ^{125}I and adapted to probe the ChTX receptor in target tissues of interest through the establishment of a ligand binding assay [38]. The interaction of $[^{125}I]ChTX$ with maxi-K channels was investigated in two different smooth muscle membrane preparations; purified sarcolemmal membrane vesicles isolated from either bovine aortic [39] or tracheal smooth muscle [40]. Incubation of aortic sarcolemmal membranes with $[^{125}I]ChTX$ results in time- and concentration-dependent association of toxin with membranes. Under conditions of low ionic strength, $[^{125}I]ChTX$ binds a single class of sites with a K_d of 100 pM and a B_{max} of 0.5 pmoles/mg protein. This membrane preparation consists of a 50/50 mixture of outside-out and inside-out vesicles. Hence, addition of low concentrations of digitonin to disrupt the osmotic permeability barrier of the vesicles increases the B_{max} of $[^{125}I]ChTX$ binding two-fold as expected. The binding reaction is a freely reversible bimolecular process; the K_d calculated from the ratio of dissociation and association rate constants is equivalent to toxin affinity measured under equilibrium conditions.

$[^{125}I]ChTX$ binding in aortic sarcolemma displays many properties that would be predicted for toxin binding to the maxi-K channel [38]. The binding reaction is inhibited by a number of different monovalent and divalent cations which are known to interact with sites located along the ion conduction pathway of the channel [41–43]. For example, K^+, Cs^+, Ba^{2+}, and Ca^{2+} inhibit $[^{125}I]ChTX$ binding with K_i values equivalent to the concentrations needed for affecting maxi-K channels in functional experiments. The quaternary ammonium ion, tetraethyl-ammonium (TEA), is known to bind in the external vestibule of the maxi-K channel, and at low concentrations, is a relatively selective inhibitor of this channel [44]. This agent inhibits $[^{125}I]ChTX$ binding with a K_i that is equivalent to its potency as an inhibitor of channel activity in patch clamp experiments, and the mechanism of inhibition appears to be competitive. These data are consistent with electrophysiological results showing that TEA affects the on, but not the off, rate of ChTX binding through a competitive mechanism [45]. The ionic strength of the media also affects [^{125}I]ChTX binding, as it does in parallel electrophysiological experiments. The toxin interaction is enhanced as ionic strength is lowered and increasing ionic strength by raising either Na^+ or Li^+ concentration results in a very steep inhibition curve for toxin binding which is due solely to decreasing the rate of [^{125}I]ChTX association. The rate constant for [^{125}I]ChTX association in aortic sarcolemma as measured in a medium consisting of 20 mM NaCl, 20 mM Tris-HCl, pH 7.4, is $6.8 \times 10^7 \, M^{-1} \, sec^{-1}$, which is a value greater than that expected for free diffusion of a peptide the size of ChTX in solution. This fast rate constant is likely to be due to an electrostatic attraction between ChTX and the maxi-K channel that was previously observed in single channel recordings. When considered together, these results strongly indicate that the receptors characterized for [^{125}I]ChTX in bovine aortic sarcolemma are directly associated with maxi-K channels. Almost identical data have been obtained in binding studies with purified sarcolemmal membranes derived from bovine tracheal smooth muscle [40]. Moreover, similar results have been obtained for [^{125}I]ChTX binding to sarcolemmal membranes isolated from porcine uterine smooth muscle (M.L. Garcia and E. Stefani, unpublished observations) and rabbit skeletal muscle (M.L. Garcia and S. Fleischer, unpublished observations), two other tissues known to contain high levels of maxi-K channel activity.

In addition to its interaction with the maxi-K channel, ChTX is known to inhibit other types of K^+ channels, including small conductance Ca^{2+}-activated K^+ channels [46] and voltage-gated K^+ channels of the $K_v1.3$ type [47]. In the latter case, [^{125}I]ChTX has been used to characterize toxin binding to $K_v1.3$ in rat brain synaptic plasma membrane vesicles [48] and human T lymphocytes [49]. Although ChTX blocks relatively few types of K^+ channels, because it is not absolutely selective for one type of channel, it was necessary to identify even more specific peptide probes for the maxi-K channel. To this end, several other crude scorpion venom extracts were tested for their ability to inhibit [^{125}I]ChTX binding to aortic smooth muscle membranes. Two of the venoms which inhibited toxin binding were further analyzed. Fractionation of venom from the old world scorpion *Buthus tamulus* yielded a peptide that was purified to homogeneity, characterized both biochemically and electrophysiologically and which was named iberiotoxin (IbTX) [50]. IbTX is a 37 amino acid peptide that is 68% homologous with ChTX (Figure 2). The most remarkable difference between IbTX

Charybdotoxin

Iberiotoxin

Figure 2. Amino acid sequence homology between toxins. The primary amino acid sequences of charybdotoxin and iberiotoxin are shown using the single letter code for the amino acids. Identical residues are enclosed by boxes.

and ChTX is the presence of four more acidic amino acid residues in IbTX than in ChTX; this gives IbTX an overall less positive net charge. Despite the high degree of homology between the two toxins, IbTX appears to be a highly selective inhibitor for the maxi-K channel. This peptide does not inhibit [^{125}I]ChTX binding to $K_v1.3$ channels in either synaptic membranes [48] or lymphocyte plasma membranes [49], or block channel activity after expression of $K_v1.3$ in *Xenopus* oocytes [51]. Furthermore, it does not inhibit the ChTX-sensitive small conductance Ca^{2+}-activated K^+ channels present in human T-lymphocytes [52]. In addition, IbTX does not appear to be a strict competitive inhibitor of [^{125}I]ChTX binding to maxi-K channels in smooth muscle sarcolemma. The mechanism by which IbTX inhibits the maxi-K channel is similar to that of ChTX, except that the channel blocking kinetics are much different; the silent periods in single channel recordings which represent toxin bound to the pore are much longer with IbTX than with ChTX [53, 54]. This indicates that IbTX dissociates much slower from the maxi-K channel than ChTX. IbTX has been produced in biologically active form by both solid phase synthesis and by expression as a fusion protein in *E. coil*. Interestingly, the three dimensional structure of IbTX as determined in 2D NMR studies displays a backbone configuration which is nearly identical to that observed for ChTX [55]. IbTX is the most potent and selective peptidyl inhibitor of the maxi-K channel that has been described to date.

Two newly discovered toxins limbatoxin (LbTX) [56] and kaliotoxin [57] also displayed nanomolar affinity for maxi-K channels. These toxins have been purified, sequenced and are presently being used to study maxi-K channel architecture. The continuing discovery of new peptidyl inhibitors of the maxi-K channel, the synthesis of chimeric peptides [51] and site-directed mutagenesis studies are important ways for elucidating structure-activity relationships which will not only define those toxin residues that are important for interacting with the channel, but may also lead to identifying complementary regions comprising the architecture of the maxi-K channel pore.

3. Molecular Characteristics of the Maxi-K Channel

Given the existence of various peptidyl probes for the maxi-K channel, it should be possible to use such agents to obtain further information about the molecular structure of the channel itself. As a first approach, chemical modifications of ChTX were attempted in order to produce derivatives which could be covalently incorporated into the ChTX receptor. The modification of lysine ε amino groups with a potential photoaffinity crosslinking reagent (*e.g.*, derivatization with N-succimidyl-6(4′-azido-2′-nitrophenylamino)hexanoate) resulted in toxin molecule adducts which had lost their biological activity [58]. In retrospect, this result is not unexpected since Lys_{27} is critical for the ChTX-maxi-K channel interaction, and this residue is easily derivatized due to its exposed location on the surface of the toxin molecule. Modification of the C-terminus of ChTX using dicyclohexyl-carbodiimide (DCCD) coupling with a different photoaffinity crosslinking reagent was also unsuccessful, perhaps because in the structure of the peptide deduced from NMR studies the C-terminus is not freely exposed to solution. It was possible, however, to construct a biologically active biotinylated derivative of ChTX by solid-phase synthesis where biotin is linked to the N-terminus of the toxin through an alkyl spacer arm. Although this probe is not appropriate for crosslinking studies and was found not to be suitable in an affinity chromatography matrix for purification of the maxi-K channel, it was useful for *in situ* labeling of channel proteins at the neuromuscular junction and demonstrating spatial colocalization of the maxi-K channel with Ca^{2+} channels in this preparation [59].

Because of lack of success in incorporating a reactive functionality into the ChTX molecule, the strategy of using a bifunctional crosslinking reagent was considered [60]. In order to eliminate nonspecific interactions of ChTX with extraneous membrane proteins that might occur because of the highly charged nature of the toxin, special conditions were employed for these experiments. [^{125}I]ChTX was bound to either aortic or tracheal sarcolemmal membranes, the vesicles were isolated by centrifugation and then resuspended in media of high ionic strength to eliminate nonspecific binding of toxin. The bifunctional crosslinking reagent, disuccimidylsuberate, was then added to form a covalent bridge between [^{125}I]ChTX and its receptor. In analysis of either membrane preparation by SDS-PAGE, [^{125}I]ChTX was found to be specifically incorporated into a single membrane protein that displays an apparent molecular weight of approximately 35 kDa. The indication is that the crosslinking of [^{125}I]ChTX to this protein is specific because the reaction is blocked by nonradiolabeled ChTX, as well as by other agents that inhibit the binding reaction (*e.g.*, IbTX, TEA and K^+). This pattern strongly suggests that the 35 kDa protein identified in crosslinking studies with aortic and tracheal sarcolemma displays the pharmacological properties expected of the ChTX receptor.

The maxi-K channel was purified to homogeneity from bovine tracheal smooth muscle by a combination of conventional chromatographic techniques and sucrose density gradient centrifugation [61]. After extracting the receptor with digitonin, it was subjected to a seven step purification scheme consisting of ion exchange, lectin affinity and hydroxylapatite column chromatographies, as well as to two sucrose density gradient centrifugation steps. At the end of this rather protracted two week procedure, fractions with the highest specific activity of [^{125}I]ChTX binding were enriched approximately 2000-fold over the initial digitonin solubilized extract to a specific activity of approximately 1 nmol [^{125}I]ChTX bound/mg protein. A final yield of 1% of the initial ChTX binding activity was detected in the peak fractions, and 32 μg of apparently homogeneous protein was isolated from 10 grams of sarcolemmal membrane vesicle protein. Silver staining after SDS-PAGE of the material from the last step of the purification demonstrates that [^{125}I]ChTX binding activity is correlated with a major component in the preparation that displays an apparent MW of 62 kDa. However, when these same fractions were alkylated with [^{125}I]Bolton-Hunter reagent, a protein modifying reagent which forms a covalent adduct with the ε amino group of lysine residues, two protein species were detected; a 62 kDa (α) and a 32 kDa (β) subunit. These proteins are in a 1:1 stoichiometry and comigrate exactly with binding activity. The β subunit is heavily glycosylated and this may account for its poor staining properties after SDS-PAGE. Deglycosylation studies demonstrate that the β subunit possesses two N-linked sugar chains and that it is the protein to which [^{125}I]ChTX is covalently incorporated with the bifunctional crosslinking reagent used. After covalent attachment of ChTX, the β subunit migrates at slightly higher MW due to the additional size of the incorporated peptide. Indentical results, in terms of subunit size and composition, have been obtained after purification of the ChTX receptor to homogeneity from bovine aorta (K. Giangiacomo, M. Garcia-Calvo, H.G. Knaus, M.L. Garcia, G.J. Kaczorowski and O.B. McManus, unpublished observations).

Binding of [^{125}I]ChTX to the purified tracheal ChTX receptor displays the same pharmacological properties that have been observed for the membrane bound and digitonin-solubilized receptor [61]. Thus, toxin binding is modulated by both peptidyl (*i.e.*, ChTX, IbTX, LbTX) and small molecule (*i.e.*, TEA, K^+, Ba^{2+}, and Cs^+) effectors of the binding reaction. Most importantly, when the purified ChTX receptor was reconstituted into liposomes that were then fused with planar lipid bilayers, maxi-K channel activity was readily observed in single channel recordings. These channels displayed a conductance of 235 picosemens (ps) in 150 mM KCl and were highly selective for K^+ over Cl^-. The channels were inhibited by ChTX and the toxin displayed the same blocking pattern as had previously been observed with the native channel. In addition, the open probability of these channels was increased by depolarizing membrane potentials and by raising Ca^{2+} concentrations at the intracellular face of the channel. Similar functional data were obtained by reconstituting the purified ChTX receptor from aorta into artificial lipid bilayers. Taken together, these results strongly suggest that the ChTX receptor purified from either trachea or aorta is composed of two subunits and that these proteins are able to functionally reconstitute maxi-K channel activity.

To further analyze the subunit composition of the maxi-K channel in smooth muscle, the target size of the ChTX receptor was determined by radiation inactivation analysis [62]. Exposure of either bovine aortic or tracheal sarcolemmal membranes to high energy irradiation results in disappearance of [^{125}I]ChTX binding activity as a monoexponential function of radiation dose, from which a molecular means of 90 kDa can be calculated. The effect of radiation on toxin binding is to decrease the number of functional ChTX receptors without affecting the affinity of receptors for toxin. This indicates that radiation is destroying rather than altering the ChTX binding site. As a control, the radiation inactivation technique was used to determine the target sizes of the α_1 subunit of the L-type Ca^{2+} channel and of 5' nuclcotidase in the same membrane preparations. The molecular masses determined for these entities were in excellent agreement with those expected from previous studies, validating the techniques used here. Thus, the ChTX receptor is likely to be associated with a single α-β heterodimer complex of 90 kDa. In analogy with other K^+ channels, it is tempting to speculate that association of four 90 kDa heterodimers may be required for functional activity. A molecular mass for the intact maxi-K channel of approximately 400 kDa would be consistent with the size of the purified ChTX receptor after sucrose density gradient centrifugation.

The structure of the α subunit of the maxi-K channel was determined using the purified tracheal ChTX receptor [63]. After subjecting this preparation to SDS-PAGE, the 62 kDa protein was electroeluted from the gel and treated with trypsin to effect proteolytic degradation. A number of fragments were subsequently isolated by microbore C_{18} reversed phase HPLC and microsequenced using Edman degradation techniques. Amino acid sequence information obtained from seven fragments revealed the existence of very high sequence homology with the Ca^{2+} activated K^+ channel, *slowpoke*, cloned from *Drosophila* [11], and with the recently identified maxi-K channel, *mSlo*, cloned from mouse brain and skeletal muscle [13]. Both of these proteins when expressed in *Xenopus* oocytes yield Ca^{2+}-activated K^+ channel currents, albeit with different properties. The putative topology of both proteins also resembles that of other K^+ channels in that they have six transmembrane domains and a pore region. However, the Ca^{2+}-activated K^+ channels have a large C-terminal domain which may form additional transmembrane regions. Therefore, the α subunit of the purified ChTX receptor from tracheal smooth muscle is the pore forming structure of the maxi-K channel in this tissue. Interestingly, the molecular mass of the α subunit, 62 kDa, is much lower than the deduced mass of *mSlo* (*i.e.*, 135 kDa). Possible explanations for this descrepancy include tissue specific differences in channel expression, post-translational modification of the extensive carboxyl-terminal region of *mSlo*, or proteolytic degradation of the α subunit of the ChTX receptor during purification. Although the latter possibility has not yet been ruled out, the purified ChTX receptor forms fully functional maxi-K channels. Further investigation will be needed to resolve the discrepancy between the expected and detected size of the tracheal smooth muscle maxi-K pore forming subunit.

To obtain further evidence that the α and β subunits of the maxi-K channel are specifically associated with each other in a complex, a sequence-directed antibody was raised against a synthetic peptide corresponding to one of the identified fragments of the α subunit of the tracheal channel [63]. This antibody was capable of specifically immunoprecipitating not only the denatured α subunit which had been covalently labeled with [^{125}I]Bolton-Hunter reagent, but also the labeled α and β subunit complex when the immunoprecipitation was carried out under non-denaturing conditions. These data demonstrate that a specific non-covalent association exists between the α and β subunits, a finding which would be predicted given the fact that both subunits remain associated through seven steps of purification.

The structure of the β subunit of the maxi-K channel was also determined for the protein isolated from bovine tracheal smooth muscle [64]. the β subunit was isolated by electroelution following SDS-PAGE of the purified ChTX receptor preparation. After digestion with V_8 endoproteinase Glu-C, proteolytic fragments were isolated on a C_4 reversed phase HPLC column and subjected to microsequencing using Edman degradation techniques. The amino acid sequence of one of these fragments was used to design an oligonucleotide probe which was then used to screen a cDNA library constructed from bovine tracheal smooth muscle. Using this procedure, a cDNA encoding the β subunit protein was isolated. This cDNA encodes a protein of 191 amino acids that contains two hydrophobic (putative transmembrane) regions, and a large extracellular domain. Two glycosylation sites for N-linked sugars are present in this domain, consistent with biochemical evidence obtained from studies of the purified subunit. In addition, there are 3 Lys residues present which are potential sites for crosslinking with ChTX. A protein kinase A recognition sequence is present on the short stretch of N-terminal residues which are located at the cytoplasmic surface. The cloned β subunit from bovine aorta is identical in sequence with the tracheal protein, and neither of these proteins display homology with subunits of other known ion channels.

Antibodies were raised against amino acid sequences present in the putative extracellular domain of the β subunit [64]. These antisera, under denaturing conditions, specifically immunoprecipitate the [^{125}I] Bolton-Hunter labeled β subunit, as well as the subunit to which [^{125}I]ChTX had been crosslinked with the bifunctional reagent. However, under non-denaturing conditions, the anti-β antibodies were able to immunoprecipitate a complex consisting of both α and β subunits. These latter results are completely consistent with data described above in which anti-α antibodies could also immunoprecipitate the α-β complex. When considered together, these data demonstrate that this is the first β subunit of a K^+ channel to be cloned to date, that the maxi-K channel consists of a multimer of α and β subunits, and that the β subunit is the protein to which ChTX becomes specifically and covalently incorporated into during crosslinking procedures. Since ChTX binds to the pore (α subunit) of the maxi-K channel, the β subunit must be in close proximity (less than 12 Å) to the pore forming subunit given that the bifunctional crosslinking reagent, disuccimidylsuberate, can bridge the distance between the two proteins. In order to determine potential regulatory functions of the maxi-K channel's β subunit, coexpression studies with the α subunit must be accomplished.

4. Molecular Pharmacology of the Maxi-K Channel

Although a number of peptidyl inhibitors of the maxi-K channel have been discovered and characterized, very few small organic molecules, besides TEA, were identified as modulators of this channel. As one approach to developing the molecular pharmacology of the maxi-K channel, [^{125}I]ChTX binding has been used to screen for potential channel effectors. This procedure is useful for assaying both synthetic molecules and natural product extracts. Using this strategy, either competitive or allosteric modulators of ChTX binding can be detected. However, such agents must be analyzed by electrophysiological techniques to determine their functional effect on channel activity. By employing [^{125}I]ChTX binding, both potent maxi-K channel agonists and antagonists have been discovered.

Desmodium adscendens is a medicinal herb used in Ghana as a treatment for asthma, dysmenorrhea and other diseases associated with disfunction of smooth muscle contraction [65]. Extracts from this plant can inhibit contractions of guinea pig ileum caused by electrical field stimulation [66], or contraction of sensitized guinea pig airways smooth muscle induced by antigen, arachidonic acid or leukotriene D_4 [67, 68].

Although the plant contains many different constituents that can contribute to the relaxation of smooth muscle tone (*e.g.*, inhibitors of arachidonic acid metabolism), these extracts also inhibit the binding of [^{125}I]ChTX to maxi-K channels in bovine tracheal smooth muscle sarcolemma [69]. The ChTX binding assay was used to purify three components from this plant that affect the toxin binding reaction, and these structures were determined by NMR and mass spectroscopy to be known glycosylated triterpenoid molecules; dehydrosoyasaponin I (DHS-I), soyasaponin I and soyasaponin III.

The most potent of these compounds, DHS-I, is a partial inhibitor of [^{125}I]ChTX binding ($K_i = 120$ nM, 62% maximal inhibition; [69]). The other two active extract components, soyasaponin I and soyasaponin III, inhibit ChTX binding with K_i's of 6 and 1 µM, respectively. Interestingly, the aglycone, soyasapogenol B, is inactive in binding at concentrations up to 100 µM, indicating that both sugar and terpenoid moieties are required for activity. Inhibition of [^{125}I]ChTX binding as analyzed in a Scatchard representation is due to a decrease in the maximum number of toxin binding sites, with only a small decrease in toxin affinity. Given the partial inhibitory effects produced by DHS-I and the other soyasaponins on [^{125}I]ChTX binding, it is likely that these agents modulate the interaction of ChTX with its receptor by an allosteric mechanism. Consistent with this notion, DHS-I increases the rate of [^{125}I]ChTX dissociation from the maxi-K channel in kinetic experiments.

In single channel recordings of the tracheal maxi-K channel reconstituted in planar lipid bilayers, DHS-I reversibly increases the open probability of the channel [69]. Strikingly, this only occurs when DHS-I is added at the intracellular face of the channel, and activation is manifested at concentrations of compound as low as 10 nM. In marked contrast, extracellular application of DHS-I at concentrations as high as 500 nM have no effect whatsoever on maxi-K channel activity. Since ChTX binds in the extracellular mouth of the channel, and DHS-I is only effective at the channel's intracellular surface, this is another indication that the glycosylated triterpenes are allosteric modulators of toxin binding. In further confirmation of this hypothesis, addition of DHS-I to the intracellular side of the bilayer in electrophysiological experiments caused a 2.3-fold increase in the rate of ChTX dissociation from its external binding site on the channel. As a measure of specificity, DHS-I had no effect on several other types of K^+ channels or membrane transporters. Thus, the glycosylated triterpene structural series is the first example of a high affinity agonist of the maxi-K channel, and, when compared with K^+ channel openers which affect other types of K^+ channels (*e.g.*, the cromakalim series of ATP-dependent K^+ channel activators), this natural product is the most potent K^+ channel opener discovered to date.

The soyasaponin components of *Desmodium adscendens* may explain some of the therapeutic utility of this medicinal herb in asthma and dysmenorrhea. The ability of DHS-I to open maxi-K channels could explain, at least in part, the antispasmogenic and spasmolytic effects of the plant extract observed *in vitro*, and may contribute to the known *in vivo* therapeutic effects. Activation of maxi-K channels would be expected to cause membrane hyperpolarization, to suppress electrical activity and to promote smooth muscle relaxation [70]. However, the intracellular site of action of these compounds, taken together with their relatively membrane impermeant nature, would suggest that the *in vivo* effects of the soyasaponins should have a slow onset, and be much weaker than the *in vitro* effects measured in binding and electrophysiological experiments. This may correlate with the fact that the medicinal herb must be given chronically to elicit its therapeutic effects. Perhaps the soyasaponins are metabolized *in vivo* to other species which are active intracellulary. In any event, the discovery of DHS-I and its congeners as maxi-K channel activators suggests that there is an agonist binding site on the channel at which other structurally dissimilar agents could bind.

The first structural series of potent nonpetidyl maxi-K channel blockers was also discovered in the [^{125}I]ChTX binding assay. Tremorigenic indole alkaloids [71] produce neurological disorders (*e.g.*, staggers syndrome) in ruminants. The mechanism of action of these fungal mycotoxins is unclear, but may be related to their known ability to potentiate the release of certain neurotransmitters [72–74]. To determine whether these effects are due to inhibition of K$^+$ channels, the interaction of various indole diterpenes with the maxi-K channel was monitored [75]. In this structural series, paspalitrem A, paspalitrem C, aflatrem, penitrem A and paspalinine inhibit binding of [^{125}I]ChTX to maxi-K channels in bovine aortic sarcolemmal membranes. Inhibition of toxin binding by these compounds occurs with a defined rank order of potency, and inhibition can be either partial or complete depending on the compound tested. In marked contrast, three structurally related compounds, paxilline, verruculogen and paspalicine enhance [^{125}I]ChTX binding to membranes in a concentration-dependent fashion. Consistent with their pattern of effects on toxin binding, covalent incorporation of [^{125}I]ChTX into the β subunit of the maxi-K channel with the bifunctional crosslinking reagent was blocked by compounds that inhibit [^{125}I]ChTX binding and enhanced by compounds that stimulate [^{125}I]ChTX binding. These results are an indication that the interaction of the indole alkaloids with the maxi-K channel occurs through a specific process.

Modulation of [^{125}I]ChTX binding by the indole diterpenes is due to an allosteric mechanism [75]. In a Scatchard analysis, stimulators of toxin binding caused an apparent increase in toxin affinity, while inhibitors of toxin binding caused a marked reduction in the maximum density of receptor sites. In kinetic experiments monitoring ChTX dissociation, stimulators of $[^{125}I]$ChTX binding decrease the rate of toxin dissociation, whereas inhibitors of binding have mixed effects; some markedly stimulate toxin dissociation, while others produce no significant effects. These data imply that the indole alkaloids are likely to bind to a unique site on the maxi-K channel where they can affect ChTX binding in the mouth of the channel by either positive or negative heterotropic interactions. Whether all of these agents share a common site on the channel is unknown, but the stimulation of $[^{125}I]$ChTX binding caused by either paxilline or verruculogen is prevented by aflatrem. Such data suggest that the indole diterpenes might compete for the same receptor.

Despite their different effects on binding of $[^{125}I]$ChTX to maxi-K channels, all of the compounds investigated potently inhibit maxi-K channels in electrophysiological experiments [75]. This was demonstrated by monitoring single channel activity in either excised membrane patches from bovine aortic smooth muscle cells, or after reconstituting the channel into bilayers. Channel inhibition by these compounds occurs with IC_{50}'s in the 1–10 nM range, and intracellular application of the indole alkaloids appeared to produce more potent channel block than did extracellular application. Moreover, the patterns of channel block produced by the various agents appear distinct. For example, paxilline causes the appearance of long silent periods devoid of channel activity, much like the pattern produced by ChTX, while paspalitrem C causes numerous brief (< 100 msec) interruptions in channel activity. The ability of these indole alkaloids to block maxi-K channels appears related to the state of the channel because when internal Ca^{2+} is elevated and the channel open probability is high, these compounds are weaker inhibitors. Interestingly, paspalicine, a des-hydroxy analog of paspalinine lacking tremorigenic activity also potently blocked maxi-K channels. These data, when considered together with the results from binding experiments, suggest a very specific interaction of the indole diterpenes with the maxi-K channel.

As a further indication of the specificity of these compounds, their effects on a number of different voltage-gated and Ca^{2+}-activated K^+ channels were examined [75]. At 100 nM, paspalitrem C had no effect on the three types of small conductance Ca^{2+}-activated K^+ channels in human T-lymphocytes, or on the $K_v1.3$ voltage-gated K^+ channel in these cells. None of the compounds affected $[^{125}I]$ChTX binding to voltage-gated K^+ channels in brain synaptic plasma membranes. Paxilline was a weak blocker of the delayed rectified K^+ channel in pancreatic β cells at 2 μM, but had no significant effect on L-type Ca^{2+} or neuronal voltage-gated Na^+ channels at 10 μM. In addition, simple chemical modifications of paxilline destroyed its ability to inhibit maxi-K channels, indicating that a defined structure-activity relationship exists for the interaction of these compounds with this channel. Together, the data suggest that the indole diterpenes are the most potent and selective nonpeptidyl inhibitors of the maxi-K channel identified to date. Some of their pharmacological properties could be explained by inhibition of maxi-K channels, although tremorigenicity appears not to be strictly related to channel block.

In addition to the agonist and antagonist series of maxi-K channel modulators which have been identified in $[^{125}I]$ChTX binding studies, several other small organic molecules have been described which affect the activity of this channel. Although the cromakalim series of ATP-dependent K^+ channel openers are not effective as direct acting agonists of the maxi-K channel [69], one compound in this structural class, BRL55834, is now reported in patch clamp experiments to promote opening of both ATP-dependent and maxi-K channels in bovine airways smooth muscle cells [76]. *In vivo*, this compound displays some selectivity as a bronchodilator when compared with its ability to lower arterial blood pressure, and this profile might be related to its effects on airway maxi-K channels [77, 78]. Recently, a new structural series of substituted benzimidazolones have been reported to be direct acting agonists of the maxi-K channel [79]. A member of this group, NS1619, reversibly activated maxi-K channels in bovine aortic smooth muscle at micromolar concentrations by shifting the activation curve towards negative membrane potentials [80]. However, questions have been raised regarding the selectivity of such molecules because another maxi-K channel agonist in this series, NS004, displays significant Ca^{2+} entry blocker activity [81]. Therefore, pharmacological experiments performed with benzimidazolones may be complicated due to their potential functional effects on a number of different ion channels. Lastly, the novel imidazole pyrazine derivative, SCA 40, has been reported to be an agonist of the maxi-K channel in guinea pig trachealis based on a pharmacological profile in which smooth muscle relaxation induced by this compound is blocked by either elevated K^+ or ChTX [82]. However, no electrophysiological evidence has been presented to demonstrate that this compound is a direct acting maxi-K channel agonist. Given its pharmacological effects, SCA 40 could also be promoting the opening of maxi-K channels indirectly by acting on a pathway which regulates channel activity. The structures of several maxi-K channel modulators are shown in Figure 3.

5. Pharmacological Modulation of the Maxi-K Channel (Functional Studies)

The first measurements of potassium channels in airways smooth muscle cells were made by McCann and Welsh in canine tracheal myocytes

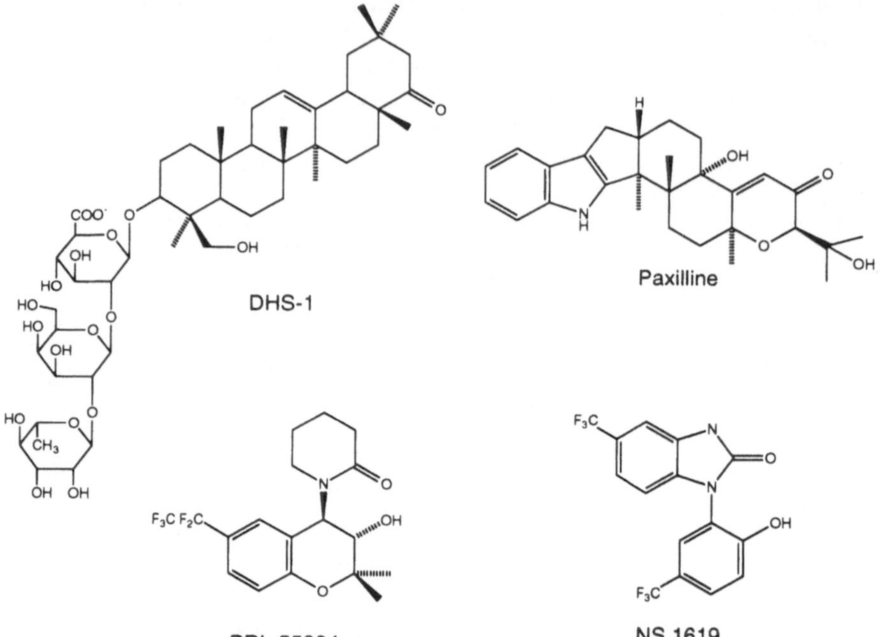

Figure 3. Structures of maxi-K channel modulators. Various compounds are depicted which have been shown to function as maxi-K channel modulators. See text for a description of their activities.

[83]. They used patch clamp recording techniques to demonstrate the presence of large conductance, Ca^{2+}-sensitive potassium channels in this tissue. These observations have since been confirmed in several laboratories using a variety of species including dog [84–86] pig [86–89] guinea pig [90] and rabbit [91–93]. The maxi-K channel in these preparations has a conductance of 120–290 pS and its open state probability is markedly increased by elevations in cytosolic Ca^{2+} and by increases in pH [92]. These channels are functionally distinct from other potassium channels such as the apamin-sensitive small conductance Ca^{2+}-activated K^+ channel, K_{ATP}, and voltage-gated K^+ channels since they are not blocked by apamin [94], glibenclamide [86, 95] or 4-aminopyridine [84, 86], but are blocked by IbTX, ChTX [86, 96, 97] and low concentrations of TEA [86]. Recent observations that ChTX [98] and IbTX [99, 100] can preferentially inhibit the relaxant responses of β-agonists in airways smooth muscle has prompted renewed interest in the role of K^+ channel opening in mediating the bronchodilator actions of these and other relaxants. Augmentation of potassium channel activity by β-agonists is a commonly observed feature of their action [101,

102] and is consistent with the suggestion that maxi-K$^+$ channels are in part responsible for their relaxant action on airways smooth muscle [91, 98].

Results from recent studies using ChTX and IbTX, [98, 100], indicate that these channels provide a pathway that can control the ionic homeostasis and hence the intrinsic contractile state of guinea pig airways smooth muscle. Similar observations have been made with ChTX on rat vascular smooth muscle [17], guinea pig airway smooth muscle [98, 99], human bronchial smooth muscle [103] and guinea pig vascular, bladder and gut smooth muscle [16]. High potassium channel conductance in airways smooth muscle has been suggested to account for the electrical quiescence of this tissue [83, 84, 104]. TEA and 4-AP, both K$^+$ channel blockers, have been known for some time to initiate action potentials and increase muscle tone in bovine [104] canine [84] and guinea pig [105] tracheal smooth muscle. The activity of K$^+$ channels in airways smooth muscle, therefore, may play a critical role in regulating intrinsic muscle tone. Recent studies with ChTX and IbTX provide support for the hypothesis that maxi-K channels in particular may be important regulators of intrinsic tone in guinea pig airways [98, 100, 106]. Whether this is unique to guinea pig trachealis and is solely related to direct effect on muscle is not known. Species differences are a possibility since a number of investigators have reported that in bovine and canine trachealis, these channels are unlikely to be open under resting conditions [83, 96, 89].

Recent findings from a number of laboratories corroborate the hypothesis that maxi-K channels are involved, in part, in mediating relaxation of guinea-pig trachea induced by β-adrenoceptor agonists. β-Agonists may require functional maxi-K channels in order to produce complete reversal of carbachol-induced tone in guinea-pig trachealis. The involvement of these IbTX-sensitive maxi-K channels may extend to a component of the relaxant response of other relaxants such as aminophylline and dibutyryl cAMP. IbTX-sensitive maxi-K channels, do not appear to be contributing to the response produced by known agonists of other types of K$^+$ channels such as cromakalim [17, 98, 100]. This hypothesis is supported by parallel observations in depolarized tissues. In the presence of depolarizing concentrations of KCl (hypertonic addition), the relaxant effects of isoprenaline and salbutamol were inhibited [98]. Similar, but less dramatic results, were obtained with K$^+$ rich media with the same osmolarity as Krebs' solution [100]. These findings are consistent with previous electrophysiological results which demonstrated that the β-agonist, isoprenaline, induced hyperpolarization of guinea-pig trachea [101]. It has been reported that isoprenaline aminophylline and dibutyryl cAMP produce hyperpolarization in guinea-pig tracheal preparations which is not considered essential for relaxation of spontaneous tone or histamine-induced tone [109].

Regardless, the membrane potential changes evoked in airways smooth muscle by β-agonists provide evidence to suggest that this class of bronchodilators promote the opening of membrane associated K^+-channels. In canine and bovine trachealis for example, isoprenaline causes hyperpolarization of the cell membrane [104, 110–114]. In guinea pig trachealis, adrenaline, isoprenaline and salbutamol suppress the spontaneous mechanical activity of the tissue and also suppress spontaneous electrical slow wave activity [101, 102, 115–117]. Hyperpolarization induced by β-agonists in tracheal smooth muscle is consistent with an increase in membrane conductance that involves the opening of K channels [101, 111, 112, 115]. Direct evidence to support this contention comes from studies which demonstrated that β-agonist could promote the efflux of $^{86}Rb^+$ from bovine trachealis [118]. Moreover, studies with selective blockers of β_1 and β_2 adrenoceptors provided support for the notion that both receptor subtypes could hyperpolarize guinea pig tracheal cell membranes and increase $^{86}Rb^+$ efflux and thus by inference open K^+ channels [116]. Somewhat inconsistent with this hypothesis are the results with salmeterol which failed to hyperpolarize guinea pig trachealis [116]. Unpublished data (T.R. Jones, L. Charette and C. Misquitta) demonstrated that low concentrations of IbTX blocked salmeterol reversal of carbachol tone. Alternative mechanisms have been suggested to explain why ChTX and high K^+ [99, 117] antagonize β agonist-induced relaxation. For example, a portion of the antagonism of β agonists by ChTX or K^+ rich medium may reflect functional antagonism due to activation of Ca^{2+} influx through L-type Ca^{2+} channels. This cannot explain the inability of IbTX to inhibit relaxation to cromakalim or pinacidil [100], nor can it explain the reduced blockade of other relaxants such as aminophylline and dibutyryl cAMP [100]. Huang et al. [99] reported that either $CdCl_2$ or nifedipine reversed the antagonism of salbutamol by IbTX. However, functional studies of this sort have their limitation, and may not be used alone to determine the degree of agonist coupling to maxi-K channels. Most functional studies, however, do concur that the interactions of IbTX with bronchodilators can be grouped into three general categories. Group I consists of compounds such as the β agonists salbutamol and isoprenaline, and the cyclic GMP elevator, sodium nitroprusside, which are non-competitively inhibited by IbTX. Group II contains compounds like aminophylline, dibutyryl cAMP, adenosine and forskolin which are competitively shifted to the right 4–7 fold. Group III contains compounds such as cromakalim and pinacidil which are not antagonized and may actually be potentiated by maxi-K channel blockers [98, 100]. Compounds within group II may include compounds that are modulated as a result of functional antagonism by IbTX and other stimuli that depolarize smooth muscle cell membranes. Obviously, the compounds in group I are of particular interest since relaxation to these agents can be completely blocked by maxi-K channel blockade. The finding that β agonists and NO donors such as sodium nitroprusside and glyceryl trinitrate fall into the same group is particularly interesting since β-receptors and NO-mediated relaxation mechanisms are physiologically relevant regulators of airway calibre [98, 100, 119]. Moreover the role of soluble and particulate guanylyl cyclase in these cyclic nucleotide-dependent pathways remains to be determined.

Patch clamp studies have suggested that β-agonists can promote K^+ channel opening in rabbit tracheal cells [91] through cyclic AMP-dependent protein kinase A mechanism and more recently through a mechanism not involving cAMP [97]. The traditional phosphorylation-dependent coupling of β-receptors and maxi-K channels was supported by results with a phosphatase inhibitor. In addition, it now appears that this is not the only mechanism by which β-receptors couple to maxi-K channels. Recent experiments have demonstrated that maxi-K channels are activated by isoprenaline in outside-out patches which are unlikely to allow phosphorylation to occur [97]. Moreover, the α-subunit of the stimulatory G-protein, G_s, directly activates maxi-K channels in [97] inside-out patches and this activation is not affected by an inhibitor of cAMP-dependent protein kinase. It this appears that G_s linked receptors can regulate maxi-K channels directly through membrane-delimited activation.

Other membrane ion channels have been reported to be regulated directly by G proteins independent of secondary transduction messages. Recent studies in other laboratories now confirm the hypothesis that maxi-K channels can indeed be regulated through membrane delimited (presumably direct) stimulatory action of G proteins on ion channels. Moreover, channel opening can be inhibited by the action of the muscarinic agonist methacholine through a pertussis-toxin sensitive G protein. These results support the hypothesis that maxi-K channels can be regulated by stimulatory and inhibitory G proteins. This regulation is similar to adrenergic/cholinergic control of adenylyl cyclase and suggests that hormone regulation of contractility in airways smooth muscle may be under cAMP-dependent and cAMP-independent control. The physiological implication of cAMP-independent control of channel activity is not known, however such a pathway may be important since many studies have failed to demonstrate a positive correlation between β agonist induced relaxation and elevations in cAMP. This is particularly evident following activation of β receptors with low doses of agonist. Such a pathway may be of particular importance in normal physiological regulation of airway calibre where very low circulating levels of catecholamines are able to maintain homeostasis. It is interesting to speculate that asthmatics may well have a defect in β receptor coupling to maxi-K channels.

Parallel studies on isolated smooth muscle cells from bovine trachea showed that activation of β_2-adrenoceptors can stimulate L-type calcium currents in cells, independent of cAMP kinase [120]. This effect on I_{Ca} parallels the effect of β-adrenoreceptor stimulation in the heart, except the effect in heart appears to be a cAMP-dependent process. Recent studies in tracheal cells suggest that β-adrenoceptors regulate I_{Ca} directly without activation of cAMP kinase, and thus, without phosphorylation of the α_1-subunit of the calcium channel in tracheal smooth muscle [97]. These results at first would appear to be inconsistent with the well known fact that stimulation of the airway β adrenoceptors induces relaxation. However, activation of Ca^{2+} entry through L-type Ca^{2+} channels in response to β agonists may provide the necessary Ca^{2+} signal for initially turning on Ca^{2+}-dependent maxi-K channels. Our results demonstrate that β-adrenoceptor activation, in addition to elevating cAMP, decreasing myosin light chain kinase phosphorylation, and decreasing intracellular free Ca^{2+}, could lead to a hyperpolarization and thus closure of L-type Ca^{2+} channels. The end result would be relaxation but, the question remains; which pathway(s) is physiologically and pharmacologically operative? Moreover, the degree to which these pathways cross talk remains to be determined. This becomes critical since muscarinic/adrenergic control of adenylyl cyclase [121–123] seems to parallel muscarinic/adrenergic control of maxi-K channels. Functional studies implicating maxi-K channels in β_2 adrenoceptor mediated relaxation were the most convincing in the presence of cholinergic receptor activation [98, 100]. Studies in which tone was increased with other spasmogens (e.g., LTD_4 or intrinsic tone) support the hypothesis that muscarinic receptor activation seems to unmask this interaction. It seems likely that β receptor coupling to maxi-K^+ channels may be more than supportive [116] when relaxation is induced in the presence of cholinergic tone. A schematic representation of the involvement of K^+-channels in cyclic nucleotide dependent and independent control of airways smooth muscle contractile mechanism is displayed in Figure 4.

Neuronal plasma membranes are also known to contain a variety of functional K^+ channels which are thought to play important roles in modulating both myogenic and neurogenic responsiveness within the lungs [124]. In general, K^+ channels act to control the excitability of nerve cells by regulating the membrane potential which in turn can affect the amount of neurotransmitter released from nerve terminals. Compared to Na^+ and Ca^{2+} channels, relatively little work has been performed on the pharmacology of K^+ channels in nerve cells. TEA was the first K^+ channel blocking drug to be used; however, it is a weak inhibitor and is nonspecific. Aminopyridines, such as 4-AP are known to be somewhat more selective in that they only block voltage-dependent K^+ channels in neuronal membranes. Blockade of K^+ currents at

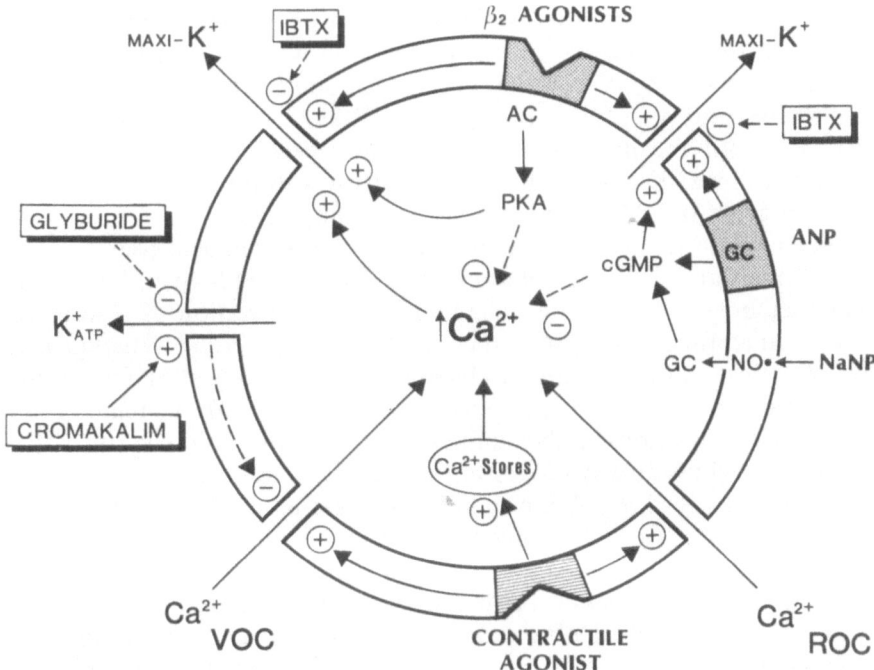

Figure 4. Potential sites of interaction between K-channel modulators and relaxation/contractile mechanisms in airways smooth muscle.
Abbreviations: **Maxi-K$^+$** – high conductance Ca^{2+}-activated K$^+$ channel; **IbTX** – iberiotoxin; **GC** – guanylyl cyclase; **ANP** – atrial natriuretic peptide; **NaNP** – sodium nitroprusside; **NO** – nitric oxide; **Ca^{2+}$_{ROC}$** – receptor operated Ca^{2+} channel; **Ca^{2+}$_{VOC}$** – voltage operated Ca^{2+}-channel; **K$^+$$_{ATP}$** – adenosine triphosphate dependent K$^+$-channel; **AC** – adenylyl cyclase; **PKA** – protein kinase A; (+) – stimulation; (−) – inhibition. Arrows within membrane depict membrane delimited activation of maxi-K channel independent of cyclic nucleotides.

motor nerve terminals by 4-AP prolongs depolarization and increases the release of acetylcholine. Relatively little has been published about ATP-regulated channels in nerve cells, but pharmacological studies with cromakalim and glybenclamide [124] suggest a role for these channels in modulating excitatory NANC and cholinergic neurotransmission within the lungs. It has also been proposed that maxi-K channels are involved in activation of certain presynaptic receptors in both guinea pig and human airway nerve teminals. For example, inhibition of eNANC responses and cholinergic responses in guinea pig airways smooth muscle by the μ opioid agonist, DAMGO, involves activation of maxi-K channels, since IbTX reverses this effect. Earlier work with ChTX was consistent with this report [103], however some caution is required when using charybdotoxin as a specific marker for maxi-K channels. Recent studies by Baker *et al.*, [125] however, failed to provide evidence for IbTX sensitive maxi-K channels in cholinergic nerves. Preliminary results from our laboratory indicate that the phasic contractile effects of IbTX on isolated guinea pig trachea treated with indomethacin can be partly reversed by a combination of atropine, the NK_2 antagonist, SR-48968 and the NK_1 antagonist, CP-99,994 (T.R. Jones and L. Charette, unpublished observations). This raises the possibility that the actions of maxi-K channel blockers in airways smooth muscle may have a neurogenic in addition to a myogenic component.

In vivo pharmacological studies suggest a physiological role for maxi-K channels and other K^+ channels sensitive to ChTX (*e.g.*, K_v and low conductance Ca^{2+}-activated channels) in the regulation of resting cardiovascular function in anaesthetized guinea pigs [126]. Whether these *in vivo* observations relate to blockade of endogenous NO-mediated relaxation is not known. A role for maxi-K channels in the regulation of pulmonary function in this animal model seems unlikely since there were no obvious airway effects following intravenous administrations of IbTX or ChTX [126]. A minor role for maxi-K channels in the mediation of the *in vivo* β agonist activity of exogenously administered salbutamol versus NKA is suggested but the significance of these findings will have to await further studies with selective maxi-K channel agonists. Unpublished observations (P. Masson, D. Foulon and T.R. Jones) showed that IbTX potentiated bronchoconstriction evoked by both NKA and substance P suggesting a possible role for maxi-K-channels in the modulation of neurokinin action in the lung consistent with the *in vitro* observations.

6. Therapeutic Opportunities

The therapeutic impact of maxi-K channel modulators in airways diseases will ultimately depend upon the demonstration that they offer improvements over existing therapies and exhibit fewer adverse effects. For example, currently employed bronchodilator drugs such as β_2 adrenoceptor agonists have beneficial effects, but these effects do not extend beyond smooth muscle relaxation. It will be important to identify properties of maxi-K channel activators which will be more than simple bronchodilators. This may well be realized since recent studies in guinea pig airways suggest that these channels may play an important role in down modulating neuronally mediated bronchoconstriction and airways inflammation. The hallmark of asthma is hyperreactivity (twitchy airways), so an understanding of the role played by maxi-K^+ channels in modulation of myogenic, neurogenic and inflammatory responses in the lungs could provide much needed fundamental information about this disease and its progression. More importantly, molecular probes should soon be available to determine whether significant changes occur in the number and/or location of these channels in lung tissues under pathophysiological conditions.

References

1. Jan LY, Jan YN. Structural elements involved in specific K^+ channel functions. Ann Rev Physiol 1992; 54: 537–55.
2. Pongs O. Molecular biology of voltage-dependent potassium channels. Physiol Rev 1992; 72: S69–S88.
3. Catterall, WA. Structure and function of voltage-sensitive ion channels. Science 1988; 242: 50–3.
4. Pongs O. Molecular basis of potassium channel diversity. Pfluger's Arch 1989; 414: 71–5.
5. Tempel BL, Papazian DM, Schwarz TL, Jan YN. Sequence of a probable potassium channel component encoded at the *Shaker* locus of *Drosophila*. Science 1987; 237: 770–5.
6. Pongs O. Structure-function studies on the pore of potassium channels. J Membrane Biol 1993; 136: 1–8.
7. Christie M.J, North RA, Osborne PB, Douglass J, Adelman JP. Heteropolymeric potassium channels expressed in *Xenopus* oocytes from cloned subunits. Neuron 1990; 4: 405–11.
8. Isacoff EY, Jan YN, Jan LY. Evidence for the formation of heteromultimeric potassium channels in *Xenopus* oocytes. Nature 1990; 345: 530–4.
9. Ruppersberg JP, Schröeter KH, Sakmann B, Stocker M, Sewing S, Pongs O. Heteromultimeric channels formed by rat brain potassium-channel proteins. Nature 1990; 345: 535–7.
10. MacKinnon R. Determination of the subunit stoichiometry of a voltage-activated potassium channel. Nature 1991; 350: 232–5.
11. Atkinson NS, Robertson GA, Ganetzky B. A component of calcium-activated potassium channels encoded by the *Drosophila slo* locus. Science 1991; 253: 551–5.
12. Adelman JP, Shen K-Z, Kavanaugh MP, Warren RA, Wu YN, Lagrutta A, et al. Calcium-activated potassium channels expressed from cloned complimentary DNA's. Neuron 1992; 9: 209–16.
13. Butler A, Tsunoda S, McCobb DP, Wei A, Salkoff L. *mSlo*, a complex mouse gene encoding "maxi" calcium-activated potassium channels. Science 1993; 261: 221–4.
14. Petersen OH, Maruyama Y. Calcium-activated potassium channels and their role in secretion. Nature 1984; 307: 693–6.
15. Latorre R, Oberhauser A, Labarca P, Alvarez O. Varieties of calcium-activated potassium channels. Ann Rev Physiol 1989; 51: 385–99.
16. Suarez-Kurtz G, Garcia ML, Kaczorowski GJ. Effects of charybdotoxin and iberiotoxin on the spontaneous motility and tonus of different guinea pig smooth muscle tissues. J Pharmacol Exp Ther 1991; 259: 439–43.
17. Winquist RJ, Heaney LA, Wallace AA, Baskin EP, Stein RB, Garcia ML, et al., Glyburide blocks the relaxation response to BRL34915 (cromakalim), minoxidil sulfate and diazoxide in vascular smooth muscle. J Pharmacol Exp Ther 1989; 248: 149–56.
18. Longman SD, Hamilton TC. Potassium channel activator drugs: mechanism of action, pharmacological properties and therapeutic potential. Med Res Rev 1992; 12: 73–148.
19. Underwood S, Raeburn D. ATP-Sensitive potassium channels. In: Raeburn D, Giembycz MA: editors. Airways Smooth Muscle: Peptide Receptors, Ion Channels and Signal Transduction. Basel: Birkhäuser, 1995.
20. Garcia ML, Galvez A, Garcia-Calvo M, King VF, Vazquez J, Kaczorowski GJ. Use of toxins to study potassium channels. J Bioenerg Biomembranes 1991; 23: 615–46.
21. Miller C, Moczydlowski E, Latorre R, Phillips M. Charybdotoxin, a protein inhibitor of single Ca^{2+}-activated K^+ channels from mammalian skeletal muscle. Nature 1985; 313: 316–18.
22. Gimenez-Gallego G, Navia MA, Reuben JP, Katz GM, Kaczorowski GJ, Garcia ML. Purification, sequence, and model structure of charybdotoxin, a potent selective inhibitor of calcium-activated potassium channels. Proc Natl Acad Sci 1988; 85: 3329–33.

23. Lucchesi K, Ravindran A, Young H, Moczydlowski E. Analysis of the blocking activity of charybdotoxin homologs and iodinated derivatives against Ca^{2+}-activated K^+ channels. J Membrane Biol 1989; 109: 269–81.
24. Strong PN, Weir SN, Beech DJ, Hiestand P, Kocher HP. Effects of potassium channel toxins form *Leiurus quinquestriatus hebraeus* venom on responses to cromakalim in rabbit blood vessels. Br J Pharmacol 1989; 98: 817–26.
25. Schweitz H, Bidard JN, Maes P, Lazdunski M. Charybdotoxin is a new member of the K^+ channel toxin family that includes dendrotoxin I and mast cell degranulating peptide. Biochemistry 1989; 28: 9708–14.
26. Anderson CS, MacKinnon R, Smith C, Miller C. Charybdotoxin block of single Ca^{2+}-activated K^+ channels. Effects of channel gating, voltage and ionic strength. J Gen Physiol 1988; 91: 317–33.
27. MacKinnon R, Miller, C. Functional modification of a Ca^{2+}-activated K^+ channel by trimethyloxonium. Biochemistry 1989; 28: 8087–92.
28. Sugg EE, Garcia ML, Reuben JP, Patchett AA, Kaczorowski GJ. Synthesis and structural characterization of charybdotoxin, a potent peptidyl inhibitor of the high conductance Ca^{2+}-activated K^+ channel. J Biol Chem 1990; 265: 18745–8.
29. Lambert P, Kuroda M, Chino N, Watanabe TX, Kimura T, Sakakibara S. Solution synthesis of charybdotoxin (ChTX), a K^+ channel blocker. Biochem Biophys Res Comm 1990; 170: 684–90.
30. Bontems F, Roumestand C, Boyot P, Gilquin B, Doljansky Y, Menez A, *et al.* Three-dimensional structure of natural charybdotoxin in aqueous solution by ^{1}H-NMR. Eur J Biochem 1991; 196: 19–28.
31. Bontems F, Gilquin B, Roumestand C, Menez A, Toma F. Analysis of side chain organization on a refined model of charybdotoxin; structural and functional implications. Biochemistry 1992; 31: 7756–64.
32. Park C-S, Hausdorff SF, Miller C. Design, synthesis, and functional expression of a gene for charybdotoxin, a peptide blocker of K^+ channels. Proc Natl Acad Sci 1991; 88: 2046–50.
33. Park C-S, Miller C. Mapping function to structure in a channel-blocking peptide: electrostatic mutants of charybdotoxin. Biochemistry 1992; 31: 7749–55.
34. Park C-S, Miller C. Interaction of charybdotoxin with permeant ions inside the pore of a K^+ channel. Neuron 1992; 9: 307–13.
35. MacKinnon R, Miller C. Mechanism of charybdotoxin block of the high-conductance Ca^{2+}-activated K^+ channel. J Gen Physiol 1988; 91: 335–49.
36. Stampe P, Kolmakova-Partensky L, Miller C. Mapping hydrophobic residues of the interaction surface of charybdotoxin. Biophys J 1992; 62: 8–9.
37. Stampe P, Kolmakova-Partensky L, Miller C. Intimations of K^+ channel structure from a complete functional map of the molecular surface of charybdotoxin. Biochemistry 1994; 33: 443–50.
38. Vazquez J, Feigenbaum P, Katz G, King VF, Reuben JP, Roy-Contancin L, *et al.* Characterization of high affinity binding sites for charybdotoxin in sarcolemmal membranes from bovine aortic smooth muscle. Evidence for a direct association with the high conductance calcium-activated potassium channel. J Biol Chem 1989; 264: 20902–9.
39. Slaughter RS, Shevell JL, Felix JP, Garcia ML, Kaczorowski GJ. High levels of sodium-calcium exchange in vascular smooth muscle sarcolemmal membrane vesicles. Biochemistry 1989; 28: 3995–4002.
40. Slaughter RS, Kaczorowski GJ, Garcia ML. Charybdotoxin binds with high affinity to a single class of sites in bovine tracheal smooth muscle sarcolemmal membrane vesicles. J Cell Biol 1988; 107: 143a.
41. Vergara C, Latorre R. Kinetics of Ca^{2+}-activated K^+ channels from rabbit muscle incorporated into planar bilayers: evidence for a Ca^{2+} and Ba^{2+} blockade. J Gen Physiol 1983; 82: 543–68.
42. Cecchi X, Wolff D, Alvarez O, Latorre R. Mechanisms of Cs^+ blockade in a Ca^{2+}-activated K^+ channel from smooth muscle. Biophys J 1987; 52: 707–16.
43. Neyton J, Miller C. Potassium blocks barium permeation through a calcium-activated potassium channel. J Gen Physiol 1988; 92: 549–67.
44. Villarroel A, Alvarez O, Oberhauser A, Latorre R. Probing a Ca^{2+}-activated K^+ channel with quaternary ammonium ions. Pfluger's Arch 1988; 413: 118–26.

45. Miller C. Competition for block of a Ca^{2+}-activated K^+ channel by charybdotoxin and tetraethylammonium. Neuron 1988; 1: 1003–6.

46. Hermann A, Erxleben C. Charybdotoxin selectively blocks small Ca-activated K channels in *Aplysia* neurons. J Gen Physiol 1987; 90: 27–47.

47. Sands SB, Lewis RS, Cahalan MD. Charybdotoxin blocks voltage-gated K^+ channels in human and murine T lymphocytes. J Gen Physiol 1989; 93: 1061–74.

48. Vazquez J, Feigenbaum P, King VF, Kaczorowski GJ, Garcia ML. Characterization of high affinity binding sites for charybdotoxin in synaptic plasma membranes from rat brain. J Biol Chem 1990; 265: 15564–71.

49. Deutsch C, Price M, Lee S, King VF, Garcia ML. Characterization of high affinity binding sites for charybdotoxin in human T-lymphocytes. J Biol Chem 1991; 266: 3668–74.

50. Galvez A, Gimenez-Gallego G, Reuben JP, Roy-Contancin L, Feigenbaum P, Kaczorowski GJ, et al. Purification and characterization of a unique, potent, peptidyl probe for the high conductance calcium-activated potassium channel from venom of the Scorpion *Buthus tamulus*. J Biol Chem 1990; 265: 11083–90.

51. Giangiacomo K, Sugg EE, Garcia-Calvo M, Leonard RJ, McManus OB, Kaczorowski GJ, et al. Synthetic charybdotoxin-iberiotoxin chimeric peptides define toxin binding sites on calcium-activated and voltage-dependent potassium channels. Biochemistry 1993; 32: 2363–70.

52. Leonard RJ, Garcia ML, Slaughter RS, Reuben JP. Selective blockers of voltage-gated K^+ channels depolarize human T lymphocytes: mechanism of the antiproliferative effect of charybdotoxin. Proc Natl Acad Sci USA 1992; 89: 10094–8.

53. Giangiacomo KM, Garcia ML, McManus OB. Mechanism of iberiotoxin block of the large-conductance calcium-activated potassium channel from bovine aortic smooth muscle. Biochemistry 1992; 31: 6719–27.

54. Candia S, Garcia ML, Latorre R. Mode of action of iberiotoxin, a potent blocker of the large conductance Ca^{2+}-activated K^+ channel. Biophys J 1992; 63: 583–90.

55. Johnson BA, Sugg EE. Determination of the three-dimensional structure of iberiotoxin in solution by ^1H nuclear magnetic resonance spectroscopy. Biochemistry 1992; 31: 8151–9.

56. Novick J, Leonard RJ, King VF, Schmalhofer W, Kaczorowski GJ, Garcia ML. Purification and characterization of two novel peptidyl toxins directed against K^+ channels from venom of new world scorpions. Biophys J 1991; 59: 78a.

57. Crest M, Jacquet G, Gola M, Zerrouk H, Benslimane A, Rochat H, Mansuelle, P, et al. Kaliotoxin, a novel peptidyl inhibitor of neuronal BK-type Ca^{2+}-activated K^+ channels characterized from *Androctonus mauretanicus mauretanicus* venom. J Biol Chem 1992; 267: 1640–7.

58. Garcia-Calvo M, Kaczorowski GJ, Garcia ML. Molecular characterization of the charybdotoxin-sensitive, high conductance, calcium-activated potassium channel. In: Glossmann H, Striessnig J. editors. Molecular and cellular biology of pharmacological targets. New York: Plenum Press 1993; 41–59 (Methods in Pharmacology; Vol. 7).

59. Robitalle R, Garcia ML, Kaczorowski GJ, Charlton MP. Functional co-localization of calcium and calcium-gated potassium channels in control of transmitter release. Neuron 1993; 11: 645–55.

60. Garcia-Calvo M, Vazquez J, Smith M, Kaczorowski GJ, Garcia ML. Characterization of the solubilized charybdotoxin receptor from bovine aortic smooth muscle. Biochemistry 1991; 30: 11157–64.

61. Garcia-Calvo M, Knaus H-G, McManus OB, Giangiacomo KM, Kaczorowski GJ, Garcia ML. Purification and reconstitution of the high-conductance, calcium-activated potassium channel from tracheal smooth muscle. J Biol Chem 1994; 269: 676–82.

62. Garcia-Calvo M, Knaus HG, Garcia ML, Kaczorowski GJ, Kempner E. Functional size of the charybdotoxin receptor in smooth muscle. Proc Natl Acad Sci 1994; 91: 4718–22.

63. Knaus H-G, Garcia-Calvo M, Kaczorowski GJ, Garcia ML. Subunit composition of the H16H conductance calcium-activated potassium channel from smooth muscle, a representative of the *mSlo* and *slowpoke* family of potassium channels. J Biol Chem 1994; 269: 3921–4.

64. Knaus H-G, Folander K, Garcia-Calvo M, Garcia ML, Kaczorowski GJ, Smith M, Swanson, R. Primary sequence and immunological characterization of the β subunit of the high conductance calcium-activated potassium channel from smooth muscle. J Biol Chem 1994; 269: 17274–8.

65. Ampofo O. Plants that heal. World Health 1977; 26: 28–33.
66. Addy ME. Several chromatographically distinct fractions of *Desmodium adscendens* inhibit smooth muscle contractions. Int J Crude Drug Res 1989; 27: 81–91.
67. Addy ME, Burka JF. Effect of *Desmodium adscendens* fractions on antigen- and arachidonic acid-induced contractions of guinea pig airways. Can J Physiol Pharmacol 1988; 66: 820–5.
68. Addy ME, Burka JF. Effect of *Desmodium adscendens* fraction F1 (DAF1) on tone and agonist-induced contractions of guinea pig airway smooth muscle. Phytother Res 1989; 3: 85–90.
69. McManus OB, Harris GH, Giangiacomo KM, Feigenbaum P, Reuben JP, Addy ME, *et al*. An activator of calcium-dependent potassium channels isolated from a medicinal herb. Biochemistry 1993; 32: 6128–33.
70. Quast U. Do the K^+ channel openers relax smooth muscle by opening K^+ channels? Trends Pharmacol Sci 1993; 14: 332–7.
71. Cole RJ, Cox RH. Handbook of toxic fungal metabolites, New York. Academic Press 1981: 355–509.
72. Norris PJ, Smith CCT, DeBelleroche J, Bradford HF, Mantle PG, Thomas AJ, *et al*. Actions of tremorgenic fungal toxins on neurotransmitter release. J Neurochem 1980; 34: 33–42.
73. Selala MI, Laekeman GM, Loenders B, Musuka A, Herman AG, Schepens P. *In vitro* effects of tremorgenic mycotoxins. J Nat Prod 1991; 54: 207–12.
74. Yao Y, Peter AB, Baur R, Sigel E. The tremorgen aflatrem is a positive allosteric modulator of the γ-aminobutyric acid$_A$ receptor channel expressed in *Xenopus* oocytes. Mol Pharmacol 1989; 35: 319–23.
75. Knaus H-G, McManus OB, Lee SH, Schmalhofer WA, Garcia-Calvo M, Helms LMH, *et al*. Tremorgenic indole alkaloids potently inhibit smooth muscle high-conductance Ca^{2+}-activated K^+ channels. Biochemistry 1994; 33: 5819–28.
76. Ward JPT, Taylor SG, Collier ML. The novel benzopyranol potassium channel activator BRL55834 activates two potassium channels in bovine airway smooth muscle. Br J Pharmacol 1992; 107: 49P.
77. Bowring NE, Arch JRS, Buckle DR, Taylor JF. Comparison of the airways relaxant and hypotensive potencies of the potassium channel activators BRL55834 and levcromakalim (BRL38227) *in vivo* in guinea-pigs and rats. Br J Pharmacol 1993; 109: 1133–9.
78. Buckle DR. Prospects for potassium channel activators in the treatment of airways obstruction. Pulmonary Pharmacol 1993; 6: 161–9.
79. Olesen SP, inventor. Neurosearch AS, assignee. Benzimidazole derivatives, their preparation and use, European patent application, EP04778191A, 1992.
80. Olesen S-P, Munch E, Moldt P, Drejer, J. Selective activation of Ca^{2+}-dependent K^+ channels by novel benzimidazolone. Eur J Pharmacol 1994; 251: 53–9.
81. Sargent CA, Grover GJ, Antonaccio MJ, McCullough JR. The cardioprotective, vasorelaxant and electrophysiological profile of the large conductance calcium-activated potassium channel opener NS-004. J Pharmacol Exp Ther 1993; 266: 1422–9.
82. Laurent F, Michel A, Bonnet PA, Chapat JP, Boucard M. Evaluation of the relaxant effects of SCA40, a novel charybdotoxin-sensitive potassium channel opener, in guinea-pig isolated trachealis. Br J Pharmacol 1993; 108: 622–6.
83. McCann JD, Welsh MJ. Calcium-activated potassium channels in canine airway smooth muscle. J Physiol 1986; 372: 113–27.
84. Muraki K, Imaizumi, Y, Kawai T and Watanabe M. Effects of tetraethylammonium and 4-aminopyridine on outward currents and excitability in canine tracheal smooth muscle. Br J Pharmacol 1990; 100: 507–15.
85. Stockbridge LL, French AS, Man SFP. Subconductance states in calcium-activated potassium channels from canine airway smooth muscle. Biochim Biophys Acta 1991; 1064: 212–8.
86. Boyle JP, Tomasic M, Kotlikoff MI. Delayed rectifier potassium channels in canine and porcine airway smooth muscle cells. J Physiol 1992; 447: 329–50.
87. Huang HM, Dwyer TM, Farley JM. Patch-clamp recording of single Ca^{2+}-activated K^+ channels in tracheal smooth muscle from swine. Biophys J 1987; 51: 50A.
88. Saunders H-M, Farley JM. Pharmacological properties of potassium current in swine tracheal smooth muscle. J Pharmacol Exp Ther 1992; 260: 1038–44.

89. Green KA, Foster RW, Small, RC. A patch clamp study of K^+-channel activity in bovine isolated tracheal smooth muscle cells. Br J Pharmacol 1991; 102: 871–8.

90. Murray MA, Berry JL, Cook SJ, Foster RW, Green KA, Small RC. Guinea pig isolated trachealis: the effects of charybdotoxin on mechanical activity, membrane potential changes and the activity of plasmalemmal K^+ channels. Br J Pharmacol 1991; 103: 1814–8.

91. Kume H, Takagi A, Tokuno H, Tomita T. Regulation of Ca^{2+}-dependent K^+ channel activity in tracheal myocytes by phosphorylation. Nature 1989; 341: 152–4.

92. Kume H, Takagi K, Satake T, Tokuno H, Tomita T. Effects of intracellular pH on calcium-activated potassium channels in rabbit tracheal smooth muscle cells. J Physiol 1990; 424: 445–57.

93. Groschner K, Silberberg SD, Gelband CH, van Breemen C. Ca^{2+}-activated K^+ channels in airway smooth muscle are inhibited by cytoplasmic adenosine triphosphate. Pflüger's Arch 1991; 417: 517–22.

94. Cook NS, Quast U. Potassium channel pharmacology. In: Cook, NS, editor. Potassium channels: structure, classification, function and therapeutic potential. Chichester: Academic Press. 1990; 181–231.

95. Langton PD, Nelson MT, Huang Y, Standen NB. Block of calcium-activated potassium channels in mammalian arterial myocytes by tetraethylammonium ions. Am J Physiol 1991; 260: H927–34.

96. Kume H, Kotlikoff MI. Muscarinic inhibition of single K_{ca} channels in smooth muscle cells by a pertussis-sensitive G protein. Am J Physiol 1991; 261: C1204–9.

97. Kume H, Graziano MP, Kotlikoff MI. Stimulatory and inhibitory regulation of calcium-activated potassium channels by guanine nucleotide-binding proteins. Proc Natl Acad Sci 1992; 89: 11051–5.

98. Jones TR, Charette L, Garcia M, Kaczorowski GJ. Selective inhibition of relaxation of guinea-pig trachea by charybdotoxin, a potent Ca^{2+}-activated K^+ channel inhibitor. J Pharmacol Exp Ther 1990; 255: 697–706.

99. Huang JC, Garcia ML, Reuben JP, Kaczorowski GJ. Inhibition of β-adrenoceptor agonist relaxation of airway smooth muscle by Ca^{2+}-activated K^+ channel blockers. Eur J Pharmacol 1993; 235: 37–43.

100. Jones TR, Charette L, Garcia ML, Kaczorowski G. Interaction of iberiotoxin with β-adrenoceptor agonists and sodium nitroprusside on guinea pig trachea. J Appl Physiol 1993; 74: 1879–84.

101. Allen SL, Beech DJ, Foster RW, Morgan GP, Small RC. Electrophysiological and other aspects of the relaxant action of isoprenaline in guinea-pig isolated trachealis. Br J Pharmacol 1985; 86: 843–54.

102. Honda K, Satake T, Takagi K, Tomita T. Effects of relaxants on electrical and mechanical activities in the guinea-pig tracheal muscle. Br J Pharmacol 1986; 87: 665–71.

103. Miura M, Belvisi MG, Stretton CD, Yacoub MH, Barnes P. Role of potassium channels in bronchodilator responses in human airways. Am Rev Resp Dis 1992; 146: 132–6.

104. Kirkpatrick CT. Tracheobronchial smooth muscle. In: Bulbring E, Brading AF, Jones AW, Tomita T, editors. Smooth muscle: An assessment of current knowledge. London: Edward Arnold. 1981; 385–95.

105. Kannan MS, Jager LP, Daniel EE, Garfield RE. Effects of 4-aminopyridine and tetraethylammonium chloride on the electrical activity and cable properties of canine tracheal smooth muscle. J Pharmacol Exp Ther 1983; 227: 706–15.

106. Foster RW, Small RC, Weston AH. Evidence that the spasmogenic action of tetraethyl-ammonium in guinea pig trachealis is both direct and dependent on the cellular influx of calcium ion. Br J Pharmacol 1983; 79: 255–63.

107. Small RC, Berry JL, Foster RW, Green KA, Murray MA. The pharmacology of potassium-channel modulators in airway smooth muscle: relevance to airways disease. In: Weston AH, Hamilton TC, editors. Potassium channel modulators: pharmacological, molecular and clinical aspects. Oxford: Blackwell Scientific Publications, 1992: 422–61.

108. Small RC, Foster RW, Boyle JP. K^+ channel opening as a mechanism for relaxing airways smooth muscle. Agents Actions 1988; 23: 89–94.

109. Small RC, Foster RW. Electrophysiology of the airway smooth muscle cell. In: Barnes PJ, Rodger IW, Thomson NC, editors. Asthma: basic mechanisms and clinical management. London: Academic Press, 1988: 35–56.

110. Suzuki H, Morita K, Kuriyama H. Innervation and properties of the smooth muscle of the dog trachea. Jap J Physiol 1976; 26: 303–20.
111. Ito Y, Tajima K. Dual effects of catecholamines on pre- and post-junctional membranes in the dog trachea. Br J Pharmacol 1982; 75: 433–40.
112. Fujiwara T, Sumimoto K, Itoh T, Suzuki H, Kurijama H. Relaxing actions of procaterol, β_2-adrenoceptor stimulant, on smooth muscle cells of the dog trachea. Br J Pharmacol 1988; 93: 199–209.
113. Ito Y. Pre- and post-junctional actions of procaterol, a β_2-adrenoceptor stimulant on dog tracheal tissue. Br J Pharmacol 1988; 95: 268–74.
114. Cameron AR, Johnston CF, Kirkpatrick CT, Kirkpatrick MCA. The quest for the inhibitory neurotransmitter in bovine tracheal smooth muscle. Q J Exp Physiol 1983; 68: 413–26.
115. Small RC. Electrical slow waves and tone of guinea-pig isolated trachealis muscle: effects of drugs and temperature changes. Br J Pharmacol 1982; 77: 45–54.
116. Cook SJ, Small RC, Berry JL, Chiu P, Downing SJ, Foster RW. β-Adrenoceptor subtypes and the opening of plasmalemmal K^+ channels in trachealis muscle: electrophysiological and mechanical studies in guinea-pig tissue. Br J Pharmacol 1993; 109: 1140–8.
117. Small RC, Chiu P, Cook SJ, Foster RW, Isaac L. β-Adrenoceptor agonists in bronchial asthma: role of K^+ channel opening in mediating their bronchodilator effects. Clin Exp Allergy 1993; 23: 802–11.
118. Chiu P, Cook SJ, Small RC, Berry JL, Carpenter JR, Downing SJ et al. β-Adrenoceptor subtypes and the opening of plasmalemmal K^+ channels in bovine trachealis muscle: studies of mechanical activity and ion fluxes. Br J Pharmacol 1993; 109: 1149–56.
119. Hamaguchi M, Ishibashi T, Imai S. Involvement of charybdotoxin-sensitive K^+ channel in the relaxation of bovine tracheal smooth muscle by glyceryl trinitrate and sodium nitroprusside. J Pharmacol Exp Ther 1992; 262: 263–70.
120. Welling A, Felbel J, Peper K, Hofman F. Hormonal regulation of calcium current in freshly isolated airway smooth muscle cells. Am J Physiol 1992; 262: L351–9.
121. Wills-Karp M, Gilmour MI. Increased cholinergic antagonism underlies impaired beta-adrenergic response in ovalbumin-sensitized guinea pigs. J Appl Physiol 1993; 74: 2729–35.
122. Fernandes LB, Fryer AD, Hirshman CA. M_2 muscarinic receptors inhibit isproterenol-induced relaxation of canine airway smooth muscle. J Pharmacol Exp Ther 1992; 262: 119–26.
123. Torphy TJ, Zheng C, Peterson SM, Fiscus RR, Rinard GA, Mayer SE. The inhibitory effect of methacholine on drug-induced relaxation, cyclic AMP accumulation and cyclic AMP-dependent protein kinase activation in canine tracheal smooth muscle. J Pharmacol Exp Ther 1985; 233: 409–17.
124. Ichinose M, Barnes JP. A potassium channel activator modulates both excitatory noncholinergic and cholinergic neurotransmission in guinea pig airways. J Pharmacol Exp Ther 1990; 252: 1207–12.
125. Baker DG, Don HF, Rodger IW, Brown JK. Muscarinic cholinergic and alpha-adrenergic inhibition of acetylcholine (Ach) release in guinea-pig trachea-role of iberiotoxin-sensitive K^+ channels. Am Rev Resp Dis 1993; 147: 504.
126. Masson P, Jones TR. Cardiovascular and respiratory effects of Ca^{2+}-activated K^+ channel antagonists, charybdotoxin and iberiotoxin in anesthetized guinea pigs. Proc Can Fed Biol Soc 1992; 35: 241.

Airways Smooth Muscle: Peptide Receptors, Ion Channels and Signal Transduction
ed. by D. Raeburn and M. A. Giembycz
© 1995 Birkhäuser Verlag Basel/Switzerland

CHAPTER 9
Adenosine Triphosphate-Activated Potassium Channels

Stephen L. Underwood and David Raeburn*

Rhône-Poulenc Rorer Ltd., Dagenham Research Centre, Dagenham, Essex, UK

1   Introduction
2   Potassium Channels in Airways Smooth Muscle
2.1 Delayed Rectifier (K_{DR}) Channel
2.2 Large Conductance, Ca^{2+}-Activated (BK_{Ca}) Channel, Maxi-K Channel
2.3 Small Conductance, Ca^{2+}-Activated (SK_{Ca}) Channel
2.4 ATP-Sensitive (K_{ATP}) Channel
3   Potassium Channel Openers
4   Mechanisms of Action of Potassium Channel Openers
4.1 Stimulation of K^+ Efflux
4.2 Hyperpolarization of the Cell Membrane
4.3 K^+ Concentration-Dependent Smooth Muscle Relaxation
4.4 Further Evidence of a Role for K_{ATP} Channels
4.5 Other Proposed Mechanisms of Action
5   Potassium Channel Openers and Airway Function *In Vitro*
5.1 Animal Airways Smooth Muscle
5.2 Human Airways Smooth Muscle
6   Potassium Channel Openers and Airway Function in Animals *In Vivo*
6.1 Airways Smooth Muscle
6.2 Microvascular Leakage
6.3 Airway Hyperreactivity
7   Neural Effects of Potassium Channel Openers in Animals
8   Clinical Pharmacology of Potassium Channel Openers in the Airways
9   Conclusions
     References

1. Introduction

A number of pharmacological agents including β_2-adrenoceptor agonists, phosphodiesterase (PDE) inhibitors, muscarinic cholinoceptor antagonists and calcium (Ca^{2+}) entry blockers (CEBs) are able to relax, or prevent the contraction of, airways smooth muscle. Agents which reduce contractility by opening potassium (K^+) ion channels in the smooth muscle cell plasmalemma are also now being investigated. This chapter will review the proposed mechanisms of action of potassium

*Present address: Department of Inflammation, Rhône-Poulenc Rorer, Inc., Collegeville, Philadelphia, USA.

channel openers (KCOs) and their pharmacological actions as elucidated in animal models and in human isolated tissues. Some findings
from clinical studies will be discussed and therapeutic prospects of
KCOs for the treatment of asthma will be considered.

2. Potassium Channels in Airways Smooth Muscle

Airways smooth muscle cells appear to differ in certain respects from
smooth muscle cells in other organs. They show little electrical activity,
probably as a result of membrane rectifying behaviour where membrane
depolarization produced by inward movement of calcium (Ca^{2+}) ions is
limited by the activation of an outward, hyperpolarizing movement of
K^+ through specific K^+ channels in the plasmalemma. The cells depolarize during activation in a graded fashion, and spontaneous or depo-
larization-induced action potentials are not commonly seen [1]. Thus,
K^+ channels appear to play an important role in maintaining the resting membrane potential (E_m) and limiting electrical responses to excitatory stimuli in these cells. In the presence of agents known to block K^+
channels, it becomes possible to demonstrate spontaneous electrical
activity, action potential discharge and contraction in airways smooth
muscle [1]. These electrophysiological changes may be similar to those
described in the asthmatic airway [2]. Indeed, it has been suggested that
the airway hyperreactivity associated with asthma may be a consequence of a partial blockade of K^+ channels in airways smooth muscle cells
[3]. These findings have suggested that agents which open K^+ channels
should reduce airways smooth muscle contractility, thus providing
bronchodilator action, and may also reduce airway hyperreactivity.

All cells in the body are likely to have K^+ channels, and each cell
may have several types. In airways smooth muscle, at least 4 types of
K^+ channels may be involved in regulating contractility:

2.1. Delayed Rectifier (K_{DR}) Channel

This is a calcium-insensitive channel which exhibits delayed activation
in response to a depolarizing voltage. Conductance in canine and
porcine airways smooth muscle cells has been reported as 13 pS [4]. The
channel is blocked potently by 4-aminopyridine (4-AP), is less sensitive
to tetraethylammonium (TEA) and is unaffected by the scorpion toxin
charybdotoxin or the sulphonylurea glibenclamide [1].

2.2. Large Conductance, Ca^{2+}-Activated (BK_{Ca}) Channel, Maxi-K Channel

Conductance has been reported to be in the range 120–290 pS [5].
These channels are activated by increasing concentrations of cytosolic

Ca^{2+} and are also voltage-sensitive in that depolarization of the cell membrane facilitates opening [1]. It has been suggested that these channels provide a mechanism whereby excitatory processes leading to rises in cytosolic Ca^{2+} are limited as a consequence of the hyperpolarization associated with the channel opening [1]. The channel is blocked by charybdotoxin or TEA but not by glibenclamide or 4-aminopyridine.

2.3. Small Conductance, Ca^{2+}-Activated (SK$_{Ca}$) Channel

Conductance is in the range 10–20 pS [5]. The channel is blocked selectively by the bee venom apamin. Unlike the BK$_{Ca}$ channel, there is less evidence to suggest that apamin-sensitive SK$_{Ca}$ channels are present (or serve a significant physiological function) in airways smooth muscle cells.

2.4. ATP-Sensitive (K$_{ATP}$) Channel

Channel opening is reduced by increasing cytosolic concentrations of adenosine triphosphate (ATP) and is only slightly voltage-dependent [5]. The channel is blocked by glibenclamide and by phentolamine (an action separate from its effects on α-adrenoceptors). K$_{ATP}$ channels were first described in pancreatic β cells where they are involved in the regulation of insulin secretion. Subsequently, similar channels have been demonstrated in smooth muscle cells where opening (activation) leads to a relaxant response. Conductance in bovine and rabbit airways smooth muscle cells is in the range 30–39 pS [5].

The remainder of this chapter will focus mainly on the effects of opening or closing K$_{ATP}$ channels on airways smooth muscle contractility. BK$_{Ca}$ channels are discussed in detail in Chapter 8.

3. Potassium Channel Openers

The coronary vasodilator drug nicorandil was the first agent shown to open K$^+$ channels (see review by Weston et al. [6]), although activation of guanylyl cyclase probably contributes to its cardiovascular actions. The unrelated smooth muscle relaxants pinacidil and diazoxide also have now been shown to exert at least some of their activity by opening K$^+$ channels [6]. However, cromakalim (BRL 34915) was the first specific KCO to be developed. Cromakalim is a racemic mixture and most of its biological activity has since been shown to reside in the L-enantiomer, levcromakalim (BRL 38227, previously referred to as lemakalim) [7]. A range of other compounds with differing chemical

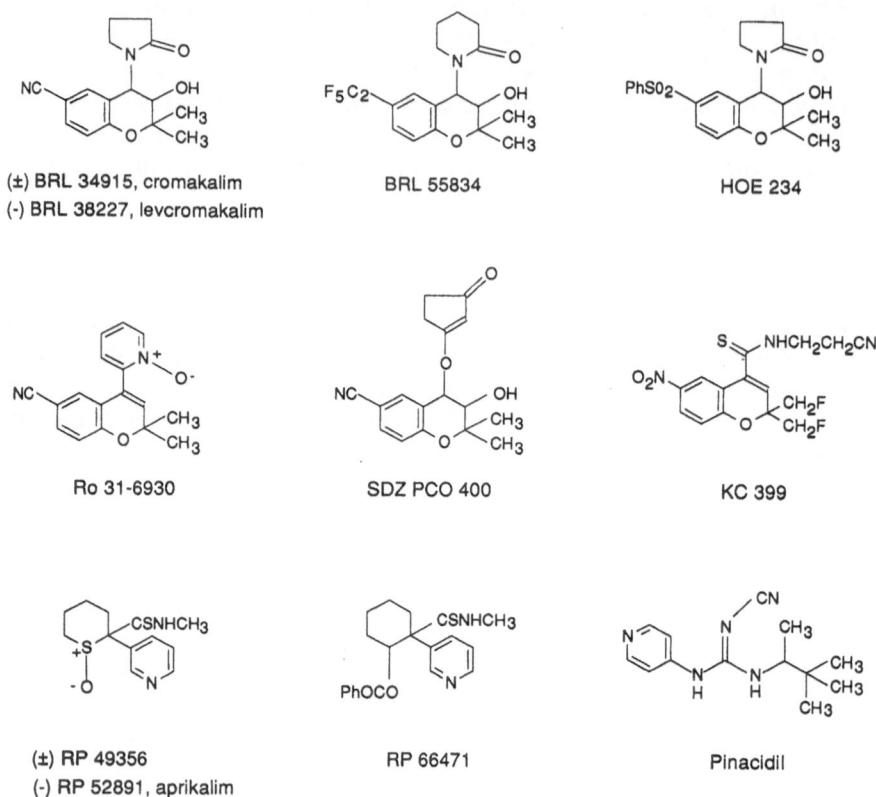

(±) BRL 34915, cromakalim
(-) BRL 38227, levcromakalim

BRL 55834

HOE 234

Ro 31-6930

SDZ PCO 400

KC 399

(±) RP 49356
(-) RP 52891, aprikalim

RP 66471

Pinacidil

Figure 1. Chemical structures of potassium channel openers mentioned in the text.

structures have now been described which relax smooth muscle by opening K^+ channels, purported to be the K_{ATP} channels. Although most of these compounds retain the benzopyran structure of cromakalim, structurally distinct tetrahydrothiopyrans such as RP 49356, its active L-enantiomer RP 52891 (aprikalim) and the more potent RP 66471 have also been developed (Figure 1).

The precise mechanisms whereby the opening of K^+ channels inhibits smooth muscle contraction are still unclear. The principal determinant for inhibition or reversal of smooth muscle contraction is a reduction in the free cytosolic Ca^{2+} concentration below a threshold value of approximately 0.1 μM [8]. This is generally achieved by several mechanisms which either inhibit Ca^{2+} influx or its mobilization from intracellular stores, or stimulate Ca^{2+} removal by promoting Ca^+ efflux or its intracellular sequestration. Because KCOs hyperpolarize the smooth muscle plasmalemma toward the K^+ equilibrium potential, there is a reduced probability that L-type voltage-operated Ca^{2+} channels in the plasmalemma will open. However, KCOs are not acting solely as inhibitors of Ca^{2+} influx since they have different properties *in vitro* to Ca^{2+} entry blockers (section 5.1).

4. Mechanisms of Action of Potassium Channel Openers

4.1. Stimulation of K^+ Efflux

Cromakalim [3, 9], levcromakalim [10] and RP 49356 [9] have each been shown to stimulate the efflux of either $^{42}K^+$ or $^{86}Rb^+$ (a marker for K^+) from smooth muscle cells of various species thereby demonstrating an action on K^+ channels. The efflux induced by cromakalim or levcromakalim is inhibited by glibenclamide, suggesting that K_{ATP} channels are being opened. Because inhibition of ion efflux is sometimes only detectable at KCO concentrations higher than those required to induce smooth muscle relaxation it has been suggested that KCOs may act preferentially on smooth muscle "pace-maker" cells from where contractions originate (see review by Buckle and Arch, [7]).

4.2. Hyperpolarization of the Cell Membrane

In airways smooth muscle *in vitro*, the relaxant actions of KCOs have been shown to be accompanied by an increase in negativity of the E_m (hyperpolarization) towards the K^+ equilibrium potential and by a reduction in spontaneous slow wave activity. The hyperpolarization induced by cromakalim or RP 49356 has been shown to be inhibited by glibenclamide [3, 11, 12], further suggesting the involvement of K_{ATP} channels.

4.3. K^+ Concentration-Dependent Smooth Muscle Relaxation

A compound which opens K^+ channels can be identified by its ability to relax smooth muscle contracted with low (< 20 mM) but not with high (> 40 mM) concentrations of K^+ (Figure 2a). This phenomenon results because, in the presence of external K^+ concentrations exceeding 40 mM, the K^+ equilibrium potential becomes less negative than the threshold for the opening of L-type voltage-operated Ca^{2+} channels. KCOs then cannot hyperpolarize the plasmalemma sufficiently to close the L-type channels and prevent Ca^{2+} influx [13]. Inhibition of airways smooth muscle spasm induced by low, but not high, concentrations of K^+ has been seen with cromakalim [3, 12, 14, 15], levcromakalim [16, 17], RP 49356 [12, 14, 18], aprikalim [17] and SDZ PCO 400 [19]. Glibenclamide inhibits the relaxant effects of KCOs (but not nifedipine) in tissues contracted with low concentrations of K^+ [15]. Further, when added at the time of peak relaxation to KCOs, glibenclamide produces

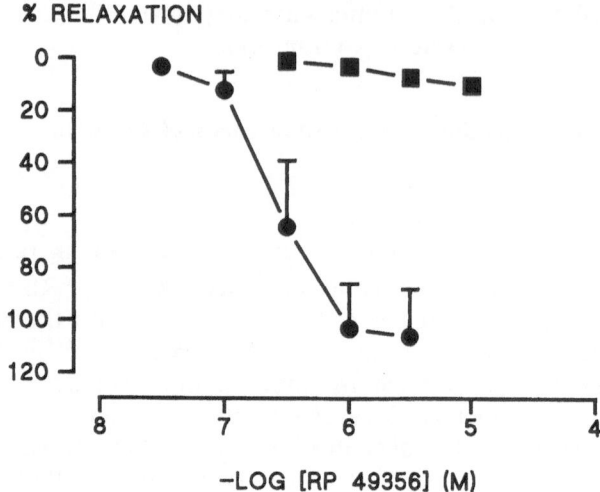

a

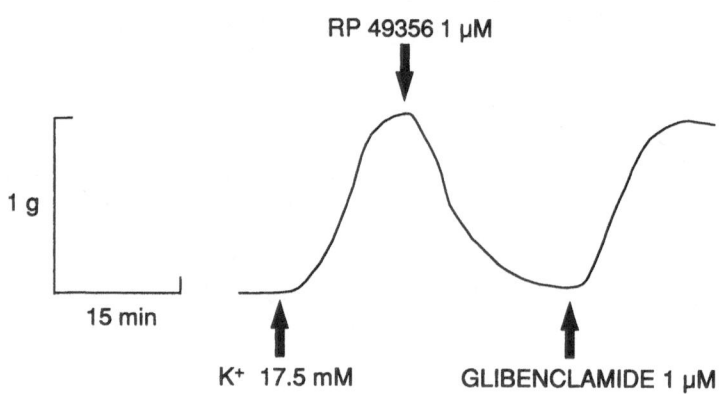

b

Figure 2. **a.** Reversal by RP 49356 of spasm induced by low (17.5 mM, ●) but not high (40 mM, ■) K$^+$ in guinea-pig isolated trachealis. Results represent mean ± s.e.m. (n = 7). Reproduced from [17]. **b.** Representative trace showing the effect of glibenclamide on relaxation of K$^+$ (17.5 mM)-induced tone by RP 49356. Reproduced from [17].

an immediate reversal of the relaxant response [15] (Figure 2b). Glibenclamide has no effect on basal tone or against K$^+$-induced tone [15] suggesting that the channel(s) blocked by the sulphonylurea is not open under these conditions and that it is the same channel opened by KCOs.

4.4. Further Evidence of a Role for K$_{ATP}$ Channels

The relaxant actions of cromakalim in airways smooth muscle are not antagonised by apamin [3] or charybdotoxin [20, 21] suggesting that

SK_{Ca} or BK_{Ca} channels are not involved in its effects. Also, cromakalim or RP 49356 had no effect on the open-state probability of BK_{Ca} channels [12]. Antagonism by glibenclamide of the actions of KCOs outlined above provides evidence that a K_{ATP} channel is the most likely channel to be activated by these agents. However, the concentration of glibenclamide required to antagonise cromakalim-like agents in smooth muscle is higher than that required to inhibit K_{ATP} channels in pancreatic β cells. This indicates that the K_{ATP} channel of smooth muscle may differ from that in insulin-secreting cells [5]. Results from recent studies suggest that, at least in rat portal vein, KCOs may open a glibenclamide-sensitive K^+ channel which is of lower conductance (17 pS) than most previously reported K_{ATP} channels. The sensitivity of this channel to ATP has not yet been established [22].

4.5. Other Proposed Mechanisms of Action

KCOs such as cromakalim and RP 49356 do not increase the cellular content of the cyclic nucleotides cyclic AMP and cyclic GMP [12, 23] indicating that they do not relax smooth muscle by inhibiting PDE activity or stimulating adenylyl or guanylyl cyclase. This suggests that, unlike the mechanism proposed for the action of isoprenaline on BK_{Ca} channels, these agents do not open K^+ channels by protein kinase-mediated phosphorylation of a channel-related protein [24]. In trachealis skinned of its plasma membranes, cromakalim did not inhibit Ca^{2+}-induced spasm thereby indicating that this agent does not act directly on the intracellular contractile machinery [3]. It seems likely that KCOs have actions which reduce cytosolic Ca^{2+} concentrations, additional to effects on Ca^{2+} influx. These are discussed below (section 5.1).

5. Potassium Channel Openers and Airway Function *In Vitro*

5.1. Animal Airways Smooth Muscle

In animal, usually guinea pig, airways smooth muscle preparations *in vitro*, cromakalim [3, 14, 15, 25, 26], levcromakalim [10, 17, 26], RP 49356 [14, 15], aprikalim [17] HOE 234 [27], Ro 31-6930 [28], KC 399 [29] and SDZ PCO 400 [19, 30] have been shown to have relaxant properties. The KCOs are generally able to relax tissues under basal tone and reverse contraction induced by low concentrations of many different spasmogens. In guinea-pig airways smooth muscle, cromakalim and levcromakalim have been shown to be markedly less potent against tone induced by cholinergic agonists than by other mediators, suggesting that the potency of KCOs in this species may to some extent depend on the spasmogen [7]. However, this may be of little relevance to the clinical situation since cromakalim and levcromakalim are effective inhibitors of cholinergic tone in the human bronchus, albeit at higher concentrations than are required to inhibit histamine-induced contraction (section 5.2).

KCOs are notably less potent at reversing contraction induced by high concentrations of spasmogens and are virtually unable to prevent the initiation of the contractile response, *i.e.* they have spasmolytic but poor antispasmogenic properties *in vitro*. These features are shared by other classes of smooth muscle relaxants including β-adrenoceptor agonists, methylxanthines and CEBs [31]. In airways smooth muscle, contractions induced by spasmogens are believed to be initiated by the rapid mobilization of Ca^{2+} from intracellular stores and maintained by the influx of Ca^{2+} through voltage-operated Ca^+ channels [15, 31, 32]. Therefore, the spasmolytic effects of KCOs may in part be explained by a hyperpolarization-induced inhibition of voltage-dependent Ca^{2+} entry. The concentration of agonist used can also influence the source of Ca^{2+} being mobilized such that responses to low agonist concentrations rely more on extracellular Ca^{2+} [31]. The lack of potency of KCOs against contractions induced by high concentrations of spasmogens provides further evidence that these agents are more effective under conditions in which contraction is due to influx of Ca^{2+} from the extracellular space.

There are, however, differences between the actions of KCOs and CEBs on smooth muscle contractility. Cromakalim or levcromakalim induces a greater relaxation of basal tone than do the CEBs nifedipine or verapamil [16, 25] and the CEBs are more potent against high K^+ than mediator-induced contractions of the trachea [15, 31]. This may suggest that KCOs induce relaxation by mechanisms additional to the inhibition of Ca^{2+} influx through L-type voltage-operated Ca^{2+} channels, possibly through an effect on intracellular Ca^{2+} stores or Ca^{2+} efflux by ion exchange mechanisms [25, 33, 34]. Indeed, levcromakalim has been shown to inhibit the loading of intracellular Ca^{2+} stores in rabbit airways smooth muscle cells [35]. Some evidence exists that RP 49356, but not cromakalim, may additionally stimulate Na^+/K^+ ATPase and thereby augment hyperpolarization [36]. Comparison of KCOs and CEBs indicates that KCOs are far more effective at relaxing airways smooth muscle and may therefore have greater therapeutic potential in the treatment of asthma.

5.2. Human Airways Smooth Muscle

Cromakalim or levcromakalim reduce resting tone in human airways smooth muscle and are potent inhibitors of contractions induced by histamine or carbachol [10, 16, 26]. The ability of these KCOs to inhibit cholinergic responses contrasts with the findings in the guinea-pig tissues *in vitro* and may be significant since, in asthma, bronchoconstriction may have a reflex parasympathetic component. In the presence of a maximal relaxation induced by verapamil, levcromakalim produced additional relaxation providing further evidence that the KCO has an action at a separate site to Ca^{2+} entry blockers [16]. As in studies using animal tissues, the relaxant effects of levcromakalim were antagonised by glibenclamide suggesting the involvement of K_{ATP} channels [10, 16].

6. Potassium Channel Openers and Airway Function in Animals *In Vivo*

6.1. Airways Smooth Muscle

In the anaesthetized guinea-pig, cromakalim, levcromakalim, BRL 55834, aprikalim, RP 66471, HOE 234, Ro 31-6930 and KC 399 each inhibit bronchoconstriction induced by either histamine or 5-hydroxy-tryptamine [25, 27, 28, 37, 38, 39, 40, 41, 42, 43]. These protective effects are seen following administration of KCOs by the oral, intravenous or inhaled (Figure 3) routes and are inhibited by glibenclamide (Figure 4). Variable effects of KCOs against cholinergic-mediated bronchoconstriction have been reported. Whereas intravenous levcromakalim and RP 66471 have been reported to inhibit bronchoconstriction induced by intravenous administration of methacholine [40], another study found intravenously administered levcromakalim to be ineffective against intravenous acetylcholine [26]. It has been suggested that these discrepancies may be due to differences in the severity of the

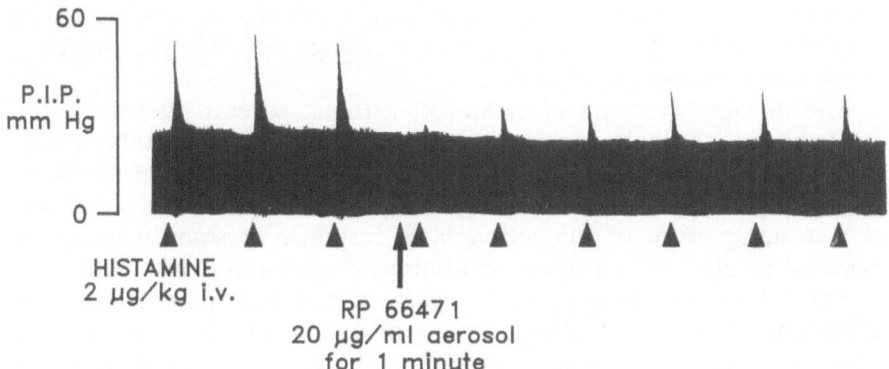

Figure 3. Representative trace showing the effect of inhaled RP 66471 on histamine-induced bronchoconstriction in the anaesthetized guinea-pig.

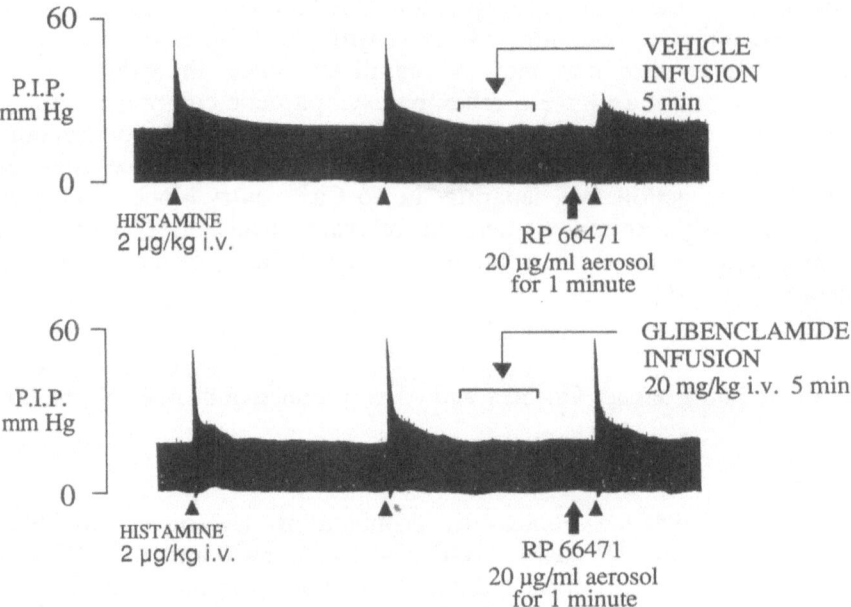

Figure 4. Representative trace to show the inhibition by glibenclamide infusion of RP 66471-induced reduction in histamine-induced bronchoconstriction in the anaesthetized guinea-pig.

bronchoconstrictor challenge, or to differences in reflex involvement between models [7].

In conscious guinea pigs, orally administered cromakalim, levcromakalim, BRL 55834, aprikalim, Ro 31-6930 and KC 399 protect against dyspnoea induced by inhaled histamine [25, 37, 38, 39, 41, 42, 43]. There is no evidence for the development of tachyphylaxis since animals pretreated with aprikalim for 7 days showed the same level of protection seen after a single dose [18]. Cromakalim and aprikalim similarly inhibit dyspnoea induced by challenge with inhaled antigen in sensitized guinea-pigs [25, 39].

This is significant since, in human asthma, several mediators may contribute to bronchoconstriction. Cromakalim does not inhibit antigen-induced mediator (histamine or leukotriene D_4) release from guinea-pig chopped lung [44], suggesting that the protective effect was due to antagonism of the accumulated response of smooth muscle to released mediators and not to inhibition of mast cell degranulation.

Taken together, the *in vivo* data clearly demonstrate that KCOs, when administered before challenge with a mediator, protect against bronchoconstriction. Although this contrasts with their weak antispasmogenic activity *in vitro*, this discrepancy is shared by β-adrenoceptor agonists and theophylline [7, 31].

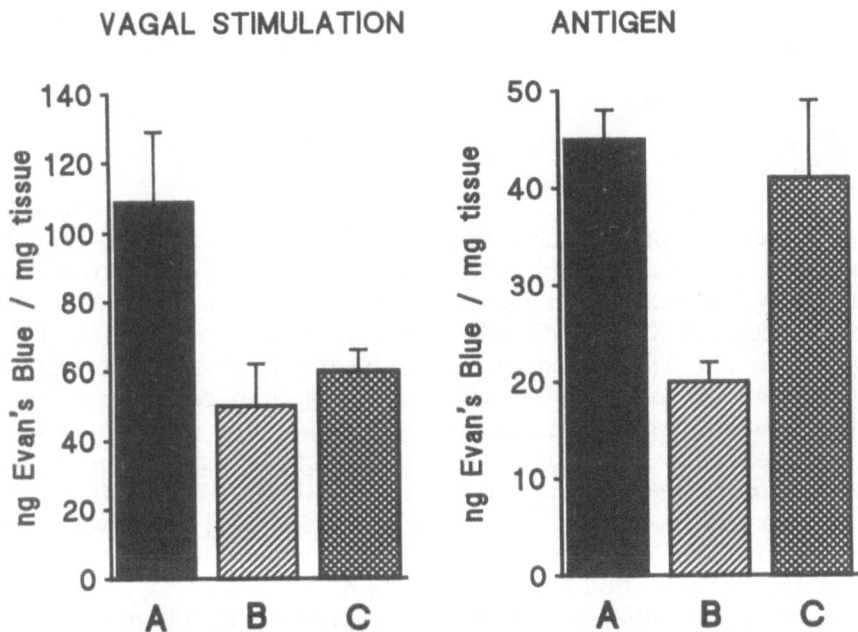

Figure 5. Comparison of the effects of inhaled cromakalim and salbutamol on microvascular leakage in intrapleural airways induced by vagal stimulation or antigen (ovalbumin) inhalation in the anaesthetized guinea-pig. Results represent mean ± s.e.m. (n = 5–7). A, vehicle control; B, salbutamol (50 µg/ml/1 minute inhalation); C, cromakalim (500 µg/ml/1 minute inhalation). Reproduced from [17].

6.2. Microvascular Leakage

An index of inflammatory changes in the lung is the presence of plasma proteins in airway tissues and in the airway lumen as a result of plasma leakage from post-capillary venules [45]. Plasma leakage can be induced in the guinea-pig lung by many exogenously administered mediators, by antigen challenge or by vagal nerve stimulation. Since leakage may be due to contraction of post capillary vascular endothelial cells in the tracheobronchial microcirculation, it has been suggested that KCOs might be able to reduce plasma extravasation by preventing this effect [46], but to date there is no experimental evidence that this happens. In the guinea-pig airways, cromakalim did not inhibit microvascular leakage induced by platelet activating factor [46], histamine [47], substance P [47] or antigen challenge (Figure 5) and levcromakalim did not inhibit leakage induced by histamine [48]. However, cromakalim was able to inhibit leakage induced by vagal nerve stimulation (Figure 5) or by bradykinin [47] suggesting that the KCO may inhibit the activation by bradykinin of excitatory non-adrenergic, non-cholinergic (e-NANC) neurones (see section 7). These findings suggest that KCOs may inhibit neurotransmission and/or the release of pro-inflammatory transmitters and point to a potential anti-inflammatory property of KCOs.

6.3. Airway Hyperreactivity

Cromakalim, Ro 31-6930 and SDZ PCO 400 each inhibit hyperreactivity induced by platelet activating factor in the guinea-pig *in vivo*. In addition, cromakalim and SDZ PCO 400 inhibit hyperreactivity induced by isoprenaline or immune complexes [7, 18, 30, 49]. The protective effects of SDZ PCO 400 were shown to be antagonized by glibenclamide. These effects were obtained using doses of the KCOs which were too low to cause bronchodilation [30] suggesting an effect at the neural level or on inflammatory cell function. Few studies have been conducted on the effects of KCOs on inflammatory cell recruitment and activation. However, neither cromakalim nor SDZ PCO 400 had an effect on antigen-induced eosinophil accumulation in the airways [30] and cromakalim has been shown to be ineffective in preventing activation and degranulation of eosinophils [46].

As indicated earlier, blockade of K^+ channels in the plasmalemma of airways smooth muscle cells may be implicated in the development of airway hyperreactivity [2, 3]. TEA-induced airways smooth muscle hyperreactivity to acetylcholine or 5-HT in the rat airways *in vitro* suggesting the involvement of BK_{Ca} channels in this model [50]. Cromakalim was able to inhibit TEA-induced hyperreactivity, but only at a high concentration which may also activate BK_{Ca} channels.

7. Neural Effects of Potassium Channel Openers in Animals

K^+ channels may have an important role in the hyperpolarization of nerves after a depolarizing stimulus and in the release of neurotransmitters [46]. This is of potential relevance to the pathogenesis of asthma since bronchoconstriction may have a reflex cholinergic component. Further, the release of tachykinins (*e.g.* substance P and neurokinin A) from e-NANC neurones may contribute to bronchoconstriction and airway inflammation [46].

Cromakalim inhibits responses of innervated guinea-pig trachealis to preganglionic vagal stimulation but not responses to exogenously applied acetylcholine *in vitro* [51, 52]. This suggests that cromakalim can modulate cholinergic neurotransmission by inhibiting the neural release of acetylcholine. The site and mechanism of action of the KCO is unclear. One study demonstrated that cromakalim does not inhibit responses to postganglionic (transmural) vagal stimulation [52], suggesting an action at the ganglionic or preganglionic sites. However, other studies have provided evidence that the KCO can inhibit postganglionic vagal stimulation [47, 53, 54]. Cromakalim, levcromakalim and aprikalim each inhibit neurally-mediated non-cholinergic contraction of guinea-pig airways smooth muscle *in vitro* but have no effect on responses to exogenous substance P or neurokinin A suggesting that KCOs also inhibit neurotransmitter release from e-NANC neurones [47, 54, 55, 56]. Levcromakalim also inhibits neurally-mediated mucus secretion by guinea-pig tracheal goblet cells with no effect on secretion induced by substance P or methacholine [57].

These *in vitro* findings have been extended by *in vivo* studies. Cromakalim, levcromakalim, RP 49356, aprikalim and RP 66471 each have a greater inhibitory effect on bronchoconstriction induced by e-NANC stimulation in the guinea-pig *in vivo* (vagus nerve stimulation in the presence of atropine) than on bronchoconstriction of a similar magnitude induced by exogenous administration of the putative mediator, substance P [39, 40, 58, 59] (Figure 6). The actions of KCOs in inhibiting cholinergic and e-NANC neurotransmission can be inhibited by glibenclamide [54, 59, 60] suggesting the involvement of neural plasmalemma K_{ATP} channels. It is not yet known whether the activation of these channels is inhibiting neurotransmitter release or affecting neurotransmission by another mechanism. Interestingly, cromakalim does not appear to suppress NANC inhibitory neurotransmission in the guinea-pig airways [54].

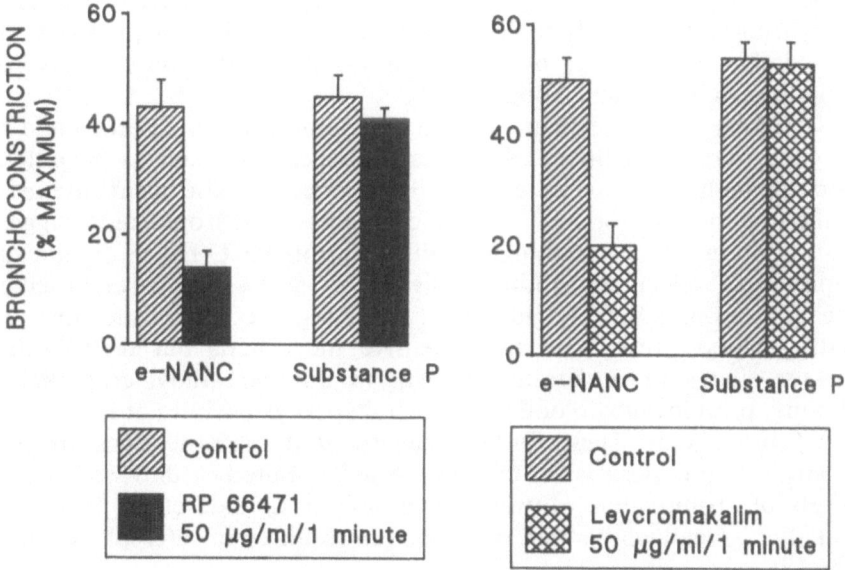

Figure 6. The effects of inhaled levcromakalim and RP 66471 on bronchoconstriction induced by e-NANC stimulation or substance P in the anaesthetized guinea-pig. Results represent mean ± s.e.m. (n = 5–7).

8. Clinical Pharmacology of Potassium Channel Openers in the Airways

Cromakalim (2 mg, p.o.) provided significant protection against a provocation concentration of histamine which caused a 40% fall in partial expiratory flow rate in normal volunteers [61]. This effect compared favourably with the protective effect of salbutamol (4 mg, p.o.). The prevention of morning dipping in nocturnal asthma is a highly desirable characteristic of new anti-asthma drugs, therefore compounds with a long duration of action enabling overnight control of symptoms are of great interest. Cromakalim and levcromakalim have plasma half-lives in man of approximately 24 hours [7]. Single (0.5 mg, p.o.) or repeat (0.25 or 0.5 mg, p.o.) doses of cromakalim reduced the early morning fall in lung function (assessed from FEV_1) when administered in the late evening [7, 62]. The predicted peak plasma concentration of cromakalim in these studies was less than that required to relax human airways smooth muscle significantly *in vitro* [26]. It has been suggested therefore that the efficacy of cromakalim may have resulted from mechanisms other than direct bronchodilation, possibly an inhibition of neural mechanisms contributing to airway hyperreactivity [13, 22] or, specifically, the vagal cholinergic bronchoconstriction that occurs at night [46].

Levcromakalim replaced cromakalim in clinical trials but failed to achieve the efficacy necessary for further development [7]. Levcromakalim (0.125, 0.25 or 0.5 mg, p.o.) did not significantly inhibit airway responsiveness to histamine or methacholine in asthmatic subjects, comparing unfavourably with salbutamol [63]. Further, after 28 days of once-daily oral dosing, only the highest dose (0.75 mg) of levcromakalim elicited bronchodilation [7].

KCOs are potent relaxants of vascular smooth muscle *in vitro* and *in vivo* and were originally developed as antihypertensive agents. Hypotension might therefore be an unwanted side effect if these agents are used in the treatment of asthma. The clinical studies with cromakalim suggest that, at doses which cause bronchodilation, this KCO has no significant hypotensive effect. In studies designed to assess the cardiovascular effects of cromakalim, single doses (1 to 2 mg, p.o.) had little effect on resting blood pressure in normotensive individuals but reduced the pressor effects of noradrenaline [64]. In another study, cromakalim (1.5 mg, p.o.) lowered blood pressure in hypertensive but not normotensive subjects [65]. These findings suggest that normotensives are less sensitive than hypertensives to the reportedly limited cardiovascular side effects of cromakalim. However, a dose-limiting side effect of levcromakalim is headache, which is believed to result from cerebral vasodilation [7].

If systemic side effects do prove to be a problem with KCOs, an inhaled preparation of a compound which is poorly absorbed from the airways may be useful. In the guinea-pig, inhaled levcromakalim can relax airways smooth muscle without achieving pharmacologically effective blood levels. Indeed, compared to intravenous dosing, inhaled administration increases selectivity for the airways by about 20-fold [38]. A marked separation of airway and vascular effects has also been demonstrated with aprikalim [18]. In order for a KCO to be a useful oral bronchodilator, it may be necessary to develop second-generation molecules with an improved intrinsic selectivity for airways smooth muscle. Recent studies suggest that, following administration by the intraduodenal, intravenous or inhaled routes to the guinea-pig or rat *in vivo*, BRL 55834 may be more airways selective than levcromakalim [41, 42]. These findings have been supported by *in vitro* studies which have demonstrated that BRL 55834 is 27-fold more potent than levcromakalim as a relaxant of histamine-induced tone in guinea-pig trachea but only 3-fold more potent as a relaxant of KCl-induced tone in the portal vein [7]. The basis of the improved selectivity is not known. Further, BRL 55834 has a longer duration of bronchodilator action suggesting a slower off-rate from its receptor [42].

9. Conclusions

KCOs are a diverse group of compounds which can relax airways smooth muscle contracted by a range of mediators. Moreover, their ability to inhibit cholinergic and e-NANC neurotransmission suggests other clinically significant actions which may extend to anti-inflammatory properties. On the cellular level, it is still not clear how the KCOs exert their inhibitory effects. There is evidence that their principal action is to open K_{ATP} channels and thereby inhibit Ca^{2+} influx, but this needs to be confirmed and additional mechanisms of action need to be clarified.

The limited clinical studies have so far yielded mixed results. Although cromakalim protected against bronchoconstriction and the early morning fall in lung function, levcromakalim failed to show the efficacy required to warrant further development. One problem with current KCOs is that they may not show adequate selectivity for airways (over vascular) smooth muscle. However, since K^+ channels are very diverse, it may be possible to develop airways-selective KCOs which will offer useful clinical benefits in the treatment of asthma.

References

1. Kotlikoff MI. Potassium channels in airway smooth muscle: a tale of two channels. Pharmacol Ther 1993; 58: 1–12.

2. Akasaka K, Konno K, Ono Y, Mue S, Abe C, Kunagai M, Ise T. Electromyographic study of bronchial smooth muscle in bronchial asthma. Tohoku J Exp Med 1975; 117: 55–59.
3. Allen SL, Boyle JP, Cortijo J, Foster RW, Morgan GP, Small RC. Electrical and mechanical effects of BRL 34915 in guinea-pig isolated trachealis. Br J Pharmacol 1986; 89: 395–405.
4. Boyle JP, Tomasic M, Kotlikoff MI. Delayed rectifier potassium channels in canine and porcine airway smooth muscle cells. J Physiol 1992; 447: 329–350.
5. Small RC, Berry JL, Cook SJ, Foster RW, Green KA, Murray MA. Potassium channels in airways. In: Pharmacology of the respiratory tract. Eds: KF Chung and PJ Barnes. Marcel Dekker Inc. 1993; 137–176.
6. Weston AH, Longmore J, Newgreen DT, Edwards G, Bray KM, Duty S. The potassium channel openers: A new class of vasorelaxants. Blood Vessels 1990; 27: 306–313.
7. Buckle DR, Arch JRS. Potassium channel activators and airways disease. Drug News and Perspectives 1993; 6: 279–288.
8. Rodger IW. Biochemistry of airway smooth muscle contraction. In: Asthma: Basic mechanisms and clinical management. Eds: PJ Barnes, IW Rodger and NC Thomson. London: Academic Press 1988; 57–79.
9. Longmore J, Bray KM, Weston AH. The contribution of Rb-permeable potassium channels to the relaxant and membrane hyperpolarizing actions of cromakalim, RP 49356 and diazoxide in bovine tracheal smooth muscle. Br J Pharmacol 1991; 102: 979–985.
10. Buckle DR, Arch JR, Bowring NE, Foster KA, Taylor JF, Taylor SG, Shaw DJ. Relaxant effects of the potassium channel activators BRL 38227 and pinacidil on guinea pig and human airway smooth muscle, and blockade of their effects by glibenclamide and BRL 31660. Pulm Pharmacol 1993; 6: 77–86.
11. Murray MA, Boyle JP, Small RC. Cromakalim-induced relaxation of guinea-pig isolated trachealis: antagonism by glibenclamide and phentolamine. Br J Pharmacol 1989; 98: 865–874.
12. Berry JL, Elliot KRF, Foster RW, Green KA, Murray MA, Small RC. Mechanical, biochemical and electrophysiological studies of RP 49356 and cromakalim in guinea pig and bovine trachealis muscle. Pulm Pharmacol 1991; 4: 91–98.
13. Small RC, Berry JL, Burka JF, Cook SJ, Foster RW, Green KA, Murray MA. Potassium channel activators and bronchial asthma. Clin Exp Allergy 1992; 22: 11–18.
14. Brown TJ, Sweetland J, Raeburn D. Comparison of the effects of RP 49356, cromakalim and nifedipine on guinea-pig trachea in vitro. Pflugers Arch 1989; 414: S188–S189.
15. Raeburn D, Brown TJ. RP 49356 and cromakalim relax airway smooth muscle in vitro by opening a sulphonylurea-sensitive K$^+$ channel: a comparison with nifedipine. J Pharmacol Exp Ther 1991; 256: 480–485.
16. Black JL, Armour CL, Johnson PR, Alouan LA, Barnes PJ. The action of a potassium channel activator, BRL 38227 (lemakalim), on human airway smooth muscle. Am Rev Respir Dis 1990; 142: 1384–1389.
17. Dumas M, Dumas J-P, Advenier C, Giudicelli J-F. Effects of three K$^+$ channel openers on airways and pulmonary circulation in the isolated guinea-pig lung. Eur J Pharmacol 1993; 239: 141–147.
18. Raeburn D, Karlsson J-A. Potassium channel openers: Airway pharmacology and clinical possibilities in asthma. Prog Drug Res 1991; 37: 161–180.
19. Small RC, Berry JL, Foster RW, Blarer S, Quast U. Analysis of the relaxant action of SDZ PCO 400 in airway smooth muscle from the ox and guinea-pig. Eur J Pharmacol 1992; 219: 81–88.
20. Jones TR, Charette L, Garcia ML, Kaczorowski GJ. Selective inhibition of relaxation of guinea-pig trachea by charybdotoxin, a potent Ca^{2+}-activated K$^+$ channel inhibitor. J Pharmacol Exp Ther 1990; 255: 697–706.
21. Murray MA, Berry JL, Cook SJ, Foster RW, Green KA, Small RC. Guinea-pig isolated trachealis: The effects of charybdoxoxin on mechanical activity, membrane potential changes and the activity of plasmalemmal K$^+$ channels. Br J Pharmacol 1991; 103: 1814–1818.
22. Buckle DR. Prospects for potassium channel activators in the treatment of airways obstruction. Pulm Pharmacol 1993; 6: 161–169.
23. Gillespie JS, Sheng H. The lack of involvement of cyclic nucleotides in the smooth muscle relaxant action of BRL 34915. Br J Pharmacol 1988; 94: 1189–1197.

24. Kume H, Takai A, Tokuno H, Tomita T. Regulation of Ca^{2+}-dependent K^+ channel activity in tracheal monocytes by phosphorylation. Nature 1989; 341:152–154.

25. Arch JRS, Buckle DR, Bumstead J, Clarke GD, Taylor JF, Taylor SG. Evaluation of the potassium channel activator cromakalim (BRL 34915) as a bronchodilator in the guinea pig: comparison with nifedipine. Br J Pharmacol 1988; 95: 763–770.

26. Taylor SG, Arch JR, Bond J, Buckle DR, Shaw DJ, Taylor JF, Ward JS. The inhibitory effects of cromakalim and its active enantiomer BRL 38227 against various agonists in guinea-pig and human airways: comparison with pinacidil and verapamil. J Pharmacol Exp Ther 1992; 261: 429–437.

27. Englert HC, Wirth K, Gehring D, Furst U, Albus U, Scholz W, Rosenkranz B, Scholkens BA. Airway pharmacology of the potassium channel opener, HOE 234, in guinea pigs: *in vitro* and *in vivo* studies. Eur J Pharmacol 1992; 210: 69–75.

28. Paciorek PM, Burden DT, Gater PR, Hawthorn YM, Spence AM, Taylor JC, Waterfall JF. Inhibition by Ro 31-6930 of agonist and allergen induced bronchoconstriction in anaesthetised guinea-pigs and cats. Pulm Pharmacol 1991; 4: 225–232.

29. Imagawa J-I, Yoshida S, Koga T, Kamei K, Nabata H. The effect of a novel benzopyran derivative, KC 399, on the isolated guinea-pig trachealis and human bronchi. Gen Pharmacol 1993; 24: 1505–1512.

30. Chapman ID, Kristerson A, Mathelin G, Schaeublin E, Mazzoni L, Bougekeur K, Murphy N, Morley J. Effects of a potassium channel opener (SDZ PCO 400) on guinea pig and human pulmonary airways. Br J Pharmacol 1992; 106: 423–429.

31. Fedan, JS, Hay DWP, Raeburn D. Ca^{2+} respiratory smooth muscle function: Is there a role for calcium entry blockers in asthma therapy? Current topics in pulmonary pharmacology and toxicology. Ed. MA Hollinger. New York: Elsevier 1987; 53–94.

32. Giembycz MA, Rodger IW. Electrophysiological and other aspects of excitation-contraction coupling and uncoupling in mammalian airway smooth muscle. Life Sci 1987; 41: 111–132.

33. Cook NS. The pharmacology of potassium channels and their therapeutic potential. Trends Pharmacol Sci 1988; 9: 21–28.

34. Quast U, Cook NS. Moving together: K^+ channel openers and ATP-sensitive K^+ channels. Trends Pharmacol Sci 1989; 10: 431–434.

35. Chopra LC, Twort CH, Ward JP. Direct action of BRL 38227 and glibenclamide on intracellular calcium stores in cultured airway smooth muscle of the rabbit. Br J Pharmacol 1992; 105: 259–260.

36. Sweetland J, Raeburn D. Effect of potassium channel openers on Ca^{2+} influx, Na^+/Ca^{2+} exchange and Na^+/K^+ pumping in guinea pig trachea *in vitro*. Br J Pharmacol 1989; 98: 882P.

37. Paciorek PM, Cowlrick IS, Perkins RS, Taylor JC, Wilkinson GF, Waterfall JF. Evaluation of the bronchodilator properties of Ro 31-6390, a novel potasssium channel opener, in the guinea-pig. Br J Pharmacol 1990; 100: 289–294.

38. Bowring NE, Buckle DR, Clarke GD, Taylor JF, Arch JR. Evaluation of the potassium channel activator BRL 38227 as an inhaled bronchodilator in the guinea-pig: contrast with nifedipine and salbutamol. Pulm Pharmacol 1991; 4: 99–105.

39. Raeburn D, Underwood SL, Lewis SA. Evaluation of the potassium channel opener RP 52891 on bronchospasm in anaesthetised, conscious guinea pigs. Thorax 1991; 46: 294P.

40. Raeburn D, Underwood SL, Lewis SA, Woodman VR, Battram CH, Sharma S, Hart TW. Evaluation of a potent, new, glibenclamide-sensitive potassium channel opener RP 66471 in guinea pig airways: a comparison with lemakalim. Fund Clin Pharmacol 1991; 5: 390P.

41. Bowring NE, Arch JR, Buckle DR, Taylor JF. Comparison of the airways relaxant and hypotensive potencies of the potassium channel activators BRL 55834 and levcromakalim (BRL 38227) *in vivo* in guinea pigs and rats. Br J Pharmacol 1993; 109: 1133-1139.

42. Arch JRS, Bowring NE, Buckle DR. Evaluation of the novel potassium channel activator BRL 55834 as an inhaled bronchodilator in guinea-pigs and rats: Comparison with levcromakalim and salbutamol. Pulm Pharmacol 1994; 7: 121–128.

43. Imagawa J-I, Kamei K, Yoshida S, Sugo I, Koga T, Nabata H. *In vivo* bronchodilator action of a novel K^+ channel opener, KC 399, in the guinea pig. J Pharmacol Exp Ther 1994; 269: 1–6.

44. Kusner EJ, Marks RL, Buckner CK, Krell RD. Effect of BRL 34915 and other potassium modulators on leukotriene D_4 and histamine release from chopped guinea-pig lung. Pharmacology 1989; 31: 144.

45. Persson CGA. Plasma exudation and asthma. Lung 1988; 166: 1–23.
46. Black JL, Barnes PJ. Potassium channels and airway function: new therapeutic prospects. Thorax 1990; 45: 213–218.
47. Martin CAE, Advenier C. Effects of cromakalim on bradykinin-, histamine- and substance P-induced airway microvascular leakage in the guinea-pig. Eur J Pharmacol 1993; 239: 119–126.
48. Kidney JC, Lotvall JP, Lei Y, Chung KF, Barnes PJ. A comparison of the inhaled K $^+$ channel openers BRL 38227 and HOE 234 against histamine-induced bronchoconstriction and microvascular leakage. Am Rev Respir Dis 1992; 145: A202.
49. Chapman ID, Mazzoni L, Morley J. Actions of SDZ PCO 400 and cromakalim on airway smooth muscle *in vivo*. Agents Actions Suppl 1991; 34: 53–62.
50. Chand N, Diamantis W, Sofia RD. Induction of non-specific airway hyperreactivity by potassium channel blockade in rat isolated trachea. Br J Pharmacol 1990; 101: 541–544.
51. Hall AK, MacLagen J. Effects of cromakalim on cholinergic neurotransmission in the guinea-pig trachea. Br J Pharmacol 1988; 95: 792P.
52. McCaig DJ, De Jonckheere B. Effect of cromakalim on bronchoconstriction evoked by cholinergic nerve stimulation in guinea-pig isolated trachea. Br J Pharmacol 1989; 98: 662–668.
53. Cooper J, MacLagen J. The effect of potassium channel opening drugs on pulmonary nerves. Agents Actions 1991; 34: 63–69.
54. Burka JF, Berry JL, Foster RW, Small RC, Watt AJ. Effects of cromakalim on neurally-mediated responses of guinea pig tracheal smooth muscle. Br J Pharmacol 1991; 104: 263–269.
55. Good DM, Hamilton TC. Effect of BRL 38227 on neurally-mediated responses in guinea-pig isolated bronchus. Br J Pharmacol 1991; 102: 336P.
56. Good DM, Clapham JC, Hamilton TC. Effects of BRL 38227 on neurally-mediated responses in the guinea-pig isolated bronchus. Br J Pharmacol 1992; 105: 933–940.
57. Kuo H-P, Rohde JAL, Barnes PJ, Rogers DF. K $^+$ channel activator inhibition of neurogenic goblet cell secretion in guinea-pig trachea. Eur J Pharmacol 1992; 215: 297–299.
58. Lewis SA, Raeburn D. Preferential prejunctional site of inhibition of non-cholinergic bronchospasm by potassium channel openers (KCOs). Br J Pharmacol 1990; 100: 474P.
59. Clapham JC, Bowring NE, Trail BK, Fuller DA, Good DM. Effects of levcromakalim and RP52891 on NANCe nerve-mediated changes in pulmonary dynamics evoked by vagal stimulation in the guinea pig. Pulm Pharmacol 1993; 6: 201–208.
60. Ichinose M, Barnes PJ. A potassium channel activator modulates both excitatory non-cholinergic and cholinergic neurotransmission in guinea pig airways. J Pharmacol Exp Ther 1990; 252: 1207–1212.
61. Baird A, Hamilton T, Richards D, Tasher T, Williams AJ. Cromakalim, a potassium channel activator, inhibits histamine-induced bronchoconstriction in healthy volunteers. Br J Clin Pharmacol 1988; 24: 114P.
62. Williams AJ, Lee TH, Cochrane GM, Hopkirk A, Vyse T, Chiew F, Lavender E, Richards DH, Owen S, Stone P, Church S, Woodcock AA. Attenuation of nocturnal asthma by cromakalim. Lancet 1990; 336: 334–336.
63. Kidney JC, Fuller RW, Worsdell YM, Lavender EA, Chung KF, Barnes PJ. Effect of oral potassium channel activator, BRL 38227, on airway function and responsiveness in asthmatic patients: comparison with oral salbutamol. Thorax 1993; 48: 130–133.
64. Nguyen PV, Davis A, Tasker TCG, Leenan FHH. Effects of BRL 34915 on pressor and chronotropic responses to iv norepinephrine, angiotensin II and isoproterenol in normal man. Cardiovascular Drugs and Therapy 1987; 1: 270.
65. Singer DRJ, Markandu ND, Miller MA, Sugden AL, MacGregor GA. Potassium channel stimulation in normal subjects and in patients with essential hypertension: An acute study with cromakalim (BRL 34915). J Hypertension 1989; 7: 294–295.

Airways Smooth Muscle: Peptide Receptors, Ion Channels and Signal Transduction
ed. by D. Raeburn and M. A. Giembycz
© 1995 Birkhäuser Verlag Basel/Switzerland

CHAPTER 10
Sodium/Potassium/Chloride Co-Transport

Alan J. Knox

Respiratory Medicine Unit, City Hospital, Nottingham, UK

1   Introduction
2   Studies of Diuretics *In Vivo*
2.1   Studies of Frusemide *In Vivo* in Asthma
2.2   Effects of Other Diuretics in Asthma
3   Possible Mechanisms of Action of Loop Diuretics in Asthma
3.1   Carbonic Anhydrase Inhibition
3.2   PGE$_2$ Production
3.3   Mediator Release from Inflammatory Cells
3.4   Effects of Neural Pathways
4   Stoichiometry and Kinetics of Na/K/Cl Co-Transport
5   Effect of Loop Diuretics on Na/K/Cl Co-Transport
6   Cellular Roles of Na/K/Cl Co-Transport
6.1   Cell Volume Regulation
6.2   Intracellular Chloride Accumulation
6.3   Smooth Muscle Contractility
7   Regulation of Na/K/Cl Co-Transport
7.1   Cyclic Adenosine 3'5' Monophosphate (Cyclic AMP)
7.2   Cyclic Guanosine 3'5' Monophosphate (Cyclic GMP)
7.3   Protein Kinase C
7.4   Calcium
7.5   ATP
8   Studies of Na/K/Cl Co-Transport in Airways Smooth Muscle
9   Studies of the Effects of Loop Diuretics on Airways Smooth Muscle *In Vitro*
9.1   Hypertonic and KCl-Induced Contraction
9.2   Neurally-Mediated Contraction
9.2.1 Cholinergic
9.2.2 Non-Adrenergic Non-Cholinergic (NANC)
9.3   Spasmogen-Induced Contraction
9.4   Protein Kinase C-Induced Contraction
10   Conclusions
      Acknowledgement
      References

1. Introduction

Considerable recent interest in the Na/K/Cl co-transport system in airways smooth muscle has been generated by studies showing that the loop diuretic frusemide, a potent inhibitor of co-transport, provides protection against several bronchoconstriction challenges in asthma (Table 1). This is particularly interesting as frusemide has the same

Table 1. Bronchoconstrictor challenges against which loop diuretics afford protection

Test	Subjects	Drug	Reference
Exercise	A	F	[1] [24]
Allergen	A	F	[2–4]
Metabisulphite	A	F,P,E	[5] [34] [16] [56]
UNDW	A	F,T	[6] [42] [35] [17]
Cold air hyperventilation	A	F	[7] [22] [20]
AMP	A	F,B	[8] [5] [14]
Hypertonic saline	A	F	[9]
Methacholine	N,A	F	[13] [25] [8]

A = asthmatic
N = normal
F = frusemide
B = bumetanide
P = piretanide
T = torasamide
E = ethacrynic acid

protective profile as two other established anti-asthma drugs, sodium cromoglycate and nedocromil sodium, suggesting that they may share a common mechanism of action.

2. Studies of Diuretics *In Vivo*

2.1. Studies of Frusemide In Vivo in Asthma

Bianco and colleagues in 1988 [1] first reported the protective effect of inhaled frusemide against exercise-induced bronchoconstriction in asthma and frusemide has subsequently been shown to protect against a wide range of challenges such as allergen [2–4], sodium metabisulphite [5], ultrasonically nebulised distilled water [6], cold air hyperventilation [7], adenosine monophosphate [8] and hypertonic saline challenge [9]. The effect of frusemide is dose related and it is only effective if given by inhalation [1]. Frusemide is relatively ineffective at protecting against histamine-[10], methacholine-[8, 11] or prostaglandin $F_{2\alpha}$-[12] induced bronchoconstriction in asthmatic subjects, although it does protect against methacholine-induced bronchoconstriction in normal subjects [13]. The fact that frusemide provides protection mainly against stimuli which act indirectly on inflammatory cells or neural pathways (exercise, allergen, distilled water, cold air) rather than stimuli which act directly on airways smooth muscle (histamine/methacholine) suggests that frusemide's effect in asthma is probably not an airways smooth muscle effect. Nevertheless, these findings have prompted several groups of investigators to examine the effects of diuretics which inhibit Na/K/Cl co-transport in airways smooth muscle (ASM).

2.2. Effects of Other Diuretics in Asthma

Several other diuretics have been looked at for their possible protective effects in asthma. Bumetanide, a more potent inhibitor of Na/K/Cl co-transport, has less effect than frusemide in protecting against adenosine monophosphate-[5, 14], metabisulphite-[5] or exercise-[15] induced bronchoconstriction. This suggests that either the effects of frusemide are not due to its effect on Na/K/Cl co-transport or that pharmacokinetic differences account for the lack of effect of bumetanide. Two other loop diuretics which are also more potent inhibitors of co-transport than frusemide, piretanide [16] and torasamide [17], also have a smaller effect than frusemide on bronchoconstriction induced by metabisulphite and ultrasonically nebulised distilled water respectively, again suggesting that co-transport inhibition is not the mechanism of action of frusemide in asthma. Amiloride, an unrelated diuretic, with effects on sodium entry channels and sodium hydrogen exchange does not inhibit the response to histamine [18], sodium metabisulphite [19] or dry air challenge in asthma [20], suggesting that inhibition of sodium entry is not responsible.

3. Possible Mechanisms of Action of Loop Diuretics in Asthma

Frusemide has several other actions apart from inhibition of Na/K/Cl co-transport which might explain its protective effect in asthma. It has effects on a number of cellular processes and there are various sites in the airway at which it might act (Table 2). These include inhibition of carbonic anhydrase, release of prostaglandin E_2, and inhibition of chloride channels either in inflammatory cells or neurones. Some of the more likely explanations for frusemide's effects are discussed below.

Table 2.

a) Potential mechanisms of action of frusemide in asthma

Na/K/Cl co-transport inhibition
Cl^- channel blockade
PGE_2 release
Carbonic anhydrase inhibition
Inhibition of Na/K ATPase

b) Possible sites of action

Airway epithelium
Airway nerves
Inflammatory cells
Vasculature
Airway smooth muscle

3.1. Carbonic Anhydrase Inhibition

Frusemide has been shown to inhibit carbonic anhydrase in choroid plexus [21]. Inhibition of carbonic anhydrase in airway epithelial cells has been suggested as a possible mechanism of action of frusemide in asthma [22]. This would seem unlikely however, as inhaled acetazolamide, a more potent carbonic anhydrase inhibitor that frusemide, had a much smaller effect than frusemide on bronchoconstriction induced by hyperventilation of cold dry air [22].

3.2. PGE_2 Production

The role of PGE_2 in mediating the effect of frusemide is also controversial. Frusemide's diuretic properties are dependent on release of PGE_2 by the kidney [23] and release of PGE_2 by airway cells, possibly airway epithelium, has been put forward as a possible mechanism of action of frusemide in asthma [24]. There are several pieces of evidence which support this hypothesis. *In vivo* indomethacin, a cyclo-oxygenase inhibitor, has been shown to inhibit the effect of frusemide on exercise induced bronchoconstriction in asthma [24] and flurbiprofen, another cyclo-oxygenase inhibitor, antagonises the effect of frusemide on methacholine-induced bronchoconstriction in normal subjects [25]. Inhaled PGE_2 shares a similar profile against bronchoconstrictor challenges in asthma inhibiting challenges to indirect stimuli [26–29], but being much less effective at inhibiting challenges by direct stimuli such as methacholine [28]. Studies *in vitro* also support the PGE_2 hypothesis. In bovine airway epithelial strips application of frusemide increased PGE_2 production [30] and frusemide increases release of PGE_2 from sensitised, human bronchial rings [31].

However there are also several pieces of evidence that argue against PGE_2 being responsible for the action of frusemide in asthma. Instillation of frusemide into nasal mucosa *in vivo* does not alter the production of PGE_2 either in normal subjects [32] or subjects with rhinitis [33]. Connor *et al* [34] showed that cyclo-oxygenase inhibition with flurbiprofen did not inhibit the protective effect of frusemide on metabisulphite-induced bronchoconstriction in asthma. Recently, Bianco and colleagues have shown that inhaled lysine acetyl salicylic acid potentiates rather than prevents the effect of frusemide on ultrasonically nebulised water-induced fog bronchoconstriction [35]. The role of PGE_2 in mediating the effect of frusemide therefore remains debatable.

3.3. Mediator Release from Inflammatory Cells

Frusemide has been shown to affect the function of several inflammatory cells in both animals and man *in vitro*. Frusemide inhibits allergen-induced release of histamine from rat peritoneal mast cells [36], eosinophil respiratory burst activity [37], superoxide generation [38] and lysosomal enzyme secretion [39] by pulmonary macrophages as well as the release of inflammatory mediators from chopped human lung fragments [40]. In contrast, in human leucocytes, frusemide had no effect on the release of histamine stimulated by anti-IgE antibody [41]. In studies *in vivo*, frusemide prevented the increased neutrophil chemotactic activity induced by fog challenge [42].

3.4. Effects of Neural Pathways

Frusemide is known to inhibit the response to low chloride cough challenge [43–45], suggesting it may have an effect on sensory nerves. It has been suggested that frusemide might be acting as a chloride channel inhibitor in these nerves, as has been demonstrated in the cornea [46] and lacrimal glands [47]. However, torasamide has a greater affinity than frusemide against chloride channels [48], but has a smaller effect than frusemide on fog challenge [17]. It would therefore seem unlikely that the effect of frusemide in asthma is due to chloride channel inhibition.

4. Stoichiometry and Kinetics of Na/K/Cl Co-Transport

The stoichiometry of the co-transporter is generally accepted to be 1:1:2 Na/K/Cl [49]. Whilst this is the case in most tissues studied, there are some notable exceptions. In some leaky epithelia Na and Cl movement is coupled as a simple Na/Cl co-transport with no concomitant movement of potassium [50]. In ferret erythrocytes [51] and squid axon [52] the ratio is 2:1:3 Na/K/Cl whilst in avian [53] and human erythrocytes [54] K fluxes are greater than Na fluxes.

There is activation of co-transport by Na, K and Cl with K_m values Na of (10–20 mmol/l), for K (1–10 mmol/l) and for Cl (60–80 mmol/l) [49]. The kinetics of the co-transporter are compliated by interactions between ions such that the affinity of one depends on the concentrations of the others. In this respect, raising Na or K increases the affinity for the other ions while raising chloride increases the affinity for the cations [49]. Some other ions are capable of substituting for Na, K or Cl in the co-transport process. Li can substitute for Na in some tissues, Rb substitutes for K and Br substitutes for Cl, although the latter substitution only supports transport at half speed [49].

5. Effects of Loop Diuretics on Na/K/Cl Co-Transport

Although the effects of different diuretics on the Na/K/Cl co-transport system in airways smooth muscle has not been studied in detail, it has been studied extensively in the kidney. The order of potency of these agents is frusemide < piretanide < torasamide < bumetanide [55]. The effect of frusemide and these related drugs is rapidly and entirely reversible. There are several requirements in the structure of these compounds for them to be active against co-transport [55]. Firstly, they must possess an anionic group such as a tetrazolate, sulphonate, carboxylate or sulphonylurea. A secondary or tertiary amine in the ortho or meta positions relative to the anionic group is required. This amino group links the anionic moiety to an apolar residue. A sulphonamide group is required in meta position to the anionic group, or, in the case of torasamide, a pyridine nitrogen. Because of their charges, hydrogen bonds and apolar moieties loop diuretics are thought to interact with the co-transporter at several sites [55].

The loop diuretics have structural similarites with chloride channel blockers and blockers of chloride/bicarbonate exchange. The fact that minor modifications of one of these compounds can lead to a block of another kind suggests that their respective membrane transport proteins may have some similar sites [55]. Ethacrynic acid is a slightly different loop diuretic which does not share the mechanism of inhibition of Na/K/Cl co-transport and is thought to act at the level of mitochondrial ATP production [55]. It is therefore of interest that ethacrynic acid has recently been shown to be at least as potent as frusemide in inhibiting bronchoconstriction induced by sodium metabisulphite [56]. This adds further weight to the hypothesis that frusemide is not acting in asthma by inhibiting Na/K/Cl co-transport.

6. Cellular Roles of Na/K/Cl Co-Transport

Na/K/Cl co-transport has several possible cellular roles but not all of these are relevant to airways smooth muscle function. In epithelial cells, co-transport is important in regulating electrolyte absorption and secretion. Possible roles which may have more relevance to airways smooth muscle are in cell volume regulation, intracellular Cl^- accumulation and determining contractility.

6.1. Cell Volume Regulation

The constancy of cell volume in several cells depends on the osmotic balance sustained by the Na/K ATPase pump in opposition to passive downhill flux through Na/K/Cl co-transport. As Na/K ATPase pumps are abundant in airways smooth muscle [57], it is likely that Na/K/Cl co-transport carries out a similar housekeeping function.

6.2. Intracellular Chloride Accumulation

Transporting chloride into the cell could be potentially important for airways smooth muscle cells. An increase in internal Cl^- would enhance HCO_3^- entry via Cl^-/HCO_3^- exchange and might therefore be an important part of the regulation of intracellular pH [49]. This hypothesis has not yet been tested in airways smooth muscle cells.

6.3. Smooth Muscle Contractility

Studies in vascular smooth muscle have suggested that Na/K/Cl co-transport may play a role in determining smooth muscle contractility. Frusemide has been shown to reduce Rb^{86} uptake and the contractile response to α adrenoceptor agonists in rat and rabbit aorta [58]. The proposed mechanism is that inhibition of Na uptake leads to a reduction in $[Na]_i$ which facilitates calcium exit by switching on Na/Ca exchange.

7. Regulation of Na/K/Cl Co-Transport

Co-transport can be regulated by several protein kinases in addition to calcium and ATP. The effect of these substances on co-transport is, however, tissue specific and the effect may vary considerably between tissues (Table 3). It is therefore impossible to extrapolate results from one cell type to another.

Table 3. Examples of regulation of Na/K/Cl co-transport

Manipulation	Effect on co-transport	Tissue	References
Elevation of cyclic AMP	Stimulation	avian erythrocytes	[59]
		shark rectal gland	[60]
	Inhibition	Flounder intestine	[53]
	No effect	MDCK cells	[61]
Elevation of cyclic GMP	Inhibition	Flounder intestine	[64]
	Stimulation	Vascular smooth muscle	[65] [66]
Activation of PKC	Inhibition	3T3 adipocytes,	[68]
		Vascular smooth muscle	[69]
	Stimulation	Hamster lung fibroblasts	[70]
Elevation in intracellular calcium	Stimulation	Rabbit tracheal epithelium	[71]
	Inhibition	Human erythrocytes	[72]
		Flounder intestine	[53]

PKC = protein kinase C

7.1. Cyclic Adenosine 3'5' Monophosphate (Cyclic AMP)

Cyclic AMP activates co-transport in a variety of tissues such as avian erythrocytes [59] and shark rectal gland [60] in response to activation of receptors coupled to adenylate cyclase. In contrast inhibition of co-transport, in response to elevations in intracellular cyclic AMP, has been described in flounder intestine [53] and cyclic AMP has no effect on co-transport in MDCK cells [61]. As airways smooth muscle has receptors such as β-adrenoceptor [62] and vasoactive intestinal peptide receptors [63] coupled to the cyclic AMP pathway, it is possible that cyclic AMP might regulate co-transport in airways smooth muscle. This has not been studied to date.

7.2. Cyclic Guanosine 3'5' Monophosphate (Cyclic GMP)

Stimulation of receptors such as atrial natriuretic peptide receptors which are coupled to cyclic GMP have been shown to inhibit co-transport in flounder intestine [64] but stimulates co-transport in vascular smooth muscle [65, 66].

7.3. Protein Kinase C

Protein kinase C (PKC) represents a heterogous group of protein kinases which are integral parts of cell signal transduction [67]. Phorbol esters which directly activate PKC have been used as probes to test the role of PKC in several tissues. Phorbol esters inhibit co-transport in 3T3 preadipocytes [68] and vascular smooth muscle [69]. In contrast phorbol esters stimulate co-transport in hamster lung fibroblasts [70].

7.4. Calcium

Like cyclic AMP effects on co-transport, calcium effects on co-transport are also tissue specific. In rabbit tracheal epithelial cells Na/Cl co-transport can be stimulated by calcium ionophore [71] suggesting that alteration in intracellular calcium may regulate co-transport activity. In these studies, α_2 adrenoceptor activation was associated with an increase in co-transport activity which was thought to be due to an increase in intracellular calcium. It is therefore possible that stimulation of receptors utilising inositol phospholipid/calcium signalling pathways in airways smooth muscle might also be capable of modifying co-transport activity. In contrast, in human erythrocytes [72] and flounder intestine [53] the opposite effect was seen and elevated calcium was found to inhibit co-transport.

7.5. ATP

Although co-transport is a passive process, there is a requirement for ATP [49]. Metabolite depletion can block co-transport in several types of cell which can be restored by repletion of ATP levels [49].

8. Studies of Na/K/Cl Co-Transport in Airways Smooth Muscle

The two most widely used methods of studying the transporter have been either to look at ion flux or to use radiolabelled bumetanide as a probe to bind to the transport system [73]. Only the former studies have been performed in airways smooth muscle. Rhoden and Douglas recently studied bumetanide-sensitive potassium uptake in guinea-pig airways smooth muscle using Rb^{86} [74]. They found that bumetanide caused a concentration-dependent inhibition of Rb^{86} uptake with a half maximal effect at 0.18 µmol. Bumetanide sensitive rubidium uptake represented 40% of total uptake into airways smooth muscle. Frusemide IC_{50} of approximately 1 µmol had a similar effect to bumetanide. The IC_{50} values are similar to that reported for frusemide and bumetanide in the kidney. The effect of dibutyryl cyclic GMP, a manoeuvre which stimulates Na/K/Cl co-transport in vascular smooth muscle was also studied. Unlike vascular smooth muscle, cyclic GMP did not alter bumetanide sensitive Rb^{86} uptake in guinea-pig airways smooth muscle.

9. Studies of the Effects of Loop Diuretics on Airways Smooth Muscle *In Vitro*

Several studies have looked at the effect of loop diuretics on the contractility of airways smooth muscle *in vitro* from several species (Table 4).

9.1. Hypertonic and KCl-Induced Contraction

In bovine airways smooth muscle frusemide had no effect on the responses to potassium chloride or hypertonic challenge [75]. The lack of effect of frusemide on hypertonic challenge was seen whether the epithelium was present or absent. Similarly in human airways smooth muscle frusemide had no effect on contraction induced by hypertonic solutions in bronchial rings with epithelium intact [75]. This suggests that the protective effects of frusemide on hypertonic challange *in vivo* are not an airways smooth muscle effect.

Table 4. Effect of loop diuretics on airways smooth muscle *in vitro*

Stimulus	Drug	Effect	Species	References
KCL	F	none	bovine	[75]
Hypertonic	F	none	bovine	[75]
Histamine	F	none	bovine	[75]
Hypertonic	F	none	human	[75]
Bradykinin	F	inhibition	guinea-pig	[79]
Capsaicin	F	inhibition	guinea-pig	[79]
Acetylcholine	F	none	guinea-pig, horse, human	[76] [77] [78]
EFS	F,B	inhibition	guinea-pig, horse, human	[76] [77] [78]
NANC	F	inhibition	guinea-pig	[79]
Toluene diisocynate	B	none	guinea-pig	[91]
Substance P	F	none	guinea-pig	[79]
Neurokinin A	F	none	guinea-pig	[79]

F = frusemide
B = bumetanide
EFS = electrical field stimulation
NANC = non adrenergic non cholinergic contraction

9.2. Neurally-Mediated Contraction

9.2.1. Cholinergic: In guinea-pig airways smooth muscle frusemide and bumetanide have been shown to inhibit the effect of electrical field stimulation [76]. This was thought to be a prejunctional effect as the response to exogenous acetylcholine was unaffected. The order of potency of these diuretics (bumetanide > frusemide) was similar to their order of potency as Na/K/Cl co-transport inhibitors, suggesting that inhibition of co-transport in cholinergic nerves might be responsible for their effect. These experiments were carried out in the presence of indomethacin and prostaglandins were therefore unlikely to be involved.

Similar results have been reported in equine [77] and human airways smooth muscle [78]. In equine airways smooth muscle, the effect was epithelium-dependent suggesting that frusemide may release a factor from epithelium which inhibits cholinergic neurotransmission. The most likely candidate for this factor is PGE_2.

9.2.2. Non-Adrenergic Non-Cholinergic (NANC): In guinea-pig airways smooth muscle frusemide inhibited contraction induced by either bradykinin or capsaicin [79]. The protective effect of frusemide on bradykinin-induced contraction was thought to be mediated via prostaglandins as it could be inhibited by indomethacin. Contraction in response to capsaicin was reduced by frusemide only in the presence of indomethacin, suggesting that either a different or an additional mechanism was involved in reducing capsaicin induced contractions, possibly a direct inhibitory effect of frusemide on C-fibres. Contractions produced by neurokinin A or substance P were not affected by frusemide.

The results from guinea-pig airways *in vitro* are consistent with studies in normal volunteers *in vivo* where frusemide inhibits capsaicin induced bronchoconstriction [80].

9.3. Spasmogen-Induced Contraction

Frusemide, at concentrations which should inhibit Na/K/Cl co-transport has no effect on histamine-induced contraction of bovine airways smooth muscle [75], acetylcholine-,substance P-or neurokinin A-induced contraction of guinea-pig airways smooth muscle [79] or acetylcholine-induced contraction of human airways smooth muscle [78]. These studies argue against a role for the Na/K/Cl co-transporter in playing a major role in spasmogen induced contraction of airways smooth muscle.

In a study of guinea-pigs at different stages in development, frusemide was found to relax airways contracted by either histamine or acetyl choline [81]. The effect diminished with maturity, however and the relaxation induced by frusemide in these preparations was greater in Hepes solution than in a bicarbonate buffered solution. This study did not look at cyclo-oxygenase inhibition on frusemide induced relaxation in airways smooth muscle and it cannot therefore be determined if prostaglandins were responsible. The significance of these findings is unclear as frusemide in not a bronchodilator in asthmatic subjects *in vivo*.

9.4. Protein Kinase C-Induced Contraction

Protein kinase C represents a heterologous group of protein kinases thought to play an essential part in the maintenance of phase of the contractile response in airways smooth muscle [82, 83]. Protein kinase C can be directly stimulated *in vitro* by the application of phorbol esters [84]. Application of phorbol esters has been shown to contract airways smooth muscle from several species [85–88]. As PKC modulates Na/K/Cl co-transport in some tissues [68–70], co-transport could potentially have a role in PKC mediated contraction. However, in bovine airways smooth muscle, frusemide had no effect on contraction induced by phorbol 12, 13 dibutyrate suggesting both that frusemide is not a PKC inhibitor and that Na/K/Cl co-transport activation does not play a role in PKC-induced contraction of airways smooth muscle [88].

In contrast, in guinea-pig airways smooth muscle phorbol esters induce a biphasic contractile and relaxant response which could be completely inhibited by frusemide [89]. The negative log of the IC_{50} in this study was much less than the IC_{50} for Na/K/Cl co-transport suggesting that co-transport inhibition is an unlikely explanation for the effect. The authors suggested that it might reflect a direct effect on PKC itself.

10. Conclusions

Taken collectively the studies *in vitro* in several species do not support a role for Na/K/Cl co-transport in determining airways smooth muscle contractility. However, in some species Na/K/Cl co-transport may play an important role in modulating neural pathways either by inhibiting cholinergic neurotransmission or by an effect on C fibres. PGE_2, produced either by epithelium in response to loop diuretics, may also have inhibitory effects on cholinergic pathways. It is of interest that in some studies frusemide can inhibit neurally mediated contractions of tissues with epithelium denuded [78, 79], and that the effect can be blocked by indomethacin. This suggests that loop diuretics may cause production of inhibitory prostaglandins from tissues other than airway epithelium. Airways smooth muscle has recently shown to be a rich source of prostaglandin E_2 [90].

Acknowledgement

I would like to thank Hilary Hughes for typing this manuscript.

References

1. Bianco S, Vaghi A, Robuschi M, Pasargiklian M. Prevention of exercise-induced bronchoconstriction by inhaled frusemide. Lancet 1988; i: 252.
2. Bianco S, Pieroni M, Fefini M, Rottoli L, Sestini P. Protective effect of inhaled furosemide on allergen-induced early and late asthmatic reactions. N Engl J Med 1989; 321: 1069–1073.
3. Robuschi M, Pieroni M, Refini M, Bianco S, Rossoni G, Magni F, et al. Prevention of antigen-induced early obstructive reaction by inhaled furosemide in (atopic) subjects with asthma and (actively sensitized) guinea-pigs. J Allergy Clin Immunol 1990; 85: 10–6.
4. Verdiani P, DiCarlo S, Baronti A, Bianco S. Effect of inhaled furosemide on the early response to antigen and subsequent change in airway reactivity in atopic patients. Thorax 1990; 45: 377–381.
5. O'Connor BJ, Chung KF, Chen Wordell YM, Fuller RW, Barnes PJ. Effect of inhaled frusemide and bumetanide on adenosine 5-monophosphate and metabisulphite-induced bronchoconstriction in asthmatics. Am Rev Respir Dis 1991; 143: 1329–1334.
6. Robushi M, Gambaro G, Spagnotto S, Vaghi A, Bianco S. Inhaled furosemide (F) is highly effective in preventing ultrasonically nebulised water (UNH$_2$O) bronchoconstriction. Pulm Pharmacol 1989; 1: 187–191.
7. Grubbe RE, Hopp R, Dave NK, Brennan B, Bewtra A, Townley R. Effect of inhaled furosemide on the bronchial response to methacholine and cold air hyperventilation challenges. J Allergy Clin Immunol 1990; 85: 881–884.

8. Polosa R, Lau LCK, Holgate ST. Inhibition of adenosine 5'-monophosphate- and methacholine-induced bronchoconstriction in asthma by inhaled frusemide. Eur Respir J 1990; 3: 665–672.
9. Rodwell LT, Anderson SD, du Toit JI, Seale JP. The effect of inhaled frusemide on airway sensitivity to inhaled 4.5% sodium chloride aerosol in asthmatic subjects. Thorax 1993; 48: 208–213.
10. Vaghi A, Robuschi M, Berni F, Bianco S. Effect of inhaled furosemide (F) on bronchial response to histamine (H) in asthmatics. Eur Respir J 1988; 1: 406S.
11. Nichol GN, Alton EWFW, Nix A, Geddes DM, Chung KF, Barnes PJ. Effect of inhaled furosemide on metabisulfite- and methacholine induced bronchoconstriction and nasal potential difference in asthmatic subjects. Am Rev Respir Dis 1990; 142: 576–580.
12. Stone RA, Yeo TC, Barnes PJ, Chung KF. Frusemide inhibits cough but not bronchoconstriction to prostaglandin $F_{2\alpha}$ in patients with asthma. Am Rev Respir Dis 1991; 143: A548.
13. Fujimura M, Sakamoto S, Kamio Y. Effect of inhaled furosemide on brochial responsiveness to methacholine. N Engl J Med 1990; 322: 935–936.
14. Polosa R, Rajakulasingam K, Prosperini G, Church MK, Holgate ST. Relative potencies and time course of changes in adenosine 5'-monophosphate airway responsiveness with inhaled furosemide and bumetanide in asthma. J Allergy Clin Immunol 1993; 92: 288–297.
15. Duggan CJ, Dixon CMS, Ind PW. The effect of furosemide and bumetanide in exercise induced asthma (EIA). Am Rev Respir Dis 1990; 141: A474.
16. Yeo CT, O'Connor BJ, Chen-Worsdell M, Barnes PJ, Chung KF. Protective effect of loop diuretics, piretanide and frusemide, against sodium metabisulphite-induced bronchoconstriction in asthma. Eur Respir J 1992; 5: 1184–1188.
17. Foresi A, Pelucchi A, Mastropasqua B, Cavigioli G, Earlesi RM, Marazzini L. Effect of inhaled furosemide and torasamide on bronchial responses to ultrasonically nebulised distilled water in asthmatic subjects. Am Rev Respir Dis 1992; 146: 364–8.
18. Knox AJ, Britton JR, Tattersfield AE. Effect of sodium-transport inhibitors on bronchial reactivity *in vivo*. Clin Sci 1990; 79: 325–330.
19. Baldwin DR, Grange KL, Pavord ID, Knox AJ. The effect of amiloride on the airway response to metasulphite in asthma: a negative report. Eur Respir J 1992; 5: 1189–1192.
20. Rodwell LT, Anderson SD, du Toit J, Seale JP. Different effects of inhaled amiloride and frusemide on airway responsiveness to dry air challenge in asthmatic subjects. Eur Respir J 1993; 8: 855–861.
21. Vogh BP, Langham MR Jr. The effect of furosemide and bumetanide on cerebrospinal fluid formation. Brain Res 1981; 221: 171–183.
22. O'Donnell WJ, Rosenberg M, Niven RW, Drazen JM, Israel E. Acetazolamide and furosemide attenuate asthma induced by hyperventilation of cold, dry air. Am Rev Respir Dis 1992; 146: 1518–1532.
23. Miyanoshita A, Terada M, Endou H. Furosemide directly stimulates prostaglandin E_2 production in the thick ascending limb of Henle's loop. J Pharmacol Exp Ther 1989; 251: 1155–9.
24. Pavord ID, Wisniewski A, Tattersfield AE. Inhaled frusemide and exercise induced asthma: evidence of a role for inhibitory prostanoids. Thorax 1992; 47: 797–800.
25. Polosa R, Rajakulasimgam K, Prosperini G, Holgate ST. Flurbiprofen abolishes the change in bronchial reactivity to methacholine with inhaled frusemide in healthy volunteers. Am Rev Respir Dis 1993; 147: A839.
26. Passargiklian M, Bianco S, Allegra L. Clinical functional and pathogenic aspects of bronchial reactivity to prostaglandins $E_{2\alpha}$, E_1 and E_2. Adv Prost Thromb Res 1976; 1: 461–75.
27. Passargiklian M, Bianco S, Allegra L, Moavero NE, Petrigni G, Robuschi M *et al.* Aspects of bronchial reactivity to prostaglandins and aspirin in asthmatic patients. Respiration 1977; 34: 79–91.
28. Pavord ID, Wisniewski A, Mathur R, Wahedna I, Knox AJ, Tattersfield AE. Effect of inhaled PGE_2 on bronchial reactivity to sodium metabisulphite and methacholine in patients with asthma. Thorax 1991; 46: 633–637.
29. Pavord ID, Wong C, Williams J, Wisniewski A, Smyth E, Tattersfield AE. Effect of inhaled prostaglandin E_2 on allergen-induced asthma. Am Rev Respir Dis 1993; 148: 87–90.

30. Pavord I, Knox A, Cole A, Tattersfield AE. Effect of frusemide on release of PGE$_2$ from bovine tracheal mucosa. Thorax 1991; 46: 751P.
31. Pavord I, Holland E, Baldwin D, Tattersfield AE, Knox AJ. Frusemide and allergen induced contractions of passively sensitised human bronchi: evidence of a role for prostaglandin E$_2$. Thorax 1992; 47: 895.
32. Mullol J, Ramis I, Prat J, Rosello-Catafau J, Xaubet A, Piera C, et al. Failure of frusemide to increase production of prostaglandin E$_2$ in human nasal mucosa in vivo. Thorax 1993; 48: 260–263.
33. Prat J, Mullol J, Ramis I, Rosello-Catafau J, Xaubet A, Nerin I, et al. Release of chemical mediators and inflammatory cell influx during early allergic reaction in the nose: Effect of furosemide. J Allergy Clin Immunol 1993; 92: 248–254.
34. O'Connor BJ, Ridge SM, Barnes PJ, Chung KF. The role of cyclo-oxygenase products in the inhibition of sodium metabisulphite-induced bronchoconstriction by frusemide in asthma. Am Rev Respir Dis 1991; 143: A210.
35. Bianco S, Vaghi A, Pieroni MG, Gambaro P, Sestini P, Berni F, et al. Potentiation of the protective effect of inhaled furosemide on the bronchial obstructive response to ultrasonically nebulised water by inhaled lysine acetylsalicylate. Eur Resp J 1991; 4 (Suppl 14): 606S.
36. Berti F, Rossoni G, Buschi A, Zuccari G, Villa LM. Inhaled furosemide prevents immunological respiratory changes and mediator release in guinea-pig. Eur Respir J 1991; 4: 445s.
37. Perkins RA, Dent G, Chung KF, Barnes PJ. Effects of anion transport inhibitors and chloride ions on eosinophil respiratory burst activity. Am Rev Respir Dis 1991; 143: A331.
38. Scoloperto M, Marini M, Brasca C, Fasoli A, Mattoli S. The protective effect of furosemide on the generation of superoxide anions by human bronchial epithelial cells and pulmonary macrophages in vivo. Pulm Pharmacol 1991; 4: 80–84.
39. Rottoli P, Rottoli L, Marino M, Sparnacci F, Vagliasindi M, Bianco S. Effect of furosemide on lysosomal enzyme secretion by alveolar macrophages. Eur Respir J 1990; 3: 307–308s.
40. Anderson SD, Wei H, Temple DM. Inhibition by furosemide of inflammatory mediators from lung fragments. N Engl J Med 1991; 324: 131.
41. Gorenberg AE, Goldbery BJ, Kaplan MS, Lad PM, Easton JG. Effects of amiloride and furosemide on histamine release from human leukocytes induced by anti-IgE antibody. J Allergy Clin Immunol 1992; 90: 691–2.
42. Moscato G, Dellabianco A, Falagiani P, Mistrello G, Rossi G, Rampulla C. Inhaled furosemide prevents both the bronchoconstriction and the increase in neutrophil chemotactic activity induced by ultrasonic "fog" of distilled water in asthmatics. Am Rev Respir Dis 1991; 143: 561–566.
43. Ventresca PG, Nichol GM, Barnes PJ, Chung KF. Inhaled furosemide inhibits cough induced by low chloride content solutions but not by Capsaicin. Am Rev Respir Dis 1990; 142: 143–146.
44. Stone RA, Barnes PJ, Chung KF. Effect of frusemide on cough responses to chloride-deficient solution in normal and mild asthmatic subjects. Eur Respir J 1993; 6: 862–867.
45. Sant'Ambrogio FB, Sant'Ambrogio G, Anderson JW. Effect of furosemide on the response of laryngeal receptors to low-chloride solutions. Eur Respir J 1993; 6: 1151–115.
46. Patarca R, Candia OA, Reinach PS. Mode of inhibition of active chloride transport in the frog cornea by furosemide. Am J Physiol 1983; 245: F660–F669.
47. Evans MG, Marty A, Tan YP, Trautmann A. Blockage of Ca-activated Cl conductance by furosemide in rat lacrimal glands. Pflügers Arch 1986; 406: 65–68.
48. Cabantchik ZI, Greger R. Chemical probes for anion transporters of mammalian cell membranes. Am J Physiol 1992; 262: C808–C827.
49. Chipperfield AR. The Na/K/Cl co-transport system. Clin Sci 1986; 71: 465–476.
50. Warnock DG, Gregor R, Dunham PB, Benhamin MA, Frizzell RA, Field M, et al. Ion transport processes in apical membranes of epithelia. Fed Proc 1984; 43: 2478–2487.
51. Ellory JC, Hall AC. Na/K/Cl co-transport stoichiometry: measurement of bumetanide-sensitive fluxes in ferret red cells. J Physiol 1984; 357: 63P.
52. Russell JM. Cation-coupled chloride influx in squid axon. Role of potassium and stoichiometry of the transport process. J Gen Physiol 1983; 81: 909–925.

53. Palfrey HC, Rao MC. Na/K/Cl co-transport and its regulation. J Exp Biol 1983; 106: 43–54.
54. Dunham PB, Stewart GW, Ellory JC. Chloride-activated passive potassium transport in human erythrocytes. Proc Nat Acad Sci 1980; 177: 1711–1715.
55. Gregor R, Wangemann P. Loop Diuretics. Renal Physiol Basel 1987; 10: 174–183.
56. Pye S, Pavord I, Wilding P, Bennett J, Knox A, Tattersfield A. Comparison of the effects of inhaled frusemide and ethacrynic acid on sodium metabisulphite induced bronchoconstriction in subjects with mild asthma. Thorax 1993; 48: 108s.
57. Knox AJ, Brown JK. Increased Na/K ATPase pump number and activity in proliferating airway smooth muscle cells. Am Rev Respir Dis 1991; 143: A607.
58. Deth RC, Payne RA, Peecher DM. Influence of furosemide on rubidium-86 uptake and alpha-adrenergic responsiveness of arterial smooth muscle. Blood Vessels 1987; 24: 321–333.
59. Haas M, Schmidt WF, McManus TJ. Catecholamine-stimulated ion transport in duck red cells. Gradient effects in electrically neutral (Na/K/Cl) co-transport. J Gen Physiol 1982; 80: 125–147.
60. Palfrey HC, Silva P, Einstein FH. Sensitivity of cAMP stimulated salt secretion in shark rectal gland to loop diuretics. Am J Physiol 1984; 246: C242–246.
61. Saier MH Jr, Boyden DA. Mechanism, regulation and physiological significance of loop diuretic-sensitive NaCl/KCl symport system in animal cells. Mol Cell Biochem 1984; 59: 11–32.
62. Hall IP, Widdop S, Townsend P, Daykin K. Control of cyclic AMP levels in primary cultures of human tracheal smooth muscle. Br J Pharmacol 1992; 107: 422–428.
63. Robberecht P, Chatelain P, De Neef P, Camus J-C, Waelbroeck M, Christophe J. Presence of vasocactive intestinal peptide receptors coupled to adenylate cyclase in rat lung membranes. Biochem Biophys Acta 1981; 678: 76–82.
64. O'Grady SM, Field M, Nash NT, Rao MC. Atrial natriuretic factor inhibits Na/K/Cl co-transport in teleost intestine. Am J Physiol 1985; 241: C531–534.
65. O'Donnell ME, Owen NE. Atrial natriuretic factor stimulates Na/K/Cl co-transport in vascular smooth muscle cells. Proc Natl Acad Sci 1986; 83: 6132–6136.
66. O'Donnell ME, Owen NE. Role of cGMP in atrial natriuretic factor stimulation of Na/K/Cl co-transport in vascular smooth muscle cells. J Biol Chem 1986; 261: 461–466.
67. Nishizuka Y. The role of protein kinase C in cell surface signal transduction and tumour promotion. Nature 1984; 308: 693–697.
68. O'Brien TG, Krzeminski K. Phorbol ester inhibits furosemide-sensitive potassium transport in BALB/c 3T3 preadipose cells. Proc Natl Acad Sci 1983; 80: 4334–4338.
69. Owen NE. Effect of TPA on ion fluxes and DNA synthesis in vascular smooth muscle cells. Biochem Res Commun 1985; 125: 500–508.
70. Paris S, Pouyssegar J. Growth factors activate the bumetanide sensitive Na/K/Cl co-transport in hamster fibroblasts. J Biol Chem 1986; 261: 6177–6183.
71. Liedtke CM. Bumetanide-sensitive NaCl uptake in rabbit tracheal epithelial cells is stimulated by neurohormones and hypertonicity. Am J Physiol 1992; 262: L621–L627.
72. Garay RP. Inhibition of the Na/K co-transport system by cyclic AMP and intracellular Ca in human red cells. Biochim Biophys Acta 1982; 688: 786–792.
73. Haas M. Properties and diversity of (Na/K/Cl) co-transports. Ann Rev Physiol 1989; 51: 443–57.
74. Rhoden KJ, Douglas JS. Bumetanide-sensitive K^+ uptake in guinea-pig airway smooth muscle. Am Rev Respir Dis 1993; 147: A52.
75. Knox AJ, Ajao P. Effect of frusemide on airway smooth muscle contractility in vitro. Thorax 1990; 45: 856–859.
76. Elwood W, Lotvall JO, Barnes PJ, Chung KF. Loop diuretics inhibit cholinergic and noncholinergic nerves in guinea-pig airways. Am Rev Respir Dis 1991; 143: 1340–1344.
77. Yu M, Wang Z, Robinson NE, Derksen FJ. The inhibitory effect of furosemide on the contractile response of equine trachealis to cholinergic nerve stimulation. Pulm Pharmacol 1992; 5: 233–238.
78. Verleden GM, Pype JL, Deneffe G, Demedts MG. Effect of loop diuretics on cholinergic neurotransmission in human airways in vitro. Thorax 1994; 49: 657–63.
79. Molimard M, Advenier C. Effect of frusemide on bradykinin- and capsaicin-induced contraction of the guinea-pig trachea. Eur Respir J 1993; 6: 434–439.

80. Karlsson J-A, Choudry NB, Zackrisson C, Fuller RW. A comparison of the effect of inhaled diuretics on airway reflexes in humans and guinea-pigs. J Appl Physiol 1992; 72: 434–438.
81. Stevens EL, Uyehara CFT, Southgate WM, Nakamura KT. Furosemide differentially relaxes airway and vascular smooth muscle in fetal, newborn, and adult guinea-pigs. Am Rev Respir Dis 1992; 146: 1192–1197.
82. Kikkawa U, Kishimoto A, Nishizuka Y. The protein kinase C family: heterogeneity and its implications. Annu Rev Biochem 1989; 58: 31–44.
83. Rasmussen H, Takuwa Y, Park S. Protein kinase C in the regulation of airway smooth muscle contraction. FASEB J 1987; 1: 177–185.
84. Kraft AS, Anderson WB. Phorbol esters increase the amount of Ca, phospholipid-dependent protein kinase associated with the plasma membrane. Nature 1983; 301: 621–623.
85. Schramm CM, Grunstein MM. Mechanisms of protein kinase C regulation of airway contractility. J Appl Physiol 1989; 66: 1935–1941.
86. Baba K, Baron C, Coburn RF. Phorbol ester effects on coupling mechanisms during cholinergic contraction of swine tracheal smooth muscle. J Physiol 1989; 412: 23–42.
87. Souhrada M, Souhrada JF. Sodium and calcium influx induced by phorbol esters in airway smooth muscle cells. Am Rev Respir Dis 1989; 139: 927–932.
88. Knox AJ, Baldwin DR, Cragoe Jr EJ, Ajao P. The effect of sodium transport and calcium channel inhibitors on phorbol ester-induced contraction of bovine airway smooth muscle. Pulm Pharmacol 1993; 6: 241–246.
89. Souhrada M, Souhrada JF. Inhibitory effect of staurosporine on protein kinase C stimulation of airway smooth muscle cells. Am Rev Respir Dis 1993; 148: 425–30.
90. Delamere F, Holland E, Patel S, Bennett J, Pavord I, Knox A. Production of PGE_2 by cultured bovine airway smooth muscle cells and its inhibition by cyclo-oxygenase inhibitors. Br J Pharmacol 1994; 111: 983–8.
91. Mapp CE, Boniotti A, Papi A, Maggi CA, Di Stefano A, Saetta M, et al. Effect of bumetanide on toluence diisocyanate induced contractions in guinea-pig airways. Thorax 1993; 48: 63–67.

Airways Smooth Muscle: Peptide Receptors, Ion Channels and Signal Transduction
ed. by D. Raeburn and M. A. Giembycz
© 1995 Birkhäuser Verlag Basel/Switzerland

CHAPTER 11
Sodium/Hydrogen Exchange

Ratna Bose

Department of Pharmacology and Therapeutics, University of Manitoba, Winnipeg, Manitoba, Canada

1   Introduction
2   Methods
3   Characteristics of the Na^+/H^+ Exchanger
3.1   Ion Selectivity and Stoichiometry
3.2   Pharmacological Inhibitors
3.3   Isoforms of the Na^+/H^+ Exchanger
3.4   Regulation
3.4.1 Intracellular Hydrogen Ion
3.4.2 Calcium-Calmodulin
3.4.3 Cyclic AMP
3.4.4 Protein Kinase C
4   Effect of pH_i on Contraction
4.1   Alkalosis
4.2   Acidosis
4.2.1 Intracellular Sites Mediating Inhibition of Contraction with Acidosis
4.2.2 Paradoxical Increase in Contraction with Acidosis
5   Potential Role in Disease States
6   Conclusions
      Acknowledgements
      References

1. Introduction

The Na^+/H^+ exchanger (formerly antiporter) is widely distributed in nature as a component of the cell membrane. The cytosolic pH of tracheal smooth muscle, similar to many other mammalian tissues, is regulated by at least three known sarcolemmal processes. These processes include 1) the Na^+/H^+ exchanger, 2) the anion transporter requiring Cl^-/HCO_3^- and 3) H^+ extruding, sodium-dependent anion transporter (Figure 1). The two sodium-dependent processes alkalinize the cells while the anion exchanger acidifies alkalotic cells. Together, these pH regulating systems maintain the cytosolic pH (intracellular pH, pH_i) at a much higher value (7.0 to 7.3) than would be predicted by thermodynamic equilibrium calculated from the Nernst equation based on the membrane potential and passive distribution of H^+ across the cell membrane. For example the resting membrane potential (E_m) of canine tracheal smooth muscle was found to be -60 ± 1 mV [1]. Based

ALKALINIZING PROCESSES

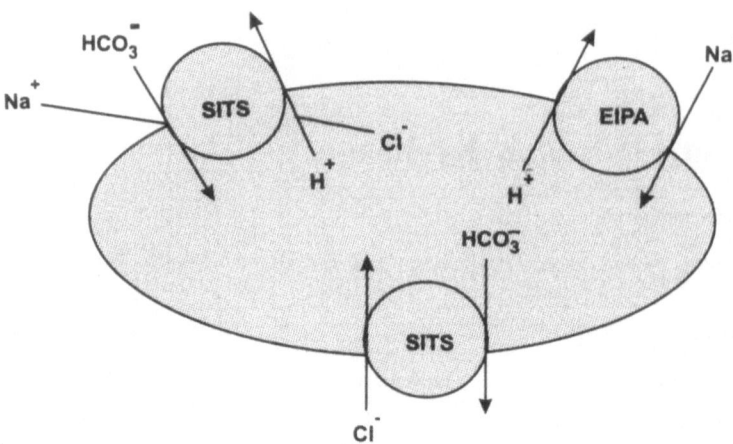

Figure 1. Three major transport pathways that regulate cytosolic pH and their inhibitors (indicated inside the circles) are shown.

on this, the calculated pH value at 37°C is expected to be 6.42, 0.6 pH units lower than the observed resting pH_i for adult canine trachealis. In sodium-free Hepes-buffered medium the pH_i of the tracheal muscle is decreased by 0.5 pH unit [2]. Thus sodium-dependent intracellular pH regulating mechanisms are important in this tissue.

In addition to these well characterized transporters, other pH regulating processes have also been described in the literature. A chloride-independent but sodium-dependent HCO_3^- exchanger has been reported to alkalinize the vas deferens smooth muscle [3]. The K^+/H^+ ATPase, that appears to be important in certain epithelial cells, has been demonstrated recently in vascular smooth muscle [4]. Inhibitors of this ATPase cause a rapid decrease in pH_i and a marked decrease in potassium content, suggesting an important role in maintaining H^+ and K^+ gradients in vascular smooth muscle. However, its role in airways smooth muscle is not known.

Cytosolic H^+ originate from: 1) metabolism, 2) displacement of protein bound hydrogen, that is released by a rapid increase in Ca^{2+} following signal transduction and 3) passive influx of H^+ or efflux of HCO_3^-. Activity of the sarcolemmal calcium pump is coupled to the influx of H^+ in certain cell types and provides an additional source of H^+ [5]. It is not clear if in an intact cell, a Ca^{2+}/H^+ exchanger activity could be generated when the sarcoplasmic or the sarcolemmal pump Ca^{2+} ATPase is forced into a reverse mode.

A downhill gradient of sodium provides the driving force for the H^+ extrusion by the Na^+/H^+ exchanger. The basic exchange process does not require chemical energy from ATP directly. However, ATP-dependent processes such as the sodium pump, are needed for maintaining the sodium gradient (see chapter 20). ATP-consuming phosphorylation processes can modulate pH regulation by the Na^+/H^+ exchanger.

The study of Na^+/H^+ exchange in airways smooth muscle is in its infancy. Therefore, in this review, in addition to the data from tracheal smooth muscle, selective relevant information from vascular smooth muscle and other tissues is also included for acquiring an overview of the exchanger and its potential role in modulating smooth muscle contraction. Alveolar epithelial cells also use a Na^+/H^+ exchanger as a major mechanism for recovery from acid loads and sodium entry from alveolar space [6, 7].

2. Methods

The techniques used to investigate Na^+/H^+ exchange include ion sensitive-microelectrodes, nuclear magnetic resonance and fluorescent dyes (discussed in [8, 9]. Studies of ^{22}Na influx, mostly done in isolated vesicles and cell cultures, have also been useful in delineating the kinetic characteristics of the exchanger [10]. In this review, methods used to obtain the data in the author's laboratory are described. Tracheal smooth muscle strips ($1mm \times 3mm$) devoid of cartilage, epithelial and adventitious tissues were tied at both ends with surgical silk and fixed to a hook on a horizontal tissue bath (Jasco TI-02) of a Jasco CAF-100 Fluorimeter (Jasco Inc, Easton, Md USA). The other end was attached to a force transducer (Grass Instruments Model FT 03). Strips were allowed to recover in a physiological medium (Krebs-Henseleit) of the following composition (mM) NaCl 140, $CaCl_2$ 2.5, KCl 4.72, $MgSO_4$ 1.2, KH_2PO_4 1.4, $NaHCO_3$ 25, Glucose 11 and adjusted to a pH of 7.4. The solution was kept at 37°C and equilibrated with a gas mixture of 95% O_2/5% CO_2. For bicarbonate-free medium, 25 mM HEPES (N-2-hydroxyethylpiperazine-N'-2-ethansulphonic acid) was substituted for $NaHCO_3$, and the solution bubbled with 100% oxygen. The strips were stimulated electrically once every 5 minutes and stretched to the length required for optimum force of contraction (Lo, usually corresponding to 0.8 to 1.5 g resting load). Tension, fluorescence signals at 500 and 440 nm excitation, ratio of the two signals, pH and temperature in the vicinity of the strip were recorded simultaneously in a 6 channel strip chart recorder (BBC-Metrawatt/Goerz). The autofluorescence was measured and then the fluorescent dye BCECF (2'7'-bis-(carboxyethyl)-5, (6)-carboxyfluorescein) was loaded by incubating with 6 µM BCECF-acetoxy-methyl ester for 1 hour at 37°C. After washing off the external dye, the fluorescence at 440 nm was observed during electrical stimulation to detect any movement artifacts; the strip was re-positioned until no artifacts were visible. Some dye leak occurred when the strips were subjected to an acid load and allowed to recover. The calibration for pH_i with adjustment for dye leak was performed according to procedures described in Yu *et al.* 1991 [11].

3. Characteristics of the Na^+/H^+ Exchanger

3.1. Ion Selectivity and Stoichiometry

Electroneutral exchange of one Na^+ for one H^+ appears to be universally true. Simultaneous measurement of Na^+ and H^+ transported by this exchanger has been accomplished in a few cell systems, for example erythrocyte and lymphocytes (reviewed in [10, 12]). The effect of extracellular sodium on the rate of the exchange process conforms to Michaelis-Menten kinetics. Only one external binding site for Na^+ and H^+ appears to be present, as determined by a Hill coefficient of one [10]. Li^+ and NH_4^+ can substitute on the external surface with rank order of selectivity as follows: $H^+ \gg Li^+ > NH4^+ \geqslant Na^+$. Other cations such as K^+, Rb^+, Cs^+, tetraethylammonium (TEA) and choline are ineffective [10].

The inhibition or activation of the Na^+/H^+ exchanger is without effect on membrane potential [13]. Unlike the Na^+/Ca^{2+} exchanger, the membrane potential has no direct effect on the Na^+/H^+ exchanger. However, after blocking the pH regulatory pathways with SITS (4-acetamido-4-isothiocyanostilbene-2,2-disulfonic acid) and amiloride, the pH_i in the mesentric artery has been shown to be increased by depolarization of the membrane [14]. The inference from such experiments was that the membrane voltage may regulate the pH_i of smooth muscle.

3.2. Pharmacological Inhibitors

At this time, amiloride and its analogues are the most widely used compounds for inhibiting the Na^+/H^+ exchanger [15]. Other agents have been used to distinguish between the different isoforms of the exchanger (discussed later). One should be aware of the nonselectivity of many of these agents, particularly amiloride. Table 1 lists the K_i for several commonly used amiloride analogues for the three major membrane processes. As is obvious, agents used for blocking the Na^+/H^+ exchanger are only relatively specific. Amiloride has higher affinity for the epithelial Na^+ channel than for the Na^+/H^+ exchanger. In most tissues these agents are effective when applied externally. Occasionally

Table 1. IC_{50} (μM) of amiloride analogues on membrane transport processes

	Na^+/H^+	$Na^+CHANNEL$	Na^+/Ca^{2+}
EIPA	0.38	> 400, 10*	130
HMA	0.16	> 400	100
MGCMA	1.35	> 400	1571
Amiloride	83.3	0.34, 10*	1100
Phenamil	> 1047	0.02	200
CBDMB	> 500	> 400	7.3

Data extracted from Kleyman and Cragoe, 1988 [15] and * from Smith and Benos, 1991 [70]. HMA, hexa methyl amiloride; CBDMB, 5-(4-chlorobenzyl)-2,4-dimethylbenzamil.

effectiveness of only internal application has been reported for the Na^+/Ca^{2+} exchanger [16]. A new benzylguanidyl derivative (Hoechst compound, HOE 694, (3-methylsulphonyl-4-piperidino-benzyl) guanidine methanesulphonate) is also being evaluated as a blocker of Na^+/H^+ exchange [17]. Other agents which possess an imidazoline or guanidyl group have also been used to characterize the isoforms of the exchanger [18].

The pharmacological effects of some of these agents on the isometric contraction and cytosolic pH have been evaluated in canine tracheal smooth muscle [11, 19, 20], and in guinea pig airways [21]. 5-(N-methyl-N-guanidocarbonyl methyl)-amiloride (MGCMA) and ethylisopnpyl-amiloride (EIPA) inhibited carbachol-induced contraction by $24 \pm 6\%$ ($n = 4$) and $28 \pm 8\%$ ($n = 9$ dogs) respectively (Figure 2). This inhibition was associated with a decrease in pH_i by 0.12 ± 0.5 unit ($n = 9$, EIPA) in canine tracheal muscle. Furthermore, acidosis was shown to increase the relaxation induced by amiloride [22]. In canine tracheal tissue, additional relaxant effects of amiloride and phenamil have been observed which are unrelated to changes in cytosolic pH. Both agents are potent epithelial Na^+ channel blockers and phenamil has no inhibitory effects on the Na^+/H^+ exchanger [19]. Allergen-(ragweed) or histamine-induced contraction and concomitant Na^+ influx in canine tracheal muscle could be blocked by phenamil and amiloride but not by MGCMA, a Na^+/H^+ exchange blocker [23]. Thus, caution is necessary in attributing an effect of amiloride to the blocking of Na^+/H^+ exchanger when intracellular pH is not measured.

Endothelin-1 is a potent constrictor of guinea pig airways smooth muscle and is believed to be released from the airway epithelium [24, 25]. Endothelin promotes proliferation the airways smooth muscle cells and its mitogenic actions are associated with the production of thromboxane [26]. The Na^+/H^+ exchange activity has been implicated in the endothelin-dependent contraction of guinea-pig tracheal muscle [21]. In contrast to the carbachol-induced contraction in canine trachea, the

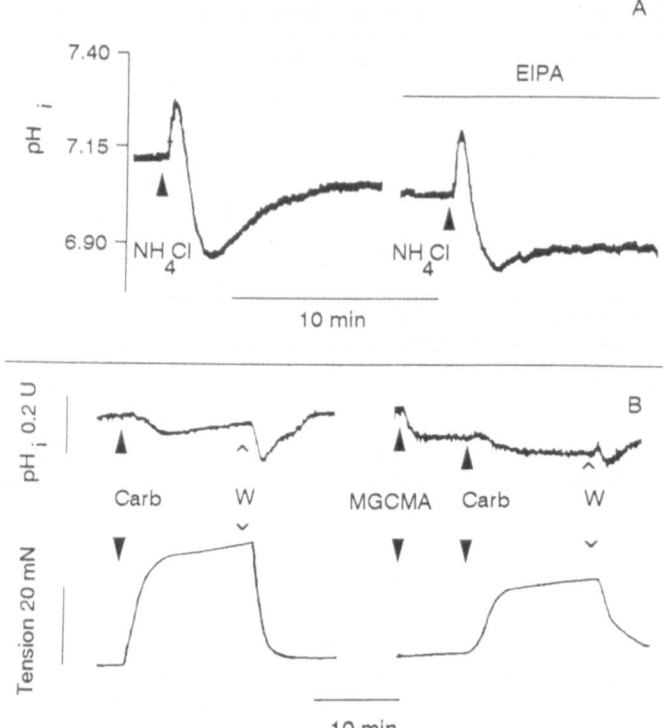

Figure 2. The effect of two Na^+/H^+ exchange blockers, EIPA and MGCMA on intracellular pH (pH_i) are shown. A. Effect on recovery from acid load induced by a brief pulse of NH_4Cl. HCO_3^- buffered physiological solution circulating at a flow rate of 25 ml/min was interrupted with 10 ml of 40 mM NH_4Cl given at the arrow. The same manoeuvre was repeated after adding EIPA (50 µM). B. The left panel shows the effect of adding 0.1 µM carbachol on intracellular pH and isometric tension [20]. W denotes washout. Right panel shows the effect of the same after the addition of MGCMA (50 µM).

endothelin-1 induced contraction of guinea-pig airways is very sensitive to Na^+/H^+ exchange blockers such as EIPA but is insensitive to amiloride. It is not clear at this time if this difference in sensitivity to amiloride analogues is due to the difference in species or represents the peculiarity of the agonist employed in the two studies.

3.3. Isoforms of the Na^+/H^+ Exchanger

Airway epithelium also has Na^+/H^+ exchange activity, which has been located at the apical surface [27]. Regulation and characteristics of this exchanger are likely to be different from those of the exchanger in smooth muscle. Epithelial cells from the intestine or kidneys contain two different types of Na^+/H^+ exchangers, one of which is located in their brush borders and the other in their basolateral membranes. Apart from their location, the isoforms appear to differ in their molecular mass, sensitivity to amiloride analogues, ion specificity and activation by cytosolic H^+, and in terms of their function [18, 28, 29].

At least four distinct gene families have been identified for the Na^+/H^+ antiporter, and the complete complementary DNA (cDNA) of two of these isoforms has been successfully transfected in Chinese hamster ovary cells [18]. Some of the functional characteristics along with other properties of these expressed molecules are summarized as follows.

The NHE-1 isoform is highly conserved and ubiquitously expressed in all tissues. It has been demonstrated in human, rat and rabbit tissues and is present in most nonepithelial cells and basolateral membrane of epithelial cells. This form of the exchanger, a 110 kD glycoprotein, has been cloned [30], and was found to be highly sensitive to amiloride and its analogues. The rank order of potency of other inhibitors is as follows cimetidine > harmaline ≥ clonidine. This isoform of the exchanger is phosporylated on a serine residue of the carboxyl terminal end in response to growth factors. The expressed exchanger displays allosteric activation with increases in cytosolic H^+ (pKa = 6.75).

The NHE-3 isoform has been localized in the apical membranes of epithelial cells, and is the most abundant isoform in the brush border membranes of kidneys. It has a molecular weight of 130 to 150 kD, a low affinity for amiloride analogues and is also inhibited by other agents with a rank order potency: clonidine > harmaline > cimetidine that differs from NHE-1. Allosteric activation with cytosolic H^+ has been shown to occur with a pKa of 6.45 for H^+. The NHE-3 isoform appears to be regulated by angiotensin II and parathyroid hormone. In addition to regulating cytosolic pH, this isoform is also involved in transepithelial electrolyte transport [29].

3.4. Regulation

3.4.1. Intracellular hydrogen ion: Cytosolic pH regulates the activity of the Na^+/H^+ antiporter in an allosteric fashion in renal cells and lymphocytes [12, 31]. This site is distinct from the site that transports H^+. This mode of regulation is physiologically important for the protection of the cell during acid loads. This second site for H^+ labelled "modifier site" controls and "sets" the threshold for activation of the exchanger [31]. At an alkaline pH above the set point the exchanger is inactive whereas below this pH set point (normally 7.0), the protonation of the modifier site greatly enhances the activity of the exchanger. Our data from unstimulated canine tracheal smooth muscle also supports this form of regulation [2]. Acid loading with an NH_4Cl pulse (after wash out of NH_4Cl, NH_3 diffuses out leaving H^+ behind) activates the exchanger while EIPA markedly reduces the rate recovery of pH_i in tracheal muscle in HCO_3-containing medium (Figure 2A) and completely abolishes recovery in HCO_3-free, Hepes-buffered medium (unpublished observation).

Agonists such as carbachol and endothelin or oncogenic transformations shift the set point for intracellular H^+ to an alkaline range [12]. A variable degree of alkalinization is observed with carbachol in canine tracheal muscle when attempts are made to mask the acidosis associated with sudden increases in cytosolic calcium and concurring contraction (Figure 3B). Frequently (26 out of 40 dogs) carbachol induced a slow recovery or alkalinization soon after the onset of contraction [20]. This recovery could be consistently blocked by EIPA or MGCMA, implying a role for the Na^+/H^+ exchanger (Figure 2B).

3.4.2. Calcium-calmodulin: The Na^+/H^+ exchanger in the brush border was shown to be inhibited by calcium-calmodulin dependent protein kinase II phosphorylation [29, 32]. The phosphorylated sites appear to be different from those for the cyclic AMP-dependent protein kinase A

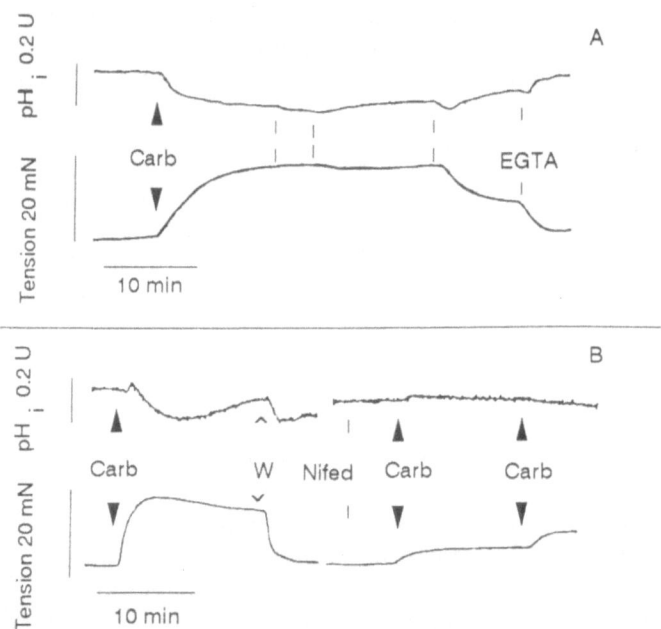

Figure 3. The effect of Ca^{2+}-chelator EGTA and calcium channel blocker, nifedipine on changes in cytosolic pH and isometric force with carbachol (0.2 μm). A. After contracting the strip with carbachol (0.1 μM), EGTA (2.5 μM) was added at each of the points indicated. B. Left panel shows the control response of adding carbachol (0.1 μM) and washout, W on intracellular pH and contractility. The right panel, the effect of adding nifedipine (10 μM), Nifed to the same strip followed by addition of carbachol (0.1 and 1.0 μM) [20].

[29]. In contrast, the exchanger in the basolateral membrane, which is expected to be similar to that in the tracheal smooth membrane, is activated by calcium-calmodulin dependent protein kinase [33].

3.4.3. Cyclic AMP: The Na^+/H^+ exchanger in several tissues has been shown to be inhibited by cyclic AMP dependent phosphorylation by protein kinase A (discussed in [12, 29]). This mechanism is also implicated in the inhibitory actions of β-adrenoceptor stimulation, forskolin, and theophylline [34].

3.4.4. Protein kinase C: Whereas protein kinase C activates the Na^+/H^+ exchanger in most tissues [35, 36] in a few it inhibits the exchanger [28]. Differences are likely due to different isoforms of the exchangers present in these tissues [28]. Protein kinase C (PKC) is activated by carbachol, which stimulates sequentially inositol phosphate hydrolysis, inositol 1,4,5 trisphosphate (IP_3) generation, calcium release from intracellular stores and constriction of treacheal smooth muscle [37]. PKC is also activated by diacylglycerol (DAG) which is produced along the IP_3. Inhibition of PKC by staurosporine inhibits the activation of the Na^+/H^+ exchanger with carbachol in dog Purkinje fibers [34]. It remains to be seen whether the same holds true for airways smooth muscle. PKC inhibition with either staurosporine or H-7 did not prevent the activation of the Na^+/H^+ exchanger by endothelin in the Purkinje fibers [34]. In rabbit tracheal smooth muscle, a phorbol ester activator of PKC, 12-deoxyphorbol 13-isobutyrate, stimulated the Na^+/K^+ pump in a manner which was independent of the activity of the Na^+/H^+ exchanger [38]. Thus the degree and mode of activation of the Na^+/H^+ exchanger by these two powerful bronchoconstrictors could be different. Interaction between the alkalinizing HCO_3^-/Cl^- exchange and the Na^+/H^+ exchanger is achieved through cyclic AMP-dependent protein kinase A and PKC in rat aortic smooth muscle cells [39].

The second messenger and transduction pathways discussed in the preceeding section can modulate the Na^+/H^+ exchanger in response to a number of hormones and neurotransmitters. Apart from the acute mode of regulation described above, long term slow regulation of the Na^+/H^+ exchanger can occur with growth factors [40] glucocorticoids, insulin, thyroxine and parathyroid hormone [12]. Na^+/H^+ exchanger messenger RNA (mRNA) is increased in the renal cortex after 5 days of metabolic acidosis but not by respiratory acidosis of the same duration [41]. Thus, the activity of exchanger is affected by acute and long term controls.

4. Effect of pH$_i$ on Contraction

Bronchoconstriction in response to hypocapnia in man was reported as early as 1968 by Sterling [42]. 5-hydroxytryptamine (5-HT)-induced contraction could be attenuated with hypercapnia [43, 44]. Respiratory acidosis was shown to depress acetylcholine-induced contraction by 30% in canine trachealis [45]. In contrast endothelin-1-induced contraction was potentiated by hypercapnia and inhibited by hypocapnia [46]. No significant effect was observed during hypocapnia when strips were stimulated with histamine and acetylcholine [46]. The authors suggested that hypocapnia released bronchodilating eicosanoids in guinea-pig airway muscle since aspirin and indomethacin reduced the actions of hypocapnia [46].

Measurement of simultaneous changes in isometric force of contraction and pH_i of tracheal and other smooth muscles following agonist addition or membrane depolarization indicate that the relationship is a complex one resulting from the interplay of several processes [20, 47, 48]. Initial rapid release of calcium and ATP hydrolysis caused a transient fall in pH_i followed by a slow recovery leading to a steady state under the prevailing conditions (Figure 2B). Thus the effects of pH manipulating agents or media should be tested either after the muscle pH_i reaches a steady state or before the agonists are applied.

4.1. Alkalosis

Alkalinization induced by decreasing pCO_2 increases the isometric force of contraction in tracheal smooth muscle when some active tone is induced by either carbachol or potassium-depolarization (Figure 4B and C). Carbachol-induced tone is potentiated more than potassium-deplorization induced tone in dogs (Table 2). In 6 dogs, 45 ± 5 and $25 \pm 8\%$ increase in force of contraction was observed with carbachol (0.2, μM, n = 15) and potassium (80 mM, n = 15) respectively for 0.4 ± 0.05 pH unit increase above the extracellular pH of 7.35 ± 0.04. Both pH_i and the increase in contractile force could be significantly attenuated with SITS (0.2 mM), but not with the Na^+/H^+ exchange

Table 2. Effect of Hypocapnic Alkalosis On Isometric force of contraction

	% Change in force of contraction	Change in pH_i
Unstimulated	0 *	0.304 ± 0.053
KCl 40 mM	57 ± 9 *	0.307 ± 0.023
Carbachol 0.1 μM	100	0.311 ± 0.045
Carb. + Verapamil 1 μm	10 ± 5 *	0.389 ± 0.085
Carb. + SITS 200 μM	57 ± 23 *	0.102 ± 0.042 *
Carb. + EIPA 10 μM	77 ± 27	0.251 ± 0.035

Alkalosis was induced by decreasing the pCO_2 from 38 ± 7 mm Hg to 15 ± 5 mm Hg, and the extracellular pH_0 from 7.4 to 7.8. Details are similar to experiments described in Figure 4. Change in force of contraction is expressed as the % change in alkalosis-induced force relative to that produced by carbachol alone. n = 4 to 11; *, p value < 0.05.

blockers EIPA and MGCMA (Table 2). Furthermore, a hypocapnic increase in contractile force could be prevented by the calcium channel blockers verapamil and nifedipine and by bathing in a calcium-free medium (Figure 4D). These results suggest that the increase in contractility is due to influx of external calcium in canine trachealis and differs from that in rat aorta cell cultures, where an increase in cytosolic calcium was attributed to intracellular calcium release from sarcoplasmic reticulum [49]. No change in contractile force, in response to a decrease in pCO_2, was observed in an unstimulated canine tracheal smooth muscle strip. These strips normally lack active tone, but acquire active tone in disease conditions, for example in ragweed-induced allergic asthma. These observations lead us to believe that the SITS-sensitive anion exchange is involved in the cytosolic alkalinization process above neutral pH. Similar conclusions concerning pH regulation by anion

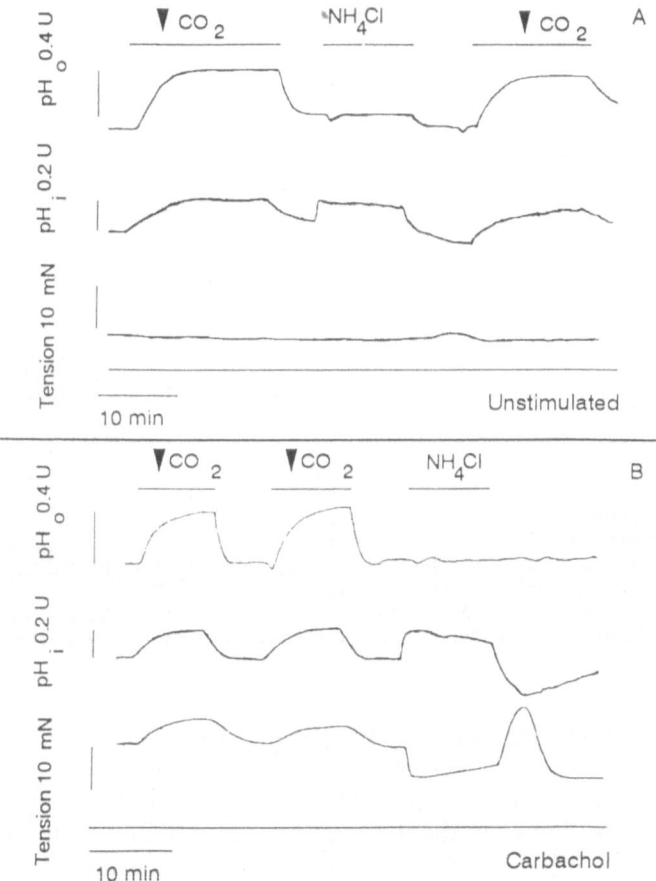

Figure 4A, B.

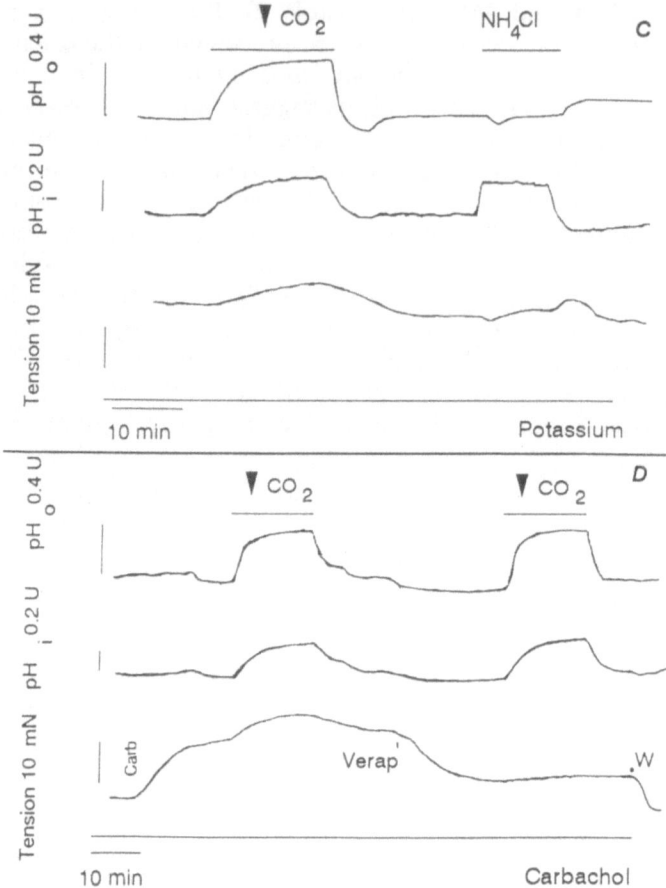

Figure 4C, D.

Figure 4. Alkalosis was induced by decreasing the pCO_2 (indicated horizontal bar) of the bathing medium from 36 to 15 mm Hg, and raising the pH to 7.8 from 7.4. Intracellular alkalosis of similar magnitude was induced by adding NH_4Cl (50 mM) to the recirculating bath, as indicated by the horizontal line. Top panel in each group denotes the extracellular pH, middle - intracellular pH and bottom- isometric force. A. Unstimulated tracheal smooth muscle; B. effect on a strip precontracted with carbachol (0.1 μM); C. tracheal strip precontracted with potassium chloride (40 mM) in the presence of atropine (1 μM); D. strip precontracted with carbachol (0.1 μM) and partly relaxed with varapamil (10 μm). W denotes washout.

exchange in response to pCO_2 changes in a smooth muscle-like cell line (BC3H-1), were reached recently by Putnam [50].

Alkalinization studies with NH_4Cl differ from the above series of experiments in that only the intracellular pH is altered without changes in extracellular pH. In three out of six different experiments with canine trachealis muscle, intracellular alkalinization with NH_4Cl (50 mM) reduced the tonic force in carbachol-contracted tissues (Figure 4B). Of the remaining three other strips there was a slight increase in tone in two and no effect was seen in the third. The variability in this response was not due to a different degrees of alkalosis and may be related to other undetermined factors. A similar inhibitory effect of NH_4Cl on noradrenaline-induced contraction of mesenteric arteries has been explained by a decrease in Ca^{2+} following repolarization of the membrane [51]. Cytosolic alkalinization increases the sensitivity of the calcium-activated potassium channels to calcium. Consequently potassium efflux is enhanced and hyperpolarization is manifest [52]. No significant increase in contractile force of either unstimulated or of potassium-depolarized muscle is observed in canine tracheal smooth muscle under these experimental conditions (Figure 4C). Prolonged incubation with high concentrations of NH_4Cl (> 40 mM) induced a small increase in tone in the resting tracheal smooth muscle, however these effects appear not to be related to cytosolic pH (unpublished observation). Guinea-pig tracheal muscle possesses an active basal tone, which was increased with the addition of NH_4Cl (5 mM) [53]. Alkalinization with NH_4Cl increases the tone in pulmonary arteries [54]. Effect of NH_4Cl in other vascular beds such as the rat aorta was qualitatively similar to our observations with tracheal smooth muscle [55]. The relative contributions of sarcoplasmic reticulum versus trans-sarcolemmal influx of calcium from the external medium varies considerably in different smooth muscles, and is also affected by the age of the animal, previous exposure to other agonists, and the characteristics of the stimuli or agonists used to increase the force of contraction.

Apart from inducing an alkalinizing effect, withdrawal of NH_4Cl leaves behind an acid load (Figure 1 and 4). Therefore it is often used to induce intracellular acidosis without changes in extracellular pH. In our studies with canine trachealis, there was only a small twitch with this protocol in unstimulated muscles during NH_4Cl washout. (Figure 4A).

4.2. Acidosis

The active force of contraction in canine tracheal smooth muscle is reduced in response to a decrease in extracellular pH, with a simultaneous drop in cytosolic H^+ concentration (Figure 5 and 6B). The change in pH_i in the two modes of contraction was not significantly different, and amounted to 0.5 to 0.6 pH unit/unit change in extracellular pH. This was greater than that observed in unstimulated tracheal muscle (0.45 to 0.55 pH unit/unit change in extracellular pH). The inhibitory effect of acidosis on isometric force was more pronounced in carbachol-contracted strips compared to potassium-depolarized muscles for a similar decrease pH_i (Figure 6) with similar magnitude of prevailing active tone. Amiloride-induced relaxation is also more pronounced

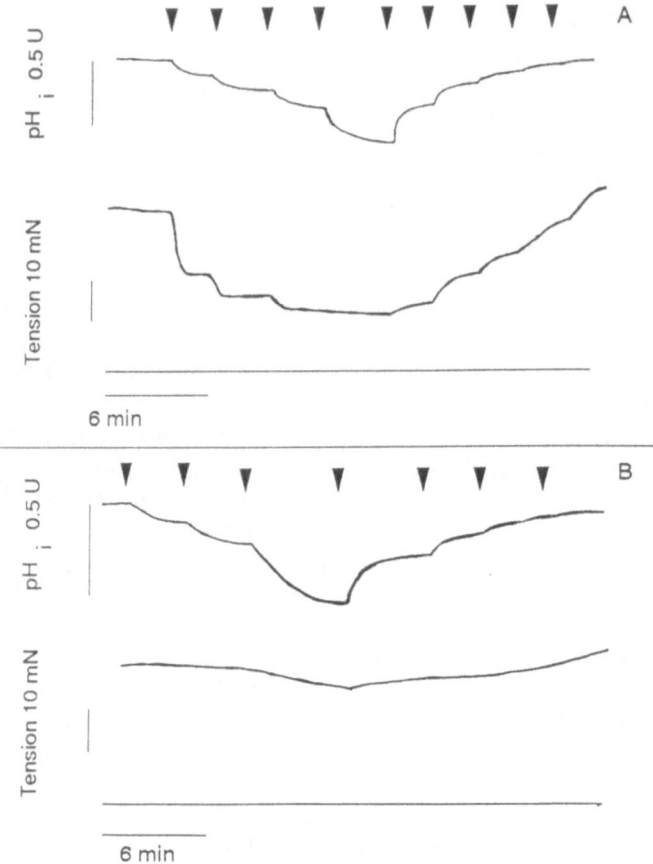

Figure 5. Effect of extracellular acidosis on precontracted tracheal smooth muscle. The pH of the bathing medium was dropped by titrating wth acetic acid and increased by adding NaOH. A carbachol (0.1 µM) was used for contraction; and in B potassium (80 mM) with atropine (1 µM) was used to contract the muscle. Similar results were obtained if propionic acid was used instead of acetic acid.

in carbachol-contracted strips [56]. One of several possible explanations could be that the cytosolic calcium concentration is greater in potassium-depolarized smooth muscle cells than in carbachol-activated strips when similar magnitudes of isometric force of contraction are generated [57]. It has been suggested that during carbachol stimulation, there is a sensitization to calcium. It is conceivable that the sensitization processes during carbachol-contraction or the utilization of different pools of calcium by different agonists could determine differences in pH sensitivity. IP_3 production is activated in response to carbachol. This pathway is activated in alkaline pH. Furthermore, optimum binding of IP_3 to its receptor occurs at an alkaline pH (pH 7.75) [58]. Therefore, this could be an additional site for mediating the pH sensitivity of carbachol contraction.

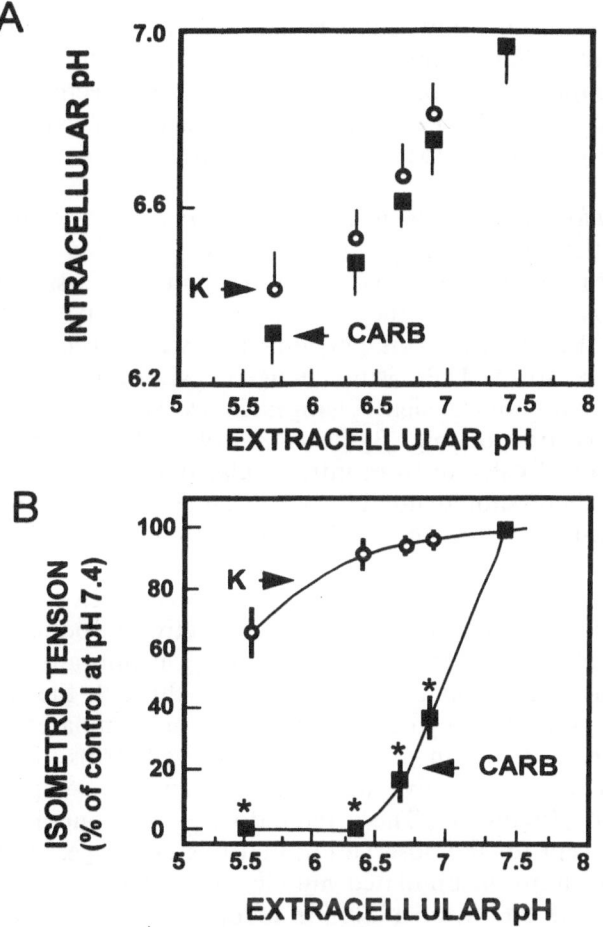

Figure 6. Summarized data of the effect of extracellular pH on intracellular pH (A) and isometric tension (B). Tracheal strips from 5 dogs were treated with media of different pHs, as described in Figure 5. * Denotes significance with a p value < 0.05 by unpaired t-test.

4.2.1. Intracellular sites mediating inhibition of contraction with acidosis: Myofibrilar proteins and sarcoplasmic reticulum are well characterized intracellular targets for increased cytosolic H^+ mediated decrease in contraction. Optimum pH for contraction of skinned smooth muscle strips is close to neutral pH and the pH-dependence of calcium sensitivity is very significant [59]. A decrease in calcium sensitivity of the myofibrils is probably due to the displacement of the Ca^{2+} from its binding sites, leading to inactivation of the ATPase and relaxation. The rate of superprecipitation of a calcium-sensitive actomyosin preparation from canine tracheal smooth muscle shows an acid-dependent decrease in activity [60]. Intracellular calcium uptake by smooth muscle sarcoplasmic reticulum increases by a factor of two across a pH range of

7.2–6.8 [61]. This could cause relaxation in strips in which the sarcoplasmic reticulum is emptied of calcium with IP_3, for example, following carbachol-stimulation, but would be less effective in potassium depolarized strips. Acid stimulation of intracellular calcium sequestration has also been proposed in the vasodilation of rat aorta [62].

4.2.2. Paradoxical increase in contraction with acidosis: While a decrease in contractile force is frequently observed in canine tracheal smooth muscle with acidosis, a transient contraction is occasionally observed following a washout of the NH_4Cl pulse in strips contracted with carbachol (Figure 4B). This phenomenon is not observed in unstimulated tracheal strips. There is no change in extracellular pH, hence these effects are due to transient changes in cytosolic pH alone. Several explanations are possible. Batlle *et al.* 1993 [49] suggested a sudden displacement of calcium from intracellular buffers by H^+. If this is the case, then there should not be such a large difference in response to NH_4Cl washout between unstimulated versus carbachol stimulated strips. Moreover, the response should also not be blunted with potassium-depolarization. A second possible explanation for this phenomenon could be the inhibition of Ca^{2+}-activated-potassium channels with intracellular acidosis [52]. The resulting membrane depolarization would lead to additional influx of calcium through the calcium channels. With very low cytosolic calcium in unstimulated muscle, and decreased K^+ current in potassium-depolarized strips, this mechanism would not be as effective and this inference is supported by our observation (Figure 4). These data do not rule out a Ca^{2+}/H^+ exchanger, except for the observation that this exchanger should induce a contraction in an unstimulated muscle strip. Blockade of the Na^+/H^+ exchanger with EIPA, or inhibition of the anion exchanger with SITS, did not prevent the cytosolic acid-induced-contraction during NH_4Cl washout. Thus, the influx of Na^+ by either of the sodium-dependent pH-regulating processes is not required for this contraction. Urinary tract smooth muscles increase their tension in response to intracellular acidosis, and this response is unaffected by the Na^+/H^+ exchange blocker MGCMA [63]. Similar results on contraction are obtained in mesentric arteries with cytosolic acidification [51].

Further investigation of the acid-induced contraction in trachealis muscle was done in experiments where intracellular acidosis was induced with the stepwise addition of sodium propionate, without any change in extracellular pH (Figure 7). An increase in isometric force was observed with decreases in cytosolic pH (Figure 7A) in stored (one day old) tracheal strips but not in freshly isolated tissues. This force development was partially ($52 \pm 8\%$) dependent on extracellular calcium and was also partly blocked by verapamil. This was somewhat different from contractions produced by extracellular alkalosis in tra-

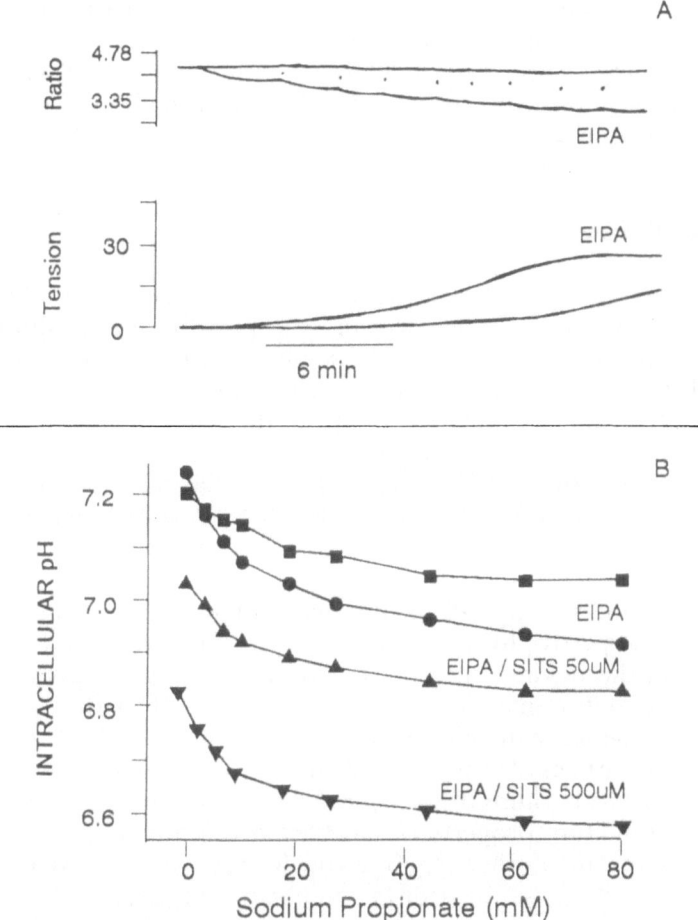

Figure 7. Effect of intracellular acidosis was tested on isometric force development by adding sodium propionate (pH 7.4) to a tracheal strip stored for 24 hours in ice-cold bathing medium. Extracellular pH did not change during the entire experiment. No force development was observed in freshly isolated tracheal smooth muscle strips although similar magnitude of acidosis was induced with sodium propionate. EIPA (upper tension trace) and SITS (not shown) enhanced the onset of tension development. A. Top tracings the ratio values for fluorescence at 500 nm and 440 nm are shown, and the calculated pH is shown in B.

cheal muscle where the contraction was almost completely blocked by verapamil (Figure 4D, Table 2). Intracellular acidosis alone without extracellular acidosis probably inhibits Ca^{2+}-dependent K^+ channels and increases the influx of calcium through calcium channels more than the acid-stimulated Ca^{2+} uptake by the sarcoplasmic reticulum, thereby generating a net increase in cytosolic calcium in the one day old strips. Intracellular acidosis inhibits voltage-gated calcium channels more effectively than extracellular H^+ in ventricular myocytes [64]. Thus, the effects of extracellular and intracellular acidosis on the calcium influx in tracheal smooth muscles are different from ventricular muscle. The calcium channels appear to be inhibited with extracellular acidosis and a net decrease in cytosolic calcium will lead to relaxation of the tracheal muscle.

5. Potential Role in Disease States

Apart from its function in modulating cytosolic pH, the Na^+/H^+ exchanger has been implicated in cell growth and multiplication. Sodium-dependent alkalosis has been observed following triggers such as fertilization or activation with transforming growth factors or with foetal calf serum [65]. Hypertrophy of the smooth muscle has been implicated in the increase in airway resistance in asthma [66]. It is conceivable that the Na^+/H^+ exchanger may play a role in initiating the process of hypertrophy. The role of the Na^+/H^+ exchanger in essential hypertension is under evaluation and the results of several studies suggest that it is an important mediator for the actions of endothelin, angiotensin and the growth factors that initiate vascular muscle hyper-plasia and hypertrophy [67]. Endothelin-1 elicits a potent mitogenic response in the bovine airways smooth muscle [26]. Recently Na^+/H^+ exchange inhibitors have been shown significantly to reduce cell prolif-eration and neointimal thickening following carotid injury [68, 69]. In one of these studies [68] the role of the Na^+/H^+ exchanger was found to be important because the structural isomer of the compound lacking exchange-inhibitory property was ineffective in preventing cell prolifera-tion. However, in another report, a role for the Na^+/H^+ exchanger was ruled out on the basis of growth inhibitory responses of EIPA on an exchanger-deficient mutant cell line [69]. It is conceivable that an adaptation to the absence of the exchanger could have occurred in these mutants. Apart from the growth promoting effects of alkalosis induced by an activated exchanger, a resetting of the modulatory site for cytosolic H^+ could occur in response to inflammatory factors. This would produce an enhanced contractile response when exposed to spasmogens. This hypothesis, however, remains to be evaluated in tracheal muscle.

6. Conclusions

Airways smooth muscle has an active Na^+/H^+ exchanger, which is able to modulate cytosolic pH. In unstimulated muscle it appears to be active when the pH_i decreases below 7.0. A sodium-dependent and SITS-sensitive anion exchanger is also active in tracheal muscle. Recovery from acid loads is possible, albeit at a much slower rate, when the Na^+/H^+ exchanger is inhibited in a bicarbonate containing physiological medium. The SITS-sensitive anion exchanger regulates intracellular pH in a more alkaline range. Although amiloride can effectively and fully attenuate active contraction in larger airways the more specific Na^+/H^+ blockers such as EIPA and MGCMA have only a modulatory effect in HCO_3^- containing medium. Moreover, the effects of intracellular acidosis alone or in conjunction with extracellular acidosis appear to be different since the interactions of the different calcium regulating process are differentially affected in intracellular or extracellular acidosis. The net effect on cytosolic calcium and thus isometric tension, is often difficult to predict. Thus, cytosolic pH, in response to the activity of the Na^+/H^+ exchanger, probably serves a modulatory role in adult airway smooth muscle and provides protection from acid loads. Further studies are required to understand the role of this important exchanger in various airway diseases.

Acknowledgements

Contributions from J. Yu, S. Sharma, A. Guia, I. Krampetz and the technical expertise of Chris Fyfe are much appreciated. Amiloride analogs were provided by Dr. E. A. Cragoe, Nagdoches, TX USA. The author's research was supported by the Medical Research Council of Canada.

References

1. Farley JM, Miles PR. Role of depolarization in acetylcholine-induced contractions of dog trachealis muscle. J Pharmacol Exper Therap 1977; 201: 199–205.
2. Krampetz IK, Bose R. Importance of the Na^+/H^+ antiporter in intracellular pH regulation of canine tracheal smooth muscle. Prog Clin Biol Res 1990; 327: 703–10.
3. Aickin CC. Intracellular pH regulation by vertebrate muscle. Ann Rev Physiol 1986; 48: 349–61.
4. McCabe RD, Young DB. Evidence of a K^+-H^+-ATPase in vascular smooth muscle cells. Am J Physiol 1992; 262: H1955–8.
5. Milanick MA. Proton fluxes associated with Ca^{2+} in human red blood cells. Am J Physiol 1990; 258: C552–62.
6. Acevedo M, Steele LW. Na^+/H^+ exchanger in isolated epithelial tracheal cells from sheep. Involvement in tracheal proton secretion. Exp Physiol 1993; 78: 383–94.
7. Nord EP, Brown SES, Crandall ED. Characterization of Na^+/H^+ antiport in type II alveolar epithelial cells. Am J Physiol 1987; 252: C490–8.
8. Grinstein S, ed. Na^+/H^+ exchange, Pub CRC press, Boca Raton, Fl, USA, 1988. (Vaughan-Jones RD, pH-selective microelectrodes, Vaughan-Jones, 3–20; Okerlund LS and Gillies RJ, Measurement of pH and Na^+ by nuclear magnetic resonance, 22–45).
9. Wray S. Smooth muscle intracellular pH: measurement, regulation and function. Am J Physiol. 1988; 254: C213–25.
10. Aronson PS. Kinetic properties of the plasma membrane Na^+/H^+ exchanger. Annu Rev Physiol 1985; 47: 545–60.
11. Yu J, Zheng J, Ong BY, Bose R. Intracellular pH measurement with fluorescent dye in canine basilar artery. Blood Vessels 1991; 28: 464–74.

12. Grinstein S, Rothstein A. Mechanism of regulation of Na^+/H^+ exchanger. J Mem Biol 1986; 90: 1–12.
13. Aickin CC, Thomas RC. An investigation of ionic mechanism of intracellular pH regulation in mouse soleus muscle fibers. J Physiol 1977; 273: 295–316.
14. Austin C, Wray S. Changes of intracellular pH in rat mesentric vascular smooth muscle with high-K^+ depolarization. J Physiol 1993; 469: 1–10.
15. Kleyman TR, Cragoe ER. Amiloride and its analogs as tools in the study of ion transport. J Mem Biol 1988; 105: 1–21.
16. Niggli E, Lederer WJ. Molecular operations of the sodium-calcium exchanger revealed by conformation currents. Nature 1991; 349: 621–24.
17. Scholz W, Albus U, Lang JH, Linz W, Martorana PA, Englert HC, Scholkens BA. Hoe 694, a new Na^+/H^+ exchange inhibitor and its effects in cardiac ischaemia. Br J Pharmacol 1993; 109: 562–8.
18. Orlowski J. Heterogeneous expression of functional properties of amiloride high affinity (NHE-1) and low affinity (NHE-3) isoforms of the rat Na^+/H^+ exchanger. J Biol Chem 1993; 268: 16369–77.
19. Bose R, Yu J, Cragoe EJ. Excitation-contraction coupling in tracheal smooth muscle studied with specific inhibitors of Na^+ channel and Na^+/H^+ antiporter. Physiologist 1989; 32: 181.
20. Bose R, Yu J, Cragoe EJ, Delaive J. Cytosolic pH changes in contracting canine trachealis smooth muscle. Prog Clin Biol Res 1990; 327: 695–702.
21. Battisini B, Filep JG, Cragoe EJ, Fournier A, Sirois P. A role for Na^+/H^+ exchange in contraction of guinea pig airways by endothelin-1 in vitro. Biochem Biophys Res Comm 1991; 175: 583–88.
22. Guia A, Bose R. Effects of extracellular acidosis on the action of smooth muscle relaxants. Proc West Pharmacol Soc 1990; 33: 77–84.
23. Yu J, Guia A, Mink S, Kepron W, Cragoe EJ, Bose R. Role of sodium in antigen-induced contraction of tracheal muscle in dogs. Respir Physiol. 1993; 91: 111–24.
24. Black PN, Ghatei MA, Bretherton-Watl D, Takahashi K, Krause T, Bloom SR. Endothelin is formed by cultured tracheal epithelial cells. Am Rev Respir Dis 1989; 139: (Suppl): A52.
25. Springall DR, Howarth PT, Counihan H, Djukanovic R, Holgate ST, Polak JM. Endothelin immunoreactivity of airway epithelium in asthamatic patients. Lancet 1991; 337: 697–701.
26. Noveral JP, Rosenberg SM, Anbar RA, Pawlowski NA, Grunstein MM. Role of endothelin-1 in regulating proliferation of cultured rabbit airway smooth muscle cells. Am J Physiol 1992; 263: L317–24.
27. Shaw AM, Steele LW, Butcher PA, Ward MR, Olver RE. Sodium-proton exchange across the apical membrane of the alveolar type II cell of fetal sheep. Biochim Biophys Acta 1990; 1028: 9–13.
28. Tse CM, Levine SA, Yun CH, Brant SR, Pouyssegur J, Montrose MH, et al. Functional characteristics of a cloned epithelial Na^+/H^+ exchanger (NHE3): resistance to amiloride and inhibition by protein kinase C. Proc Natl Acad Sci 1993; 90: 9110–4.
29. Weinman EJ, Shenolikar S. Regulation of the renal brush border membrane Na^+/H^+ exchanger. Annu Rev Physiol 1993; 55: 289–304.
30. Sardet C, Franchi A, Pouyssegur J. Molecular cloning, primary structure, and expression of human growth factor activatable Na^+/H^+ exchanger. Cell 1991; 56: 271–80.
31. Aronson PS, Nee J, Suhm MA. Modifier role of internal H^+ in activating the Na^+/H^+ exchanger in renal microvillus membrane vesicles. Nature 1982; 299: 161–63.
32. Rood RP, Emmer E, Wslok J, McCullen J, Hussain Z, et al. Regulation of rabbit ileal brush border Na^+/H^+ exchanger by ATP requiring Ca^{2+}/calmodulin mediated process. J Clin Invest 1988; 82: 1091–97.
33. Burns KD, Homma T, Harris RC. Calmodulin antagonism inhibits basolateral and activates apical Na^+/H^+ exchange in LLC-PK/C14. J Am Soc Nephrol 1990; 1: 714.
34. Wu M, Tseng Y. The modulatory effects of endothelin-1, carbachol and isoprenaline upon Na^+/H^+ exchange in dog cardiac purkinje fibers. J Physiol 1993; 471: 583–97.
35. Alvarez J, Garcia-Sancho J, Mollinedo F, Sanchez A. Intracellular Ca^{2+} potentiates Na^+/H^+ exchange and cell differentiation induced by phorbol ester in U937 cells. Eur J Biochem 1989; 183: 709–14.
36. Danthuluri NR, Berk BC, Brock TA, Cragoe EJ, Deth RC. Protein Kinase C-mediated intracellular alkalinization in rat and rabbit aortic smooth muscle cells. Eur J Pharmacol 1987; 141: 503–6.

37. Hall IP, Chilvers ER. Inositol phosphate and airway smooth muscle. Pulmonary Pharmacol, 1989; 2: 113–20.
38. Schramm CM, Grunstein MM. Mechanism of protein kinase C regulation of airway contractility. J Appl Physiol 1989; 66: 1935–41.
39. Vigne P, Breittmayer JP, Frelin C, Lazdunski M. Dual control of intracellular pH in aortic smooth muscle cells by a cyclic AMP-sensitive HCO_3^-/Cl^- antiporter and a protein kinase C-sensitive Na^+/H^+ antiporter. J Biol Chem 1988; 263: 18023–8.
40. Mitsuka M, Berk BC. Long term regulation of Na^+/H^+ exchange in vascular smooth muscle cell: role protein kinase C. Am J Physiol 1991; C562–9.
41. Krapf R, Pearce D, Lynch C, Xi XP, Reudelhuber TL, Pouysseggur J, Rector FC. Expression of rat renal Na/H antiporter mRNA levels in response to respiratory and metabolic acidosis. J Clin Invest 1991; 87: 747–51.
42. Sterling GM. The mechanism of bronchoconstriction due to hypocapnia in man. Clin. Sci. 1968; 34: 277–85.
43. Duckles SP, Raynor MD, Nadel JA. Effect of CO_2 and pH on drug-induced contraction of airway smooth muscle. J Pharmacol Exp Ther 1974; 190: 472–81.
44. Sterling GM, Holst PE, Nadel JA. Effect of CO_2 and pH on bronchoconstriction caused by serotonin vs acetylcholine. J Appl Physiol 1972; 32: 39–43.
45. Stephens NL, Mitchell RW. Mechanism of action of respiratory acidosis on tracheal smooth muscle. Am J Physiol 1974; 227: 647–51.
46. Uchida Y, Ninomiya H, Saotome M, Nonomura A, Ohtsuka M, Yanagisawa M, et al. Endothelin, a novel constrictor peptide, as a potent bronchoconstrictor. Eur J Pharmacol 1988; 154: 227–8.
47. Aalkjaer C, Cragoe EJ. Intracellular pH regulation in resting and contracting segments of rat mesentric resistance vessels. J Physiol 1988; 402: 391–410.
48. Taggart M, Wray S. Simultaneous measurement of intracellular pH and contraction in uterine smooth muscle. Pflugers Arch 1993; 423: 527–9.
49. Batlle DC, Peces R, Lapointe MS, Ye M, Daugirdas JT. Cytosolic free calcium regulation in response to acute changes in intracellular pH in vascular smooth muscle. Am J Physiol 1993; 264: C932–43.
50. Putnam RW. pH Regulatory transport systems in a smooth muscle like cell line. Am J Physiol 1990; C470–9.
51. Jensen PE, Hughes A, Boonen HCM, Aalkjaer C. Force, membrane potential, and $[Ca^{2+}]_i$ during activation of rat mesentric small arteries with norepinephrine, potassium, aluminum fluoride and phorbol ester. Effect of changes in pH_i. Circ Res 1993; 73: 314–24.
52. Kume H, Takagi K, Satake T, Tokuno H, Tomita T. Effects of intracellular pH on calcium-activated potassium channels in rabbit tracheal smooth muscle. J Physiol 1990; 424: 445–57.
53. Raeburn D. Effect of altered availability of Na^+ on guinea-pig airway smooth muscle contractility. Pulmon. Pharmacol. 1990; 3: 121–7.
54. Krampetz IK, RA. Intracellular pH: effect on pulmonary arterial smooth muscle. Am J Physiol 1991; 260: L516–21.
55. Danthuluri NR, Deth R. Effect of intracellular alkalinization on resting and agonist-induced vascular tone. Am J Physiol 1989; 256: H867–75.
56. Krampetz IK, Bose R. Relaxant effect of amiloride on canine tracheal smooth muscle. J Pharmacol Exper Therap 1988; 246: 641–48.
57. Ozaki H, Kwon SC, Tajimi M, Karaki H. Stimulants and relaxants in canine tracheal smooth muscle. Pfluger Arch 1990; 416: 351–9.
58. Chilvers ER, Challiss RA, Willcocks Al, Potter BV, Barnes PJ, Nahorski SR. Characterization of stereospecific binding sites for inositol 1,4,5-triphosphate in airway smooth muscle. Br J Pharmacol. 1990; 99: 297–302.
59. Mrwa U, Achtig I, Ruegg JC. Influences of calcium concentration and pH on tension development and ATPase activity of the arterial actomyosin contractile system. Blood Vessels 1974; 11: 277–86.
60. Bose R, Stephens NL. Mechanism of action of hypoxia. Role of contractile proteins. In: Stephens NL, editor. Biochemistry of Smooth Muscle. Univ. Park Press. 1977: 499–512.
61. Grover AK, Sampson SE. Pig coronary artery smooth muscle: Substrate and pH dependence of the two calcium pumps. Am J Physiol 1986; 20: C529–34.

62. Loutzenhiser R, Matsumoto Y, Okawa W, Epstein M. H^+ induced vasodilation of rat aorta is mediated by alterations in intracellular calcium sequestration. Circ Res 1990; 67: 426–39.
63. Fry CH, Liston TG, Cole RS. The effect of pH on urinary tract smooth muscle function. Prog Clin Biol Res 1990; 327: 717–23.
64. Irisawa H, Sato R. Intra- and extracellular actions of protons on the calcium current of isolated guinea pig ventricular cells. Circ Res. 1986; 59: 348–55.
65. Moolenaar WH. Effect of growth factors on intracellular pH regulation. Ann Rev Physiol 1986; 48: 363–76.
66. James AL, Pare PD, Hogg JC. The mechanics of airway narrowing in asthma. Am Rev Resp Dis 1989; 139: 242–6.
67. Huot SJ, Aronson PS. Na^+/H^+ exchanger and its role in essential hypertension and diabetes mellitus. Diabetes Care 1991; 14: 521–35.
68. Kranzhofer R, Schirmer J, Schomig A, von Hodenberg E, Pestel E, Metz J, et al. Suppression of neointimal thickening and smooth muscle cell proliferation after arterial injury in rat by inhibitors of Na^+/H^+ exchange. Circ Res 1993; 73: 264–8.
69. Mitsuka M, Nagae M, Berk BC. Na^+/H^+ exchange inhibitors decrease neointimal formation after rat carotid injury. Circ Res 1993; 73: 269–75.
70. Smith PR, Benos DJ. Epithelial Na^+ channels. Annu Rev Physiol 1991; 53: 509–30.

Airways Smooth Muscle: Peptide Receptors, Ion Channels and Signal Transduction
ed. by D. Raeburn and M. A. Giembycz
© 1995 Birkhäuser Verlag Basel/Switzerland

CHAPTER 12
Sodium/Potassium-Dependent Adenosine Triphosphatase

Susan J. Gunst

Department of Physiology and Biophysics, Indiana University School of Medicine, Indianapolis, Indiana, USA.

1 Introduction
2 Methods Used to Study Na$^+$/K$^+$ ATPase Activity in Smooth Muscle Tissues
3 Evidence for a Sodium Electrogenic Pump in Airways Smooth Muscle
4 Role of the Na$^+$/K$^+$ ATPase in the Relaxation of Airways Smooth Muscle
5 Role of the Na$^+$/K$^+$ ATPase in the Regulation of Airway Reactivity
Acknowledgements
References

1. Introduction

Most mammalian cells are known to have Na$^+$/K$^+$-pumps in their membranes which function to pump Na$^+$ ions out of the cell and K$^+$ ions into the cell. This creates an electrochemical gradient for the movement of extracellular Na$^+$ into the cell and intracellular K$^+$ out of the cell [1]. The pumping requires energy and is accomplished by a membrane-bound ATP hydrolyzing enzyme system, referred to as the Na$^+$/K$^+$ ATPase. The coupling ratio for the exchange of Na$^+$ and K$^+$ is not equal; it approximates three Na$^+$ ions for two K$^+$ ions for each ATP molecule hydrolyzed. Thus there is a net outward movement of positive charge which results in an electrogenic current [2]. This contributes to the negativity on the inner side of the membrane; thereby increasing the potential difference (membrane potential, E_m) across the membrane.

There is no evidence that the Na$^+$/K$^+$ ATPase in smooth muscle cells differs from that found in other mammalian cells [3, 4]. The Na$^+$ pump appears to be electrogenic in most smooth muscle tissues, including airways smooth muscle [3–5]. There is evidence that the Na$^+$ pump can generate significant membrane hyperpolarization; however, the contribution of the Na$^+$ pump to the resting E_m is controversial and difficult to assess because of the many secondary effects of pump inhibition on ion gradients [3].

The Na$^+$ gradient generated by activity of the Na$^+$/K$^+$ ATPase, as well as the Ca^{2+} gradient generated by the membrane-bound Ca^{2+}

ATPase, provide considerable potential energy to drive secondary membrane transport processes. In most tissues the gradient for Na^+ is used as a free energy source for the cotransport (symport) of substances such as sugars, amino acids, Cl^-, or K^+ plus Cl^-, as well as for the countertransport of Ca^{2+} or H^+ against their concentration gradients via passive membrane bound carriers [1, 3].

The existence of a mechanism for Na^+/Ca^{2+} exchange is now well-established in smooth muscle tissues [6–8]. This mechanism has been extensively studied in other tissues, and has generally been believed to exchange one Ca^{2+} ion for three Na^+ ions and to be reversible [3]. The direction of ion movement is determined by the ion gradients for Na^+ and Ca^{2+} as well as the membrane potential. In smooth muscle tissues, the Na^+/Ca^{2+} exchange mechanism utilizes the Na^+ gradient to extrude Ca^{2+} in the presence of an active Na^+ pump, thereby maintaining a low intracellular Ca^{2+} concentration. In the absence of a functional Na-pump, i.e. in Na^+-loaded smooth muscle, the Na^+/Ca^{2+} exchange mechanism can use the Ca^{2+} gradient to extrude Na^+, and thereby reduce intracellular Na^+ [3].

The electrochemical potential generated by the Na^+ pump is also important in maintaining the intracellular pH of smooth muscle tissues. The extrusion of H^+ in exchange for Na^+ utilizes the energy generated by active Na^+ pumping to reduce the acidity of the cell [9–11]. The intracellular pH of smooth muscle cells has been recorded in the range of 7.0 to 7.2 [9]. Assuming an extracellular pH (pH_e) of 7.4 and a resting E_m of -60 mV, a considerably more acidic value of 6.4 would be predicted on the basis of passive ion distribution [9, 10]. There is also evidence that the Na^+ gradient drives the transport of both Cl^- and HCO_3^- into smooth muscle cells by cotransport mechanisms [12]. Na^+/HCO_3^- cotransport may provide a second Na^+ dependent mechanism for acid extrusion. In addition, a $Na^+/K^+/2Cl^-$ cotransport mechanism may also modulate cellular pH by affecting the HCO_3^- concentration through HCO_3^-/Cl^- exchange [13]. However, it is not yet clear whether these latter mechanisms are universal features of all smooth muscle tissues. Their existence in airways smooth muscle has not been established. The role of these transport systems are discussed in other chapters.

Changes in the activity of the Na^+/K^+ ATPase can therefore be expected to have multiple secondary effects on cellular metabolism. The most profound effects of Na^+ pump activity on smooth muscle excitability are likely to be mediated by its effects on the membrane potential, intracellular pH, and intracellular Ca^{2+} concentration.

2. Methods Used to Study Na^+/K^+ ATPase Activity in Smooth Muscle Tissues

Assessment of the effects of Na^+/K^+ ATPase activity in smooth muscle tissues has traditionally involved measuring the mechanical or electrical effects of inactivation of the Na^+ pump [4, 14–17]. Recently, the use of ion sensitive microelectrodes and optical methods for monitoring intracellular pH and intracellular Na^+ and Ca^{2+} activity in conjunction with more traditional methods have yielded important advances in our knowledge of ion transport in smooth muscle tissues [18].

The membrane bound Na^+/K^+ ATPase is activated by Na^+ on the cytoplasmic side of the membrane and by K^+ on the extracellular side [1]. The extracellular K^+ sites are saturated at about 20 mM K^+ and the cytoplasmic Na^+ sites are saturated at about 130 mM Na^+ [1]. Thus, lowering the extracellular K^+ concentration below 20 mM reduces activity of the Na^+/K^+ ATPase, and increasing intracellular Na^+ concentration up to 130 mM (Na^+-loading) increases Na^+ pump activity. Kinetic studies of Na^+/K^+ ATPase isolated from ox brain indicate that with a normal Na^+ ion concentrations (20–30 mM intracellular, 140 mM extracellular) and normal K^+ ion concentrations (about 4 mM extracellular and 120 mM intracellular) the Na^+/K^+ ATPase operates with an activity of about 25–35% maximum [1].

In smooth muscle tissues, activity of the Na^+/K^+ ATPase is completely inhibited by replacing the normal bathing medium with a K^+-free medium [16]. This results in an accumulation of intracellular Na^+. The Na^+/K^+ ATPase can be rapidly reactivated by returning K^+ to the bathing medium after intracellular Na^+ has been allowed to accumulate. This results in a transient hyperpolarization of the cell and relaxation of the tissue, followed later by a "rebound" depolarization and recontraction (see Figure 3 for example) [17, 19].

The cardiac glycosides act specifically to inhibit activity of the Na^+/K^+ ATPase by binding to the extracellular side of the system [20]. The most widely used cardiac glycoside has been ouabain (g-strophanthin), which is the most water soluble. For enzymes from most tissues, half-maximal inhibition of Na^+/K^+ ATPase activity can be obtained at ouabain concentrations between 10^{-7} and 10^{-5} M; however, the inhibitory effects of cardiac glycosides differ among different tissues and species [1]. Vanadate in micromolar concentrations also inhibits Na^+/K^+ ATPase. However, vanadate is not as specific as the cardiac glycosides, which inhibit only the Na^+/K^+ ATPase. Vanadate also inhibits the $Ca^{2+}/ATPase$ and other ATPases.

Because the Na^+-pump is energy-dependent, its activity is also sensitive to temperature and to metabolic inhibitors. Activity of the Na^+/K^+ ATPase can be inhibited by cooling the tissues to inhibit cellular metabolism; however, cooling also inhibits other metabolic processes in the cell.

3. Evidence for a Sodium Electrogenic Pump in Airways Smooth Muscle

Evidence for a contribution of the Na-K ATPase to the resting membrane potential of airways smooth muscle was first provided by

Souhrada *et al.* [5]. In these studies, the Na^+/K^+ ATPase was inhibited, and the effects on E_m were determined by microelectrode recording from single muscle cells in bovine or guinea-pig trachealis muscle strips. The administration of ouabain (10^{-5} M) or the exposure of tracheal muscle cells to K^+-free physiological saline solution significantly decreased the resting membrane potential (E_m) in tracheal smooth muscle tissues (Figure 1). The magnitude of the decrease in membrane potential was similar using either of these interventions, which suggests that their efficacy at inhibiting the Na^+/K^+ ATPase is comparable. However, inhibition with ouabain resulted in a depolarization which was maximal by 20 minutes and which remained for at least 60 minutes. In contrast, after exposure to K^+-free medium, membrane depolarization was followed by a hyperpolarization which could not be blocked with ouabain. Studies in guinea-pig taenia coli have suggested that this hyperpolarization is due to an increase in K^+ permeability and an increase in the ratio of $[K^+]_i/[K^+]_o$ [14–16]. The decrease in E_m in tracheal muscle after

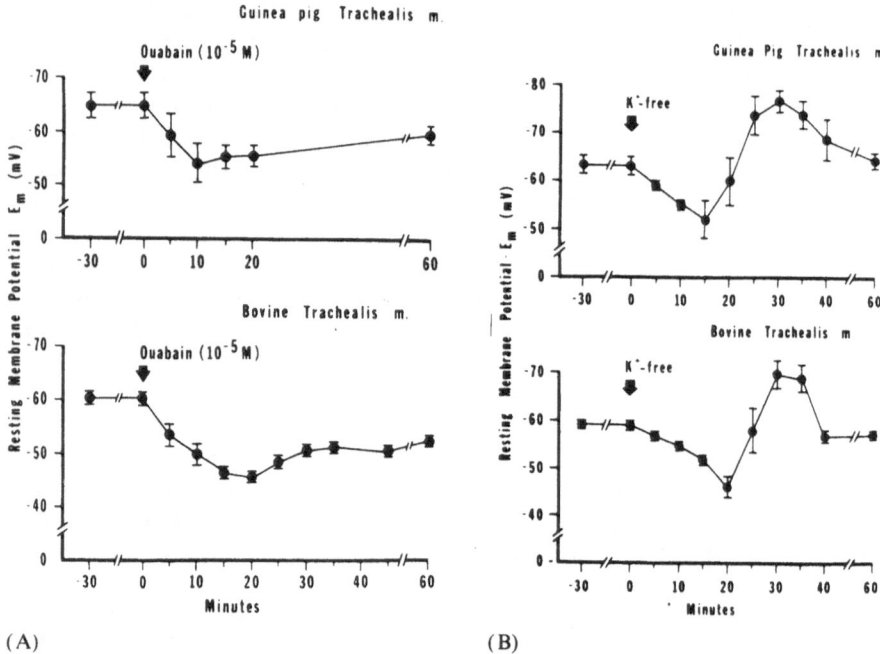

(A)   (B)

Figure 1. Effects of ouabain (10^{-5} M) (A) and K^+-free physiological salt solution (B) on the resting membrane potential of guinea-pig and bovine trachealis muscle. Inhibition of the Na^+/K^+ ATPase using ouabain (A) resulted in a significant decrease in E_m in both tissues. Inhibition of the Na^+/K^+ ATPase by exposure to K^+-free physiological saline solution resulted in an initial depolarization followed by hyperpolarization. Data points represent mean $\pm$ SE. In experiments with guinea pig tissue, 6 individual preparations from 6 animals were tested; each point represents 24–36 successful impalements. In experiments with bovine tissue 5 preparations from 5 animals were used; in these experiments each point represents 16–25 successful impalements. From Souhrada *et al* [5].

Na^+/K^+ ATPase inhibition was rapid compared with the response in tissues such as intestinal smooth muscle [14–16, 21]. Souhrada *et al.* also showed that a decrease in temperature to 22°C also led to significant depolarization in tracheal smooth muscle tissues, providing further evidence that an electrogenic Na^+/K^+ ATPase contributes to the resting membrane potential.

Experiments with Na^+ loaded airways smooth muscle also provided evidence for the presence of an active Na^+/K^+ ATPase (Figure 2) [5]. Guinea-pig and bovine tracheal smooth muscles were loaded with Na^+ by incubating them in K^+-free solution at 22° for a period of 18 hr. Subsequent exposure of the tissues to normal physiological saline solution resulted in significant hyperpolarization of both muscle preparations.

The mechanical effects of the Na^+ pump on airways smooth muscle have also been evaluated. Chideckel *et al.* [22] reported that the incubation of human airway tissues in ouabain (10^{-7} to 10^{-5} M) or in K^+-free

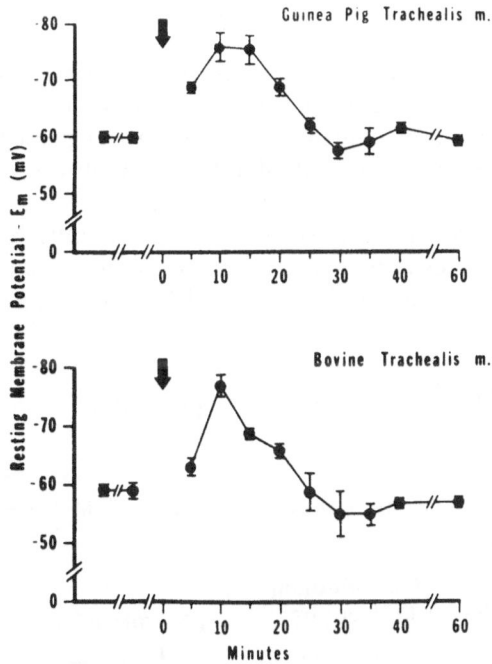

Figure 2. Effect of normal temperature (37°C) and physiological salt solution (PSS) on resting membrane potential (E_m) of guinea-pig and bovine trachealis smooth muscles previously exposed for 18 h to K^+-free PSS (4°C) ("Na^+-loaded") airways smooth muscle. Base-line E_m was measured before tissues were exposed to 4°C PSS. Exposure of tissue to 37°C normal PSS, indicated by arrows, resulted in hyperpolarization within 10 min. In experiments with guinea pigs, 5 individual preparations from 5 animals were tested; each point represents 20–25 successful impalements. In experiments with bovine tissue 5 preparations from 5 animals were used; in these experiments each point represents 16–25 successful impalements. Data represents means ± SE. From Souhrada *et al.* [5].

medium results in immediate and sustained contractions. In contrast, ouabain (10^{-5} M) caused only transient contractions in canine and guinea-pig tracheal muscle. In rabbit airway muscle ouabain was without effect. Other studies have reported more sustained contractions in response to ouabain in canine tissue [23]; however, these contractions are not substantial except at relatively high (10^{-5} M) concentrations of ouabain, and may occur only after a delay of up to 30 min. In canine tracheal muscle, K^+-free medium has no effect on resting tension. Long delays in the effect of Na^+/K^+ ATPase inhibition may indicate that the effects on tension are not the direct result of Na^+/K^+ ATPase inhibition but occur in response to secondary effects of Na^+/K^+ ATPase inhibition.

The introduction of K^+ into the bathing medium of canine tracheal muscles which have been subjected to Na^+-loading by incubation in K^+-free medium for one hour results in marked relaxation in tissues which have been contracted with either 5-hydroxytryptamine (5-HT) or acetylcholine [23] (Figure 3). The amount of relaxation is proportional to the concentration of extracellular K^+ up to 10 mM (Figure 4). When the same procedure is performed in the presence of ouabain, the addition of K^+ causes contraction rather than relaxation, indicating that the K^+-induced relaxation depends on the presence of an active Na^+/K^+ ATPase. The amount of relaxation observed in response to K^+ also depends on which agonist is used to contract the muscle initially; much more relaxation is observed in muscles contracted by 5-HT than in muscles contracted by acetylcholine, even when the initial contractile tensions are closely matched [23]. The relaxation induced by

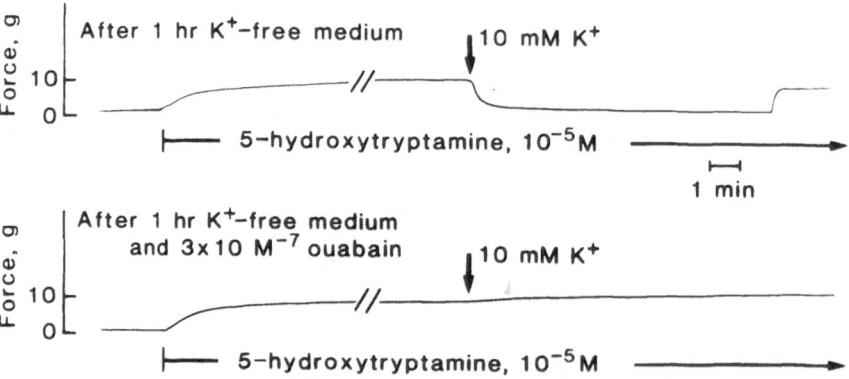

Figure 3. Effect of the activation of the Na^+/K^+ ATPase by KCl on tension in a canine tracheal smooth muscle strip. The Na^+/K^+ ATPase was inhibited for 1 hour by maintaining the tissue in K^+-free physiological salt solution. The muscle was then contracted with 5-hydroxtryptamine. Activation of the Na^+/K^+ ATPase by the addition of KCl (10 mM) to the bathing medium resulted in immediate relaxation eventually followed by recontraction. When this procedure was performed in the presence of ouabain (3×10^{-7} M), the muscle contracted in response to KCl (10 mM) rather than relaxed.

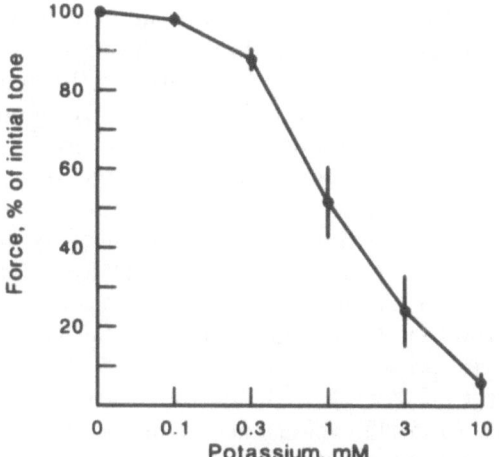

Figure 4. Concenration-response curve for potassium activation of Na$^+$/K$^+$ ATPase. Canine tracheal smooth muscle tissues were contracted with 5-hydroxytryptamine after incubation for 1 h in K$^+$-free physiological salt solution. After force reached a steady state, KCl (0.1–10 mM) was added cumulatively to the bathing medium. Near maximal relaxation was obtained in response to KCl (10 mM). From Gunst and Stropp [23].

the activation of the Na$^+$/K$^+$ ATPase is accompanied by a marked drop in intracellular Ca^{2+} (Figure 5) [24].

Studies in a number of smooth muscle tissues have demonstrated that membrane depolarization caused by inhibition of the Na$^+$/K$^+$ ATPase with ouabain may be associated with an increase in sensitivity to

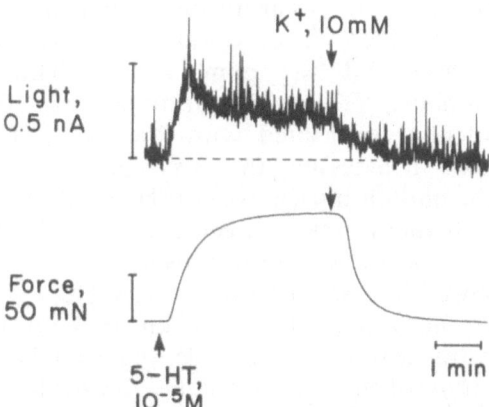

Figure 5. Effect of stimulation of Na$^+$/K$^+$ ATPase activity by K$^+$ ion on force and aequorin luminescence (light) in a canine trachealis muscle strip contracted with 5-hydroxytryptamine (10^{-5} M) after 1 hr incubation in K$^+$-free physiological saline solution. Intracellular Ca^{2+} was measured using the Ca^{2+}-indicator aequorin. Aequorin luminescence is calibrated in nanoamperes (nA) of current output from a photomultiplier tube which is positioned so as to measure light emitted by the muscle strip. Intracellular Ca^{2+} declined concurrently with the relaxation which occurred in response to KCl (10 mM). From Gunst and Bandyopadhyay [24].

pharmacologic agonists [25–28]. There is limited data available which addresses the effect of Na^+/K^+ ATPase inhibition on airways smooth muscle responsiveness or sensitivity to agonists. In guinea-pig tracheal smooth muscle, inhibition of the Na^+ pump by ouabain or K^+-free solution potentiates contractions in response to histamine [29]. In canine tracheal smooth muscle, inhibition of Na^+/K^+ ATPase activity by ouabain increases the sensitivity of the muscle to contraction by 5-HT; however the sensitivity to acetylcholine is not significantly affected, even in tissues where Na^+/K^+ ATPase inhibition elicits an increase in force [23]. As noted above, the sensitivity of Na^+-loaded tracheal smooth muscle to relaxation induced by Na^+-pump activation is also greater in muscles contracted with 5-HT [23]. The greater modulatory effect of Na^+/K^+ ATPase activity on the responses of tracheal muscles to 5-HT versus acetylcholine may reflect a greater influence of the membrane potential on coupling mechanisms during activation with 5-HT [30–32].

In guinea-pig taenia coli, inhibition of Na^+/K^+ ATPase activity with ouabain, K^+-free medium, or low extracellular Na^+ causes a marked increase in the release of prostaglandin E (PGE) [33, 34]. This effect is not the result of the changes in E_m, intracellular pH, intracellular Na^+ or intracellular Ca^{2+} which occur as a result of Na^+/K^+ ATPase inhibition. The PGE release is not affected by elevations in extracellular K^+ which depolarize the membrane by an amount similar to that caused by Na^+/K^+ ATPase inhibition, or by agents which increase mechanical activity. Rather, the effects of ouabain on PGE release appear to be directly related to Na^+/K^+ ATPase activity or to the intracellular K^+ concentration. The effect of Na^+/K^+ ATPase inhibition on PGE release might occur through an active PGE transport process linked to Na^+/K^+ ATPase activity or via a link between prostaglandin synthesis and Na^+ pump activity. Thus the possibility that an increase in tissue PGE concentration may contribute to changes in membrane potential associated with Na^+/K^+ ATPase inhibition must be considered in assessing the electrogenic effects of Na^+/K^+ ATPase activity in smooth muscle tissues. However, to date it has not been established whether a link between Na^+/K^+ ATPase activity and PGE release exists in airways smooth muscle.

There is evidence that activity of the Na^+/K^+ ATPase may be potentiated by the activation of protein kinase C in airways smooth muscle. In rabbit tracheal smooth muscle, greater relaxation occurs in response to activation of the Na^+/K^+ ATPase with K^+ in Na^+ loaded muscles that are pretreated with phorbol esters to stimulate protein kinase C [35]. This effect occurs independently of alterations in Na^+/H^+ exchange or extracellular Ca^{2+} influx. In addition, the inhibitory action of phorbol esters on agonist-induced contractions can be reversed by inhibition of the Na^+/K^+ ATPase. Similar data has been obtained in guinea-pig ileal smooth muscle [36]. However, increased activity of the Na^+/K^+ ATPase under such conditions should be rapidly self-limiting, as increased activity of the Na^+/K^+ ATPase in the absence of an increased inward leak of Na^+ will rapidly lower intracellular Na^+ and thereby reduce Na^+ pump activity. Further information will be needed to fully evaluate whether there is a direct link between Na^+/K^+ ATPase activity and protein kinase C and to determine its role in contractile activation.

4. Role of the Na^+/K^+ ATPase in the Relaxation of Airways Smooth Muscle

Studies in vascular and visceral smooth muscle tissues resulted in a hypothesis that activation of the Na^+/K^+ ATPase is directly involved in regulating the relaxation of smooth muscles by β-adrenoceptor agonists [38–41]. The increase in cyclic AMP induced by β-adrenoceptor stimulation has been proposed directly to stimulate activity of the Na^+/K^+ ATPase, causing muscle relaxation by generating the Na^+ gradient necessary to extrude Ca^{2+} via Na^+/Ca^{2+} exchange [38, 39]. Na^+/K^+ ATPase activity may also cause relaxation by hyperpolarizing the sarcolemmal membrane, thereby reducing Ca^{2+} influx through membrane-potential dependent Ca^{2+} channels [42, 43].

β-adrenoceptor activation in airways smooth muscle stimulates adenylyl cyclase resulting in increases in cyclic AMP [44–46]. Relaxation of airways smooth muscle by PGE_2 and vasoactive intestinal peptide (VIP) have also been associated with increases in cyclic AMP [47]; whereas relaxation by sodium nitroprusside (SNP) has been associated with increases in cyclic GMP [46]. Thus, if a direct link between cyclic AMP production and stimulation of the Na^+/K^+ ATPase exists, the Na^+/K^+ ATPase should be play a role in regulating the relaxation of airways smooth muscle by those agonists which stimulate adenylyl cyclase.

The effects of Na^+/K^+ ATPase inhibition on the relaxation of canine tracheal smooth muscle have been found to be similar in many respects to the effects of Na^+/K^+ ATPase inhibition on the relaxation of other smooth muscle tissues [23]. Inhibition of activity of the Na^+/K^+ ATPase by ouabain, K^+ free medium, or low temperature decreases the amount of relaxation which can be elicited by isoprenaline, PGE_2 or forskolin (an activator of adenylyl cyclase [48]) [23] (Figure 6). As in arterial muscle, tracheal muscle relaxation in response to SNP is not affected by Na^+/K^+ ATPase inhibition [23]. These observations are consistent with a hypothesis that relaxation induced by agents which stimulate cyclic AMP production is coupled to activation of the Na^+/K^+ pump [38].

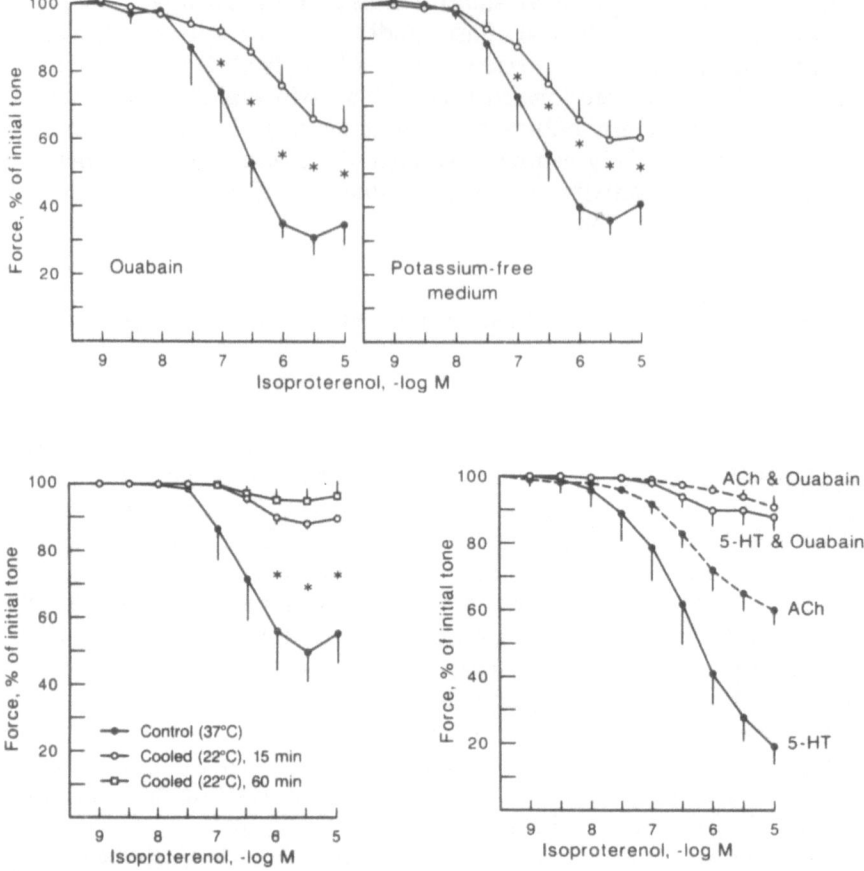

Figure 6. Effect of inhibition of the Na^+/K^+ ATPase on relaxation to isoproterenol in tissues contracted by acetylcholine (all panels) or 5-hydroxytryptamine (bottom right). Solid dots indicate relaxation in response to cumulatively increasing concentrations of isoproterenol. Open dots indicate responses to isoproterenol (isoprenaline) after inhibition of the Na^+/K^+ ATPase by: (top left) 1 hour incubation in ouabain (3×10^{-7} M); (top right) 1 hour incubation in potassium-free physiological saline solution; (bottom left) cooling to $22°C$; (bottom right) 1 hour incubation in 10^{-5} M ouabain. Values shown are means $\pm$ SEM. Asterisks (*) indicate responses at which relaxation was significantly decreased after Na^+/K^+ ATPase inhibition compared with control ($P < 0.05$; n = 6). From Gunst and Stropp [23].

However, the question of whether the Na^+/K^+ ATPase plays a direct or an indirect role in regulating relaxation is difficult to address directly. The degree to which inhibition of the Na^+/K^+ ATPase depresses relaxation of tracheal muscles in response to isoprenaline or other agents varies with the method used to inhibit the Na^+/K^+ ATPase and with the duration of Na^+/K^+ ATPase inhibition prior to relaxation [23]. Relaxation to isoprenaline can be completely inhibited in canine tracheal muscle by incubating the muscles in ouabain (10^{-5} M) or by cooling the muscles to 22°C after they are contracted with either 5-HT or by acetylcholine (Figure 6). However, if the absence of a response of Na^+ loaded tissues to extracellular K^+ is used to indicate the inhibition of Na^+/K^+ ATPase activity, a relatively short incubation period (15 min) in a much lower concentration of ouabain (3×10^{-7} M) is found to be sufficient to completely inhibit the Na^+ pump. One hour incubation in K^+-free medium also completely inhibits the Na^+ pump. Yet ouabain (3×10^{-7} M) has no effect on relaxation to isoprenaline in tracheal smooth muscle, and one hour in K^+-free medium only partially blocks relaxation to isoprenaline [23] (Figure 6). Thus conditions which fully inhibit Na^+/K^+ ATPase activity either do not affect or only partially inhibit relaxation in tracheal smooth muscle. These observations suggest that the effects of Na^+/K^+ ATPase inhibition on relaxation may be secondary to the general disruption of cellular regulatory mechanisms, rather than solely to effects on Ca^{2+} fluxes caused by the direct stimulation of the Na^+/K^+ ATPase by cyclic AMP. Long periods of Na^+/K^+ ATPase inhibition resulting in a build up of intracellular Na^+ and Ca^{2+}, intracellular acidosis, and a decline in E_m could disrupt processes essential for relaxation. Whether or not the role of the Na^+/K^+ ATPase in relaxation is a direct one, the fact that some relaxation occurs in the presence of Na^+/K^+ ATPase inhibition indicates that stimulation of Na^+/K^+ ATPase is not the only pathway by which β-adrenoceptor-induced relaxation is affected. Other pathways which are not directly dependent on the presence of a functional Na^+/K^+ ATPase must be involved [49, 50].

Relaxation which is mediated either wholly or partly by the stimulation of Na^+/K^+ ATPase would be expected to be accompanied by a reduction in intracellular Ca^{2+}, resulting from inhibition of Ca^{2+} influx through voltage-dependent Ca^{2+} channels and from the stimulation of Na^+/Ca^{2+} exchange. Thus it is noteworthy that β-adrenoceptor-induced relaxation of tracheal smooth muscle can be partially inhibited by the administration of Ca^{2+}-channel antagonists such as diltiazem or verapamil [23]. More importantly, in canine tracheal muscles loaded with the Ca^{2+}-indicator aequorin, relaxation stimulated by either isoprenaline or forskolin is accompanied by a decrease in intracellular Ca^{2+} [24] (Figure 7). Both the relaxation and decline in Ca^{2+} which accompanies it can be partially blocked by inhibition of the Na^+/K^+ ATPase with ouabain or incubation in K^+-free medium (Figure 7) [24]. Thus there is direct evidence that stimulation of the Na^+/K^+ ATPase causes a decrease in intracellular Ca^{2+} in airways smooth muscle, and that β-adrenoceptor stimulation is also accompanied by a decrease in intracellular Ca^{2+} which is partially dependent on a functional Na^+/K^+ ATPase.

Decreases in intracellular Ca^{2+} associated with relaxation in response to isoprenaline and forskolin have also been demonstrated in canine

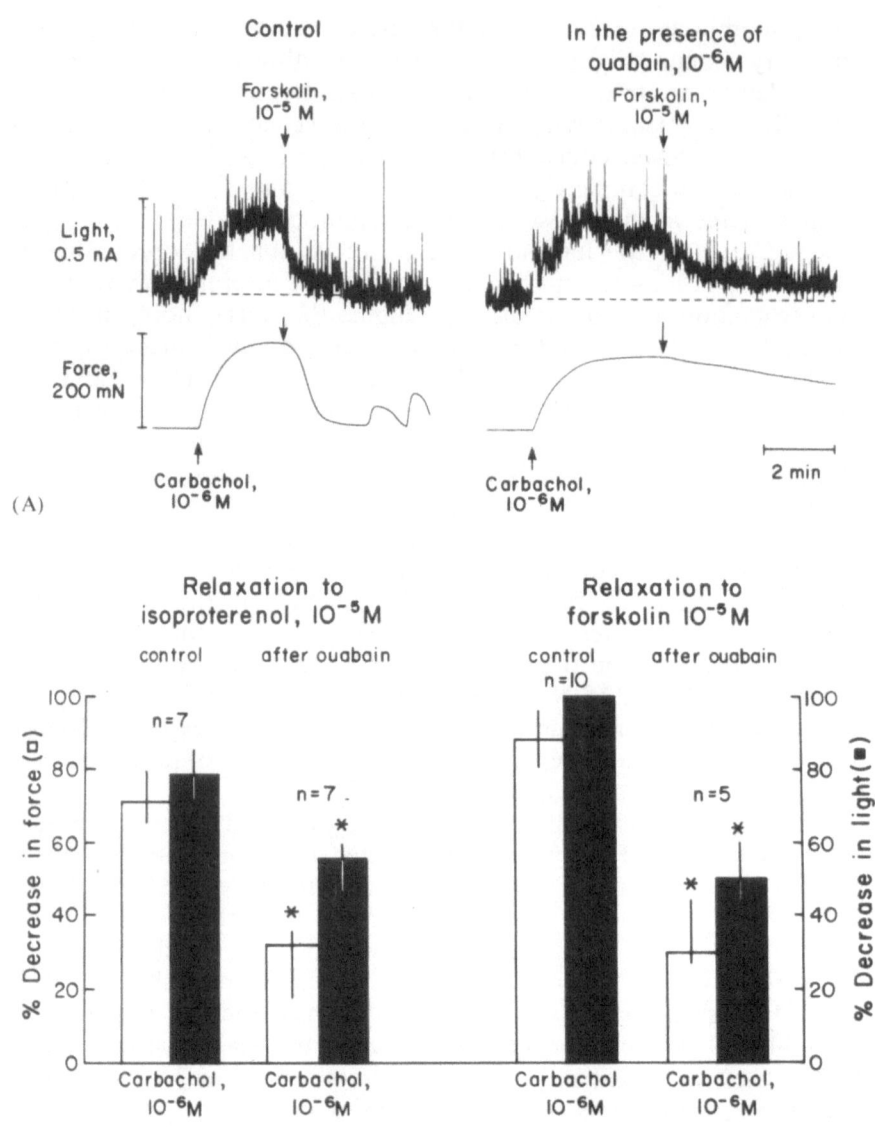

Figure 7. Effect of Na^+/K^+ ATPase inhibition on intracellular Ca^{2+} and tension during relaxation in canine tracheal smooth muscle elicited by forskolin or isoproterenol (A). Effect of forskolin on intracellular Ca^{2+} and tension in a single strip of canine trachealis muscle loaded with the Ca^{2+}-indicator aequorin and contracted with carbachol. Aequorin luminescence is calibrated in nanoamperes (nA) of current output from a photomultiplier tube which is positioned so as to measure light emitted by the muscle tissue. In the presence of ouabain (10^{-6} M), both the decline in $[Ca^{2+}]_i$ and relaxation were partially inhibited. B. Median changes in force and aequorin luminescence elicited by forskolin or isoproterenol in control muscles and in muscles treated with ouabain. Error bars indicate responses between the 25th and 75th percentile. * Indicates statistically different responses to isoproterenol or forskolin in the presence and absence of ouabain. From Gunst and Bandyopadhyay[24].

tracheal smooth muscles loaded with fura-2 [51]. Similar observations have been made in isolated smooth muscle cells from bovine trachea [52, 53]. In contrast, Takuwa et al. observed an increase in intracellular Ca^{2+} associated with the administration of forskolin or isoprenaline in intact bovine tracheal muscle strips loaded with aequorin [54]. However, the observations of Takuwa et al. were made in muscles maintained at 25°C. At this temperature activity of the Na^+/K^+ ATPase is inhibited and no relaxation occurs in response to isoprenaline [23].

5. Role of the Na^+/K^+ ATPase in the Regulation of Airway Reactivity

Early experiments by Dixon and Brodie [55] and by Weber [56] in anaesthetized cats and rabbits suggested that bronchoconstriction can be induced by large intravenous doses of digitalis. Later, Marco et al. [57] demonstrated that the intravenous administration of the cardiac glycosides digoxin and acetyl strophanthidin caused an increase in airway resistance and pulmonary compliance in anaesthetized dogs. They attributed this effect to direct constrictor effects of these drugs on bronchial smooth muscle, suggesting that Na^+/K^+ ATPase inhibition leads to the constriction of airways smooth muscle in vivo. Agrawal and Hyatt [58] subsequently reported that inhaled ouabain causes a decrease in specific airway conductance in conscious guinea-pigs. The dose producing a 30% fall in specific airway conductance in response to ouabain was highly correlated with the ED_{50} for the response to inhaled histamine, suggesting that the Na^+/K^+ ATPase was more active in the airways of the animals with a greater histamine sensitivity. Pretreatment with ouabain also significantly increased airway responsiveness to histamine. These authors proposed that the Na^+/K^+ ATPase serves a homeostatic function by preventing Na^+ and Ca^{2+} loading of the cell and that the increase in Na^+/K^+ ATPase activity was not the primary cause of airway hyperreactivity. They suggested that an increased permeability to Na^+ in the hyperractive animals was the primary cause of both the greater airway reactivity to histamine and the increased Na^+/K^+ ATPase activity in these animals.

The role of the electrogenic sodium pump in the reactivity of airways smooth muscle from sensitized animals has been investigated [29, 59–63]. Souhrada and Souhrada recorded E_m with glass microelectrodes in single cells of tracheal muscle strips from guinea pigs or rabbits sensitized with ovalbumin in vivo either actively or passively. Recordings were also made in tracheal smooth muscle preparations passively sensitized to ovalbumin in vitro [29, 59]. Under all of these conditions airways smooth muscle cells showed significant membrane hyperpolarization relative to tissue from control animals. Measurements of the

membrane potential before and after the Na^+ pump was inhibited with
ouabain showed a marked potentiation of the contribution of the
electrogenic sodium pump to the E_m of the airways smooth muscle cells
in tissues from sensitized animals. The hyperpolarization was not prevented by agents that inhibit specific mediators of anaphylaxis, indicating that it was not caused by mediators released from mast cells or other
inflammatory cells. Thus, the hyperpolarization appeared to be due to
an increased contribution of the electrogenic sodium pump to the
resting membrane potential in sensitized animals.

Further studies by Souhrada et al. have demonstrated that the
changes in airways smooth muscle cells induced by passive in vitro
sensitization can be inhibited when the muscle preparations are pretreated with amiloride or furosemide to block sodium influx [60, 61]. In
guinea-pig tracheal muscle, these effects on sodium conductance are
mediated by the occupancy of a low-affinity F_c receptor [62] (Figure 8).
Thus the binding of specific reaginic antibodies appears to cause an
increase in Na^+ influx leading to the increase in Na^+/K^+ ATPase
activity and consequently hyperpolarization. The alteration in the Na^+
gradient which occurs as a result of the increased membrane permeability for Na^+ may also lead to an increase in Ca^{2+} influx through
Na^+/Ca^{2+} exchange, and therefore be responsible for the airway hyperreactivity to mediators observed in sensitized tissue [63].

There is also evidence that response of airways smooth muscle to
challenge with specific IgG involves the activation of protein kinase C

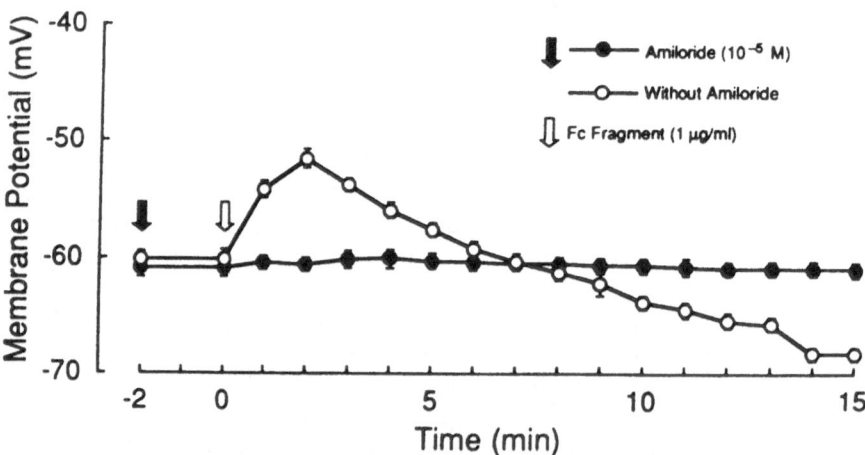

Figure 8. Effect of amiloride pretreatment on changes in membrane potential of guinea pig
tracheal smooth muscle cells. Exposure of the cells to enzymatically prepared F_c fragments of
IgG_l resulted first in depolarization of the membrane followed by hyperpolarization. F_c
fragment-induced changes in membrane potential were blocked by the administration of the
Na^+-channel blocker amiloride. From Souhrada and Souhrada [62].

[61]. If the activation of protein kinase C potentiates activity of the Na^+/K^+ ATPase [35], this could provide another mechanism by which antigen challenge could affect Na^+ pump activity.

The role played by the Na^+/K^+ ATPase in mediating the airway hyperractivity has also been investigated in human asthmatics [64]. Human asthmatics and non-asthmatic controls were challenged with ouabain and histamine. Subjects without asthma did not exhibit a consistent bronchomotor response to the inhalation of ouabain. Bronchodilation was the predominant response of patients with mild asthma and moderate airway reactivity. Bronchoconstriction occurred in patients who had severe asthma and high airway reactivity. The responses of subjects to the inhalation of ouabain were complex, probably due to complications caused by systemic or local effects of ouabain such as the stimulation of vagal reflexes; however the results did demonstrate that Na^+/K^+ ATPase activity may be altered in patients who have asthma. Further research will be necessary to determine whether an alteration in Na^+/K^+ ATPase activity is an important factor in the airway hyperreactivity characteristic of human asthmatics.

Acknowledgements

This work was supported by research grant R01 HL29289 from the National Heart, Lung, and Blood Institute, National Institutes of Health, Bethesda, MD.

References

1. Skou JC. Overview: The Na,K-Pump. Meth Enzymol 1988; 156: 1–25.
2. Glynn IM, Membrane adenosine triphosphatase and cation transport. Br Med Bull 1968; 24: 165–170.
3. Brading AF, Aickin CC. Ions, transporters, exchangers and pumps in smooth muscle membranes. Frontiers in Smooth Muscle Research. Alan R. Liss, Inc. 1990, 323–343.
4. Fleming WW. The electrogenic Na^+, K^+ pump in smooth muscle: physiologic and pharmacologic significance. Ann Rev Pharmacol Toxicol 1980; 20: 129–149.
5. Souhrada M, Souhrada JF, Cherniack RM. Evidence for a sodium electrogenic pump in airway smooth muscle. J Appl Physiol 1981; 51: 346–352.
6. Reuter H, Blaustein MP, Haeusler G. Na-Ca exchange and tension development in arterial smooth muscles. Phil Trans R Soc B 1973; 265: 87–94.
7. Pritchard K, Ashley CC. Na^+/Ca^{2+} exchange in isolated smooth muscle cells demonstrated by the fluorescent calcium indicator fura-2. FEBS Lett 1986; 195: 23–27.
8. Aaronson PI, Benham CD. Alterations in $[Ca^{2+}{}_i]$ mediated by sodium calcium exchange in smooth muscle cells isolated from the guinea pig ureter. J Physiol 1989; 416: 1–18.
9. Wray S. Smooth muscle intracellular pH: measurement, regulation, and function. Am J Physiol 1988; 254: C213–C225.
10. Aickin CC. Direct measurement of intracellular pH and buffering power in smooth muscle cells of guinea-pig vas deferens. J Physiol 1984; 349: 571–585.
11. Aickin CC. Movement of acid equivalents across the mammalian smooth muscle membrane. In: Proton Passage across Cell Membranes. CIBA Foundation Symposium, 1988; 139: 3–22.
12. Aickin CC, Brading AF. The role of chloride-bicarbonate exchange in the regulation of intracellular chloride in guinea pig vas deferens. J Physiol 1984; 349: 587–606.

13. Aickin CC, Brading AF. Effect of Na$^+$ and K$^+$ on the Cl$^-$ distribution in guinea-pig vas deferens smooth muscle: evidence for Na$^+$,K$^+$,Cl$^-$ co-transport. J Physiol 1990; 421: 13–32.

14. Casteels R. The action of ouabain on the smooth muscle cells of the guinea pig's taenia coli. J Physiol 1966; 183: 131–142.

15. Casteels R. Droogmans G, Hendricks H. Electrogenic sodium pump in smooth muscle cells of the guinea pig's taenia coli. J Physiol 1971; 217: 297–313.

16. Casteels R. Droogmans G, Hendricks H. Membrane potential of smooth muscle in K-free solution. J Physiol 1971; 217: 281–285.

17. Hendrickx H, Casteels R. Electrogenic sodium pump in arterial smooth muscle cells. Pfluegers Arch 1974; 346: 299–306.

18. Aickin CC, Brading AF. Advances in the understanding of transmembrane ionic gradients and permeabilities in smooth muscle obtained by using ion-sensitive microelectrodes. Experientia. 1985; 41: 879–887.

19. Webb RC, Bohr DF. Potassium-induced relaxation as an indicator of Na$^+$,K$^+$ATPase activity in vascular smooth muscle. Blood Vessels 1978; 15: 198–207.

20. Schwarz A, Lindenmayer GE, Allen JC. The sodium-potassium adenosine triphosphatase: pharmacological, physiological and biochemical aspects. Pharmacol Rev 1975; 27: 3–134.

21. El-Sharkaway TY, Daniel EE. Electrogenic sodium pumping in rabbit small intestinal smooth muscle. Am J Physiol 1975; 229: 1277–1286.

22. Chideckel EW, Frost JL, Mike P, Fedan JS. The effect of ouabain on tension in isolated respiratory tract smooth muscle of humans and other species. Br J Pharmacol 197; 92: 609–614.

23. Gunst SJ, Stropp JQ. Effects of Na-K adenosinetriphosphatase activity on relaxation of canine tracheal smooth muscle. J Appl Physiol 1988; 64: 635–641.

24. Gunst SJ, Bandyopadhyay S. Contracile force and intracellular Ca^{2+} during relaxation of canine tracheal smooth muscle. Am J Physiol 1989; 257: C355–C364.

25. Urquilla PR, Westfall DP, Goto K, Fleming WW. The effects of ouabain and alterations in potassium concentration on the sensitivity to drugs and the membrane potential of the smooth muscle of the guinea pig and rat vas deferens. J Pharamacol Exp Ther 1978; 207: 356–63.

26. Brender D, Strong CG, Shepard JT. Effects of acetylstrophanthidin on isolated veins of the dog. Circ Res 1970; 26: 647–55.

27. Ozawa H, Katsuragi T. Ouabain-induced potentiation of the contractions of the guinea-pig vas deferens. Eur J Pharmacol 1974; 25: 147–54.

28. Karaki H, Ozaki H, Urakawa N. Effects of ouabain and potassium-free solution on the contraction of isolated blood vessels. Eur J Pharmacol 1978; 48: 439–43.

29. Souhrada M, Souhrada JF. Potentiation of Na$^+$-electrogenic pump of airway smooth muscle by sensitization. Respir Physiol 1982; 47: 69–81.

30. Coburn RF, Yamaguchi T. Membrane potential-dependent-independent tension in the canine tracheal muscle. J Pharmacol Exp Ther 1977; 201: 276–284.

31. Coburn RF. Electromechanical coupling in canine trachealis muscle: acetylcholine contractions. Am J Physiol 1979; 236: C177–C184.

32. Farley JM, Miles PR. Role of depolarization in acetylcholine-induced contractions of dog trachealis muscle. J Pharmacol Exp Ther 1977; 201: 199–205.

33. Coburn RF. Na$^+$,K$^+$-ATPase inhibition stimulates prostaglandin release and phosphatidylinositol metabolism in smooth muscle. In: Braquet P, Garay RP, Frolich JC, Nicosia S, editors. Prostaglandins and Membrane Ion Transport. New York: Raven, 1984: 29.

34. Coburn RF, Soltoff S. Na$^+$,K$^+$-ATPase inhibition stimulates PGE release in guinea pig taenia coli. Am J Physiol 1977; 232: C191–C195.

35. Schramm CM, Grunstein MM. Mechanisms of protein kinase C regulation of airway contractility. J Appl Physiol 1989; 66: 1935–1941.

36. Sasaguri T, Watson SP. Phorbol esters inhibit smooth muscle contractions through activation of Na$^+$-K$^+$-ATPase. B J Pharm. 1990; 99(2): 237–42.

37. Limas CJ, Cohn JN. Stimulation of vascular smooth muscle sodium potassium-adenosine-triphosphatase by vasodilator. Circ Res 1974; 35: 601–607.

38. Scheid CR, Fay FS. β-Adrenergic effects on transmembrane Ca fluxes in isolated smooth muscle cells. Am J Physiol 1984; 246: C431–C438.

39. Scheid CR, Honeyman TW, Fay FS. Mechanism of β-adrenergic relaxation of smooth muscle. Nature 1979; 277: 32–36.

39. Scheid CR, Honeyman TW, Fay FS. Mechanism of β-adrenergic relaxation of smooth muscle. Nature 1979; 277: 32–36.

40. Webb RC, Bohr DF. Relaxation of vascular smooth muscle by isoproterenol, dibutyrl-cyclic AMP and theophylline. J Pharmacol Exp Ther 1981; 217: 26–35.

41. Lockette WE, Webb RC, Bohr DF. Prostaglandins and potassium relaxation in vascular smooth muscle of the rat. Circ Res 1980; 46: 714–720.

42. Somlyo AP, Somlyo AV, Smiesko V. Cyclic AMP and vascular smooth muscle. In: Greenard P, Robinson GA, editors. Advances in Nucleotide Research. New York: Raven, 1972: 175–194.

43. Marshall JM, Kroeger EA. Adrenergic influences on uterine smooth muscle. Phil Trans R Soc Biol 1973; 265: 135–148.

44. Diamond J. Role of cyclic nucleotide in control of smooth muscle contraction. In: George WJ, Ignarro LJ, editors. Advances in Cyclic Nucleotide Research. New York: Raven, 1978: 327.

45. Andersson RG, Kovesi G, Ericsson E. Beta adrenergic stimulation and cyclic AMP level in bovine tracheal muscle. Acta Pharmacol Toxicol 1978; 43: 323–334.

46. Torphy TJ, Zheng C, Peterson SM, Fiscus RR, Rinard GA, Mayer SE. Inhibitory effect of methacholine on drug-induced relaxation, cyclic AMP accumulation, and cyclic-AMP-dependent protein kinase activation in canine tracheal smooth muscle. J Pharmacol Exp Ther 1985; 233: 409–417.

47. Diamond L, Szarek JL, Gillespie MN, Altiere RJ. In vivo bronchodilator activity of vasoactive intestinal peptide in the cat. Am Rev Respir Dis 1983; 128: 827–832.

48. Seamon KB, Daley JW. Forskolin: a unique diterpene activator of cyclic AMP-generating systems. J Cyclic Nucleotide Res 1981; 7: 201–224.

49. Conti MA, Adelstein RS. The relationship between calmodulin binding and phosphorylation of smooth muscle myosin kinase by catalytic subunit of 3:5 cAMP-dependent protein kinase. J Biol Chem 1981; 256: 3178–3181.

50. Meisheri KD, Van Breeman C. Effects of β-adrenergic stimulation on calcium movements in rabbit aortic smooth muscle: relationship with cyclic AMP. J Physiol 1982; 331: 429–441.

51. Ozaki H, Kwon SC, Tajimi M, Karaki H. Changes in cytosolic Ca^{2+} and contraction induced by various stimulants and relaxants in canine tracheal smooth muscle. Pflugers Arch. 1990; 416: 351–359.

52. Taylor DA, Bowman BF, Stull JT. Cytosolic Ca^{2+} is a primary determinant for myosin phosphorylation in smooth muscle cells. J Biol Chem 1989; 264: 6207–6213.

53. Felbel J, Trockur B, Ecker W, Landgraf W, Hofmann F. Regulation of cytosolic calcium by cAMP and cGMP in freshly isolated smooth muscle cells from bovine tracheal. J Biol Chem 1988; 263: 16764–16771.

54. Takuwa Y, Takuwa N, Rasmussen H. The effects of isoproterenol on intracellular calcium concentration. J Biol Chem 1988; 263: 762–768.

55. Dixon WE, Brodie TG. Contributions to the physiology of the lungs. Part I. The bronchial muscles, their innervation and the action of drugs upon them. J Physiol 1903; 29: 97–180.

56. Weber E. Neue untersuchungen uber experimentalles asthma und die innervation der bronchialmuskeln. Arch f. Physiol, Leipz. 1914; 63–154.

57. Marco V, Park CD, Aviado DM. Bronchopulmonary effects of digitalis in the anesthetized dog. Dis Chest 1968; 54: 437–444.

58. Agrawal KP, Hyatt RE. Airway responses to inhaled ouabain and histamine in conscious guinea pigs. J Appl Physiol 1986; 60: 2089–2093.

59. Souhrada M, Souhrada JF. Immunologically induced alterations of airway smooth muscle cell membrane. Science 1984; 225: 723–725.

60. Souhrada M, Souhrada JF. Sensitization-induced sodium-influx in airway smooth muscle cells of guinea pigs. Respir Physiology 1985; 60(2): 157–68.

61. Souhrada M, Souhrada JF. The role of protein kinase-C in sensitization and antigen response of airway smooth muscle. Am Rev Respir Dis, 1989; 140: 1567–1572.

62. Souhrada M, Souhrada JF. Effect of IgG1 and its subfragments on resting membrane potential of isolated tracheal myocytes. J Appl Physiol 1993; 74: 1948–1953.

63. Souhrada M, Souhrada JF. A transient calcium influx into airway smooth muscle cells induced by sensitization. Respir Physiol 60: 1985; 157–168.

64. Agrawal KP, Reed CE, Hyatt RE, Imber WE, Krell WS. Airway responses to inhaled ouabain in subjects with and without asthma. Mayo Clin Proc 1986; 61: 778–784.

Index

Acetylcholine release 76
Acidosis 245
Adenylyl cyclase 58
β-adrenoceptor 161, 263, 265
β-agonists 171
Airway epithelium 19
Airway reactivity 90
Airway resistance 267
Airways 155
-, hyperreactivity of 100, 210
-, inflammation of 92
- secretion 91
Airways smooth muscle 16, 51, 217, 237
- Na$^+$ loaded 259
-, proliferation of 60
Alkalosis 242
Allergen-induced contraction 237
ANP analogues 118
ANP-$_A$ receptor 119
ANP-$_B$ receptor 120
ANP-$_C$ receptor 120
Aprikalim 202
Asthma 28, 164, 218, 269
Asthmatic subjects 59
Atrial natriuretic peptide 115
Atriopeptins 125
Autocrine regulation 135

Basement membrane 146
BCECF 235
Benzimidazolones 185
β subunit 178
Bleomycin 138, 141
BQ-123 11
Bradykinin 51
- receptor signal transduction 57
- B$_1$ receptors 52
- B$_2$ receptor antagonists 53
- B$_2$ receptors 52
- B$_3$ receptors 53
Brain natriuretic peptide 117
BRL 55834 213
Bronchial asthma 100
Bronchial circulation 92
Bronchial reactivity 116
Bumetanide 219

C-type natriuretic peptide 118

Ca^{2+}-activated K$^+$ channels 170
Ca^{2+}-channel antagonists 265
Ca^{2+} channels 73, 155, 202
^{45}Ca^{2+} efflux 54
Ca^{2+}, intracellular 256, 261, 265, 266
Calcitonin gene-related peptide (CGRP)
   68
Calcium antagonists 165
Calcium influx factor 161
Capsaicin-sensitive primary afferent
   neurones 67
Cardiac glycosides 257
Charybdotoxin 171
Cl$^-$ channel 73
Connective tissue growth factor (CTGF)
   141-142
Cromakalim (BRL 34915) 201
Cyclic ADP-ribose 156
Cyclic AMP 72, 263
Cyclic guanosine 5′-monophosphate
   (cGMP) 122
Cytokine 132-134, 144-145, 148-149
Cytosolic pH 233

Dehydrosoyasaponin I 182
Dihydropyridine 158
Diltiazem 164

e-NANC neurones 210
Electromechanical coupling 156
Endothelin 2
- ET-1 2
- ET-2 2
- ET-3 2
Endothelin-converting enzyme (ECE) 3
ET receptor subtypes 8
ET$_A$ receptor 9
ET$_B$ receptor 9
Ethylisopropylamiloride (EIPA) 237
Excitation-contraction coupling 156, 170
Extracellular K$^+$ 257

Ferret trachea 57
Fibroblasts 138, 141, 145-146, 148
Frusemide 217

G-proteins 171
Glibenclamide 201

Guinea-pig trachea   53

H^+ 234
HCO_3^- exchanger   234
Histamine-induced contraction   237
HOE 234   205
Human bronchus   11, 55
Hyperpolarization   170

Iberiotoxin (IbTX)   175
IFN-γ 138, 148
Indole diterpenes   183
Inositol-1,4,5 trisphosphate   156
Insulin-like growth factor (IGF-I)   149
Ion channels   162, 169
Isoprenaline   264-265
Isoproterenol   264, 266

K^+-free medium   257-258, 263, 265
K^+-induced relaxation   260
K^+ channels   73, 79, 200
K_{ATP} channels (ATP-sensitive)   201
KC 399   205

L-type VDCs   160
Levcromakalim (BRL 38227)   201
Ligand-gated channels   169
Loop diuretics   219
Lung cancer   103
Lung injury, acute   102

Macrophages   137-138, 142, 144,
   147-148
Maxi-K (BK) channel   170, 179
Membrane depolarization   261
Membrane potential (Em)   255, 258, 262,
   268
Membrane transport   255
-, cotransport   256
MGCMA(Na^+/H^+ exchange blocker)
   237
Mitogenesis   59
Mitotic cycle   133
mSlo   170, 179
Mucous glands   20
Mucous secretion   76
Myogenic activity   170

Na^+-loading   260
Na^+ loaded tissues   265
Na^+ loaded tracheal smooth muscle   262
Na^+ loaded airways smooth muscle   259
Na^+/Ca^+ exchange   256, 263, 265, 268

Na^+/H^+ antiporter   233
NANC inhibitory system   97
Neurogenic inflammation   67
Neuropeptides   89
Neutral endopeptidase   123
Nifedipine   160
Nitric oxide   78, 98
Nocturnal asthma   212
NS1619 (a new benzimidazolone)   185

Ouabain   257-258, 260-266, 268-269

Paracrine regulation   135
Particulate guanylyl cyclases   121
Passive sensitization   268
Paxilline   183
pCO_2 242
PDGF receptor   136
PGE_2 263
pH, intracellular   256
Pharmacomechanical coupling   156
Phenotype   131
Phenotypic expression   131
pH_i 233
Phosphoinositide turnover   72
Phospholipase C   54, 72
Phospholipase D   54
Phosphoramidon   125
Piretanide   219
Plasma protein extravasation   67
Platelet-derived growth factor (PDGF)
   132-134, 137-141, 143
Potassium channel openers   199
Potassium channels   169
Potassium-depolarized smooth muscle
   cells   246
Potassium-free physiological saline
   solution   264
Proliferation   59
Prostaglandin E (PGE)   262
Prostaglandin E_2 263
Prostaglandin synthesis   54
Protein kinase C   262-263, 269
Pulmonary circulation   92
Pulmonary disorders   33
Pulmonary hypertension   31, 102
Pulmonary microvascular permeability
   99
Pulmonary system   2
Pulmonary vascular smooth muscle   17

Receptor subtypes   71

Receptor-operated Ca^{2+} channel (ROC)
  157
Relaxation 261, 263-265
Ro 31-6930 205
RP 49356 202
RP 52891 (aprikalim) 202
RP 66471 202

Sarafotoxin S6c 11
Second messengers 95
Sensitized animals 267
Signal transduction mechanisms 25
SITS (4-acetamido-4-isothiocyano-
  stilbene-2,2-disulfonic acid) 236
Smooth muscle 143, 155, 170
- and fibroblast proliferation 20
- cells 146
- tone 90
Sodium nitroprusside (SNP) 263
Sodium/hydrogen exchange 233
Sodium/potassium/chloride co-transport
  217
Soluble guanylyl cyclases 122
Species-related variations of amino
  acids 70

Tachykinins 68
Torasamide 219
Tracheal epithelium 56
Tracheal smooth muscle, Na$^+$ loaded
  262
Tracheal muscle 237
Transforming growth factor-β (TGF-β)
  132, 139-140, 147
Tyrosine kinases 60

Urodilatin 117

Vanadate 257
Vasoactive intestinal peptide (VIP) 88
Verapamil 164
VIP analogues 101
VIP receptors 94
Voltage-dependent Ca^{2+} channels
  (VDC) 157
Voltage-gated calcium channels 249
Voltage-sensitive channels 169

WIN 64338 54

D. Raeburn, *Rhône-Poulenc Rorer Ltd, Dagenham, UK*
M.A. Giembycz, *Royal Brompton National Heart and Lung Institute, London, UK (Eds)*

Airways Smooth Muscle: Structure, Innervation and Neurotransmission

1994. 328 pages. Hardcover
ISBN 3-7643-5010-5

In the context of a growing incidence of respiratory disorders worldwide, particularly asthma and allergy, the importance of research into the respiratory system is increasingly recognized. The emphasis of this first volume in a new series of research monographs is on the anatomical aspects of airways smooth muscle, including its innervation and neurotransmission.

Scientists of international repute were invited to contribute chapters on anatomy, gross morphology and ultrastructure, sympathetic, parasympathetic and NANC innervation, vagal reflexes, prejunctional regulation of neurotransmission, and neural elements in airways smooth muscle.

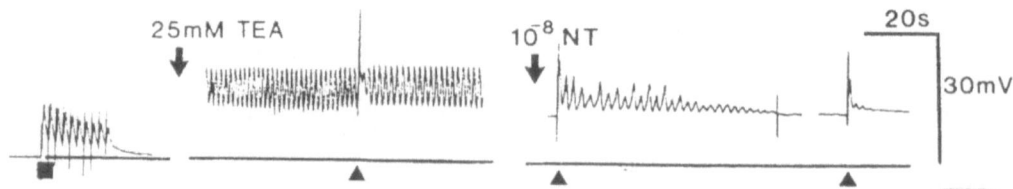

Birkhäuser Verlag • Basel • Boston • Berlin

D. Raeburn, *Rhône-Poulenc Rorer Ltd, Dagenham, UK*
M.A. Giembycz, *Royal Brompton National Heart and Lung Institute, London, UK (Eds)*

Airways Smooth Muscle: Development, and Regulation of Contractility

1994. 420 pages. Hardcover
ISBN 3-7643-5011-3

The focus of this second volume in a new series of research mono-graphs is on the growth and development of airways smooth muscle, and its regulation. It also addresses the role of nerves and other physi-ological factors responsible for regulating contractility.

Internationally acclaimed experts review the latest research data and emerging themes in the field. Aspects discussed include trophic factors and the control of smooth muscle development, cell-to-cell coupling, electrophysiology, voltage-dependent calcium channels, and the effects of ageing on contractility.

This comprehensive and up-to-date work of reference is a valuable source of information which will benefit researchers in physiology, pharmacology, anatomy and developmental biology as well as clini-cians.

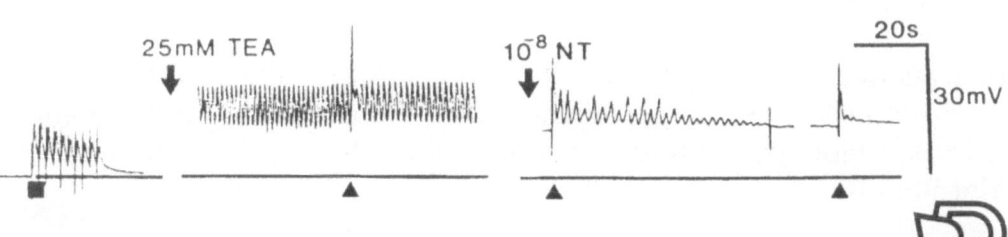

Birkhäuser Verlag • Basel • Boston • Berlin

D. Raeburn, *Rhône-Poulenc Rorer Ltd, Dagenham, UK*
M.A. Giembycz, *Royal Brompton National Heart and Lung Institute, London, UK (Eds)*

Airways Smooth Muscle:
Biochemical Control
of Contraction and Relaxation

1994. 352 pages. Hardcover
ISBN 3-7643-5043-1

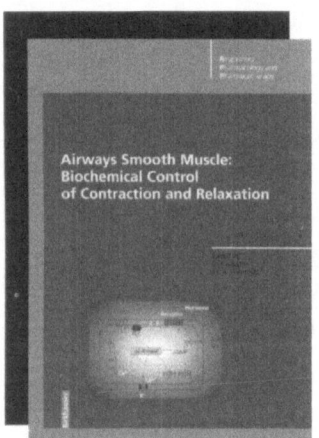

The study of airways smooth muscle has intensified over recent years against a background of a growing incidence of asthma and other respiratory disorders. Building on the previous volumes in the series, this research monograph focuses upon the biochemical regulation of contraction and relaxation of airways smooth muscle.

Written by international experts, this up-to-date reference work includes chapters on actin, myosin, diglyceride and protein kinase C, inositol polyphosphates, current theories regarding mechanisms of force generation and maintenance, G-proteins, cyclic nucleotides and properties of airways smooth muscle cells in culture.

All academic and clinical research workers in the field of airways smooth muscle physiology, biochemistry, pharmacology and cell and molecular biology will find this volume an indispensable source of information.

Birkhäuser Verlag • Basel • Boston • Berlin